# TRANSITION METAL CHEMISTRY

## Volume 7

# TRANSITION METAL CHEMISTRY

## A Series of Advances

EDITED BY
### RICHARD L. CARLIN

DEPARTMENT OF CHEMISTRY
UNIVERSITY OF ILLINOIS AT CHICAGO CIRCLE
CHICAGO, ILLINOIS

## VOLUME 7

1972
MARCEL DEKKER, INC., New York

MARCEL DEKKER, INC.
95 Madison Avenue, New York, New York 10016

LIBRARY OF CONGRESS CATALOG CARD NUMBER 65-27431

ISBN 0-8247-1082-7

PRINTED IN THE UNITED STATES OF AMERICA

# Contributors to Volume 7

S. L. Holt, *Chemistry Department, University of Wyoming, Laramie, Wyoming*

S. Mitra,* *Department of Inorganic Chemistry, University of Melbourne, Australia*

John E. Rives, *Department of Physics and Astronomy, University of Georgia, Athens, Georgia*

C. Rosenblum, *Chemistry Department, University of Wyoming, Laramie, Wyoming*

*Present address: Tata Institute of Fundamental Research, Colaba, Bombay-5, India.

v

# CONTENTS OF VOLUME 7

CONTENTS OF OTHER VOLUMES

TRANSITION METAL

CHEMISTRY

*Volume 7*

Chapter 1

# MAGNETIC PHASE TRANSITIONS
# AT LOW TEMPERATURES

John E. Rives
Department of Physics and Astronomy
University of Georgia
Athens, Georgia

1

PART 1

INTRODUCTION

## I. SCOPE

A number of reviews of various aspects of magnetic phase transitions have been published in the past few years. Most of these papers either cover special topics, or attempt to give a concise review of the entire field of magnetism.

Recently, however, there has been renewed interest in the behavior of insulating magnetic compounds which order at relatively low temperatures. In particular, the behavior of antiferromagnetic compounds in magnetic fields can be studied in detail if the compound orders at reasonably low temperatures. Presently available laboratory magnets can produce magnetic potential energies comparable to, or greater than, the internal exchange and anisotropy energies in such compounds.

Molecular field and spin wave theories present a rather complete description of the behavior of antiferromagnetic compounds in external fields. Depending on the nature, and magnitude, of the various internal interactions, a number of field induced phases are predicted which may be experimentally investigated at low temperatures.

The present discussion is limited to magnetic insulators which order in, or near, liquid helium temperatures. The main aim is to review the area of field induced phase transitions. The best known systems which order at low temperatures are the iron group transition metal halides, a number of double salts containing transition elements, and certain rare earth compounds. Because of the diversity of crystal and magnetic structure of such compounds, they offer an interesting group to study.

In Part 2 a short review of static critical point phenomena will be presented with results for both ferromagnetic, as well as antiferromagnetic compounds cited for completeness. Several excellent reviews on the topic have been published in recent months, and the reader will be referred to these works for a complete survey of the current situation in this very fascinating area of research.

Part 3 contains the discussion of the behavior of antiferromagnetic compounds in magnetic fields. In Part 5 a list of a number of common magnetic insulators is given, together with a list of some of the more important magnetic parameters which have been determined for these substances.

## II. SOME GENERAL PROPERTIES OF MAGNETIC PHASE TRANSITIONS AT LOW TEMPERATURES

### A. Specific Heat and Entropy

In recent years a large number of insulating magnetic compounds have been discovered which order at low temperatures. Both ferromagnetic and antiferromagnetic order is observed, and in some cases a more complicated situation exists where antiferromagnetic order is followed at lower temperatures by ferromagnetic order. In zero applied magnetic field, the transition from the paramagnetic to the ordered state is characterized by a somewhat anomalous behavior of some of the thermodynamic properties of the material. The specific heat, for instance, usually shows a $\lambda$-type behavior at the transition temperature. An example shown in Fig. 1 is that of antiferromagnetic $NiCl_2 \cdot 6H_2O$ reported by Robinson and Friedberg [1]. The data given is the total specific heat which is the sum of the magnetic and the lattice contributions. When the lattice contribution is subtracted it is found that a sizable magnetic contribution still persists above the transition temperature. This has been interpreted as evidence that short range order persists even in the paramagnetic state. Further evidence of this interpretation can be obtained if one calculates the entropy of magnetic order. The relationship between the magnetic contribution to the specific

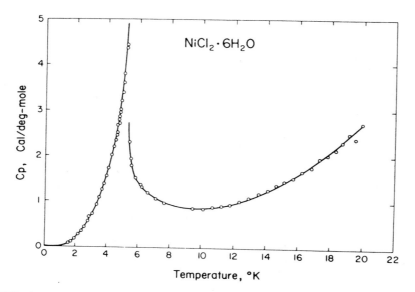

FIG. 1. Molar specific heat of $NiCl_2 \cdot 6H_2O$ as a function of temperature. (Reprinted from Ref. 1, p. 404, by courtesy of Am. Phys. Soc.)

heat, $C_M$, and the entropy of magnetic order, $S_M$, follows from the thermodynamic relation

$$C_M = T(\partial S_M / \partial T) \tag{1}$$

By integrating this we have

$$S_M = \int_0^T (C_M/T) \, dT \tag{2}$$

From statistical mechanics we have

$$S_M = k \, \log_e W_M \tag{3}$$

where $k$ is the Boltzmann constant and $W_M$ is the number of ways of achieving the same total energy for the macroscopic magnetic system by different
microscopic arrangements of the elementary moments. In the perfectly
disordered state there are $2S+1$ quantum states of different orientation
open to each ion, where $S$ is the total spin quantum number for the ion.
In the perfectly ordered state there is only one such state. Hence the
difference in entropy between the perfectly ordered and the completely
disordered states is, per mole

$$\begin{aligned} \Delta S_M &= Nk \, \log_e (2S+1) - Nk \, \log_e 1 \\ &= Nk \, \log_e (2S+1) \end{aligned} \tag{4}$$

where $N$ is Avogadro's number. Figure 2 shows the result of such a calculation for $NiCl_2 \cdot 6H_2O$ by Robinson and Friedberg [1]. It is evident from
this curve that some short range order persists well above the transition
temperature.

Modern microscopic theories indeed predict such a short range order,
whereas earlier molecular field models predicted a zero magnetic contribution to the entropy above the ordering temperature.

Although most magnetic transitions exhibit a $\lambda$-type anomoly in the
specific heat where the asymptotic limit of the specific heat is singular on
one or both sides of the transition, $GdCl_3$[2,3] appears to be an exception.
The early work of Leask, Wolf and Wyatt [2] indicated that the specific
heat was finite but had a simple discontinuity. More recent high resolution
measurements of Landau [3] shown in Fig. 3 verify that the specific heat is
finite on both sides of the transition, but that the derivatives are infinite.
The asymptotic form for the specific heat is not singular, hence the
transition appears to be of the type 2b, or diffuse, second order transition.

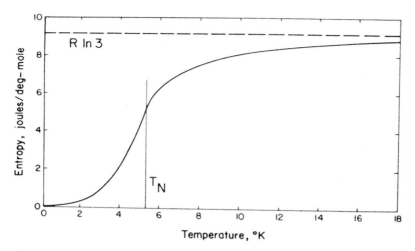

FIG. 2. Magnetic entropy as a function of temperature for $NiCl_2 \cdot 6H_2O$. (Reprinted from Ref. 1, p. 407, by courtesy of Am. Phys. Soc.)

## B. Susceptibility and Magnetization

At temperatures well above the ordering temperature the magnetic susceptibility $\chi = dM/dH$, in the limit of zero external field, usually follows a Curie-Weiss behavior. Thus, well into the paramagnetic region one has $\chi = C/(T \pm \theta)$, where C is the Curie constant for the material and $\theta$ is a constant related to the energy of interaction between the spins. The minus and plus signs indicate ferromagnetic and antiferromagnetic order, respectively. In Fig. 4 the Curie-Weiss behavior is shown for, (a) the antiferromagnetic case, (b) the ferromagnetic case, and (c) the ideal noninteracting paramagnetic case.

Below the ordering temperature there exists a spontaneous magnetization. In the ferromagnetic case this magnetization is a maximum at T=0 and drops to zero at the ordering temperature. From molecular field theory one obtains

$$M_S = M_o B_J (\mu M_S/kT) \tag{5}$$

where $B_J (\mu M_S/kT)$ is the Brilluoin function of $\mu M_S/kT$, i.e.,

$$B_J(x) = \frac{2J+1}{2J} \coth \frac{2J+1}{2J} x - \frac{1}{2J} \coth \frac{1}{2J} x \tag{6}$$

The measured spontaneous magnetization of a number of ferromagnetic materials has a temperature dependence which closely resembles Eq. (5) (See, for instance, Ref. 3).

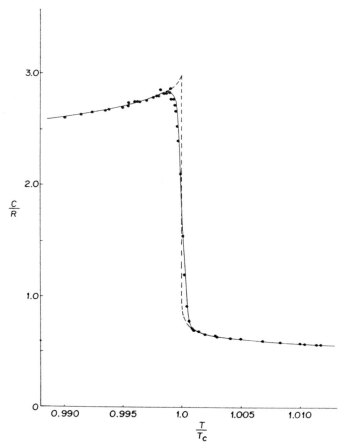

FIG. 3. Specific heat of GdCl$_3$ vs. T/T$_c$. Experimental results . . . , asymptotic unrounded critical behavior ---, rounded result from Gaussian distribution with half-width $\Gamma$ = 1.5 mK (Reprinted from Ref. 3, p. 916, by courtesy of La Societé Francaise de Physique.)

In the antiferromagnetic case the situation is complicated by the anti-parallel alignment. However, in most cases one can view this case in terms of several interpenetrating sublattices. Within each sublattice there exists a spontaneous magnetization, which according to molecular field theory, should follow Eq. 5. In the case of a simple two sublattice model the spontaneous magnetization of one sublattice is aligned antiparallel to that of the other sublattice. Anisotropy effects lead to a preferred direction for the alignment.

If the susceptibility of a single crystal is measured along this preferred

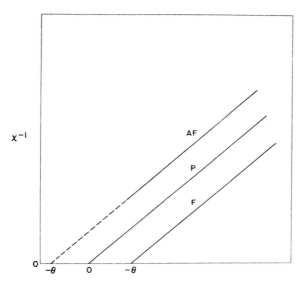

FIG. 4. Inverse magnetic susceptibility against temperature for paramagnetic (P), ferromagnetic (F), and antiferromagnetic (AF) materials
according to the Curie-Weiss theory.

direction, one finds, as the temperature is reduced through the ordering
temperature, that the susceptibility goes through a maximum (near $T_c$)
and then decreases to zero as T approaches zero. In a direction perpendicular to the preferred direction the susceptibility remains almost independent of temperature below ordering temperature.

## C.  Field Induced Phase Transitions

In the case of a simple two sublattice antiferromagnet with anisotropy
Fig. 5 shows a typical H-T phase diagram with the field along the preferred
direction. At zero field $T_N$ separates the paramagnetic state from the
ordinary antiferromagnetic state. As the field is increased from zero a
second order phase boundary, CD, results. At temperatures below $T_t$ at
a critical field $H_{cl}$ there is a first order transition to the spin flop state.
As the field is increased still further there is a second order transition
to the paramagnetic state.

It is possible to give a simple argument to determine which state is stable
at a given temperature and field by considering the free energy of the
system. For a field parallel to the preferred direction the magnetic
potential energy is given by

$$E = - (1/2) \chi_{||} H^2 \qquad (7)$$

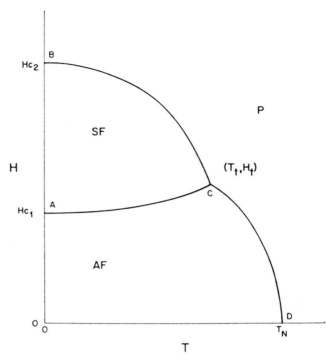

FIG. 5. Typical H-T phase diagram for a simple two-sublattice antiferromagnet with the field along the preferred axis.

If the sublattice moments are aligned perpendicular to the field the magnetic potential energy will be, neglecting anisotropy,

$$E = -(1/2) \chi_\perp H^2 \tag{8}$$

Below $T_N$, as was already pointed out, $\chi_{||} < \chi_\perp$, hence it is seen that if
there is no anisotropy the total free energy of the system will be minimized
if the sublattice moments flip over perpendicular to the external field.
This state is referred to as the spin flop state. Thus for an isotropic
system the spin flop state would be the stable state for any field greater
than zero.

The anisotropic case can be considered most easily by introducing a
simple uniaxial anisotropy energy of the form

$$E_a = K \sin^2\theta \tag{9}$$

where $\theta$ is the angle between the sublattice moments and the preferred direction. Thus Eq. (8) becomes

$$E = -1/2\chi_\perp H^2 + K \tag{10}$$

Equations (7) and (10) define a critical field

$$H_{cl}^2 = 2K/(\chi_\perp - \chi_{||}) \tag{11}$$

For fields below $H_{cl}$ the antiferromagnetic state is stable, and for fields greater than $H_{cl}$ the spin flop state is stable.

When the field is increased further to a value such that the magnetic potential energy becomes more or less equal to the interaction energy (exchange) between the spins, a second order transition to the paramagnetic state results.

It was assumed in the above example that the anisotropy energy is small compared to the exchange energy. In certain cases, such as $FeCO_3$ [5] and $FeBr_2$ [6] the anisotropy energy is larger than the exchange energy. In such cases the spin flop phase region disappears, since the external field energy overcomes the exchange energy at a field less than the critical field for the spin-flop transition.

A more detailed account of the behavior of antiferromagnetic materials in external fields will be presented in Part 3. An excellent review of the molecular field model of antiferromagnetism has been presented by Nagamiya, Yosida and Kubo [7]. Anderson and Callen [8] present a spin wave theory, using a Green-Function method, and also offer a good review of previous work.

## REFERENCES

1. W. K. Robinson and S. A. Friedberg, Phys. Rev., 117, 402 (1960).

2. M. J. M. Leask, W. P. Wolf, and A. F. G. Wyatt, Proc. Eighth Int. Conf. on Low Temperatures, London, September 1962 (Butterworth's Scientific Publications, Ltd, London 1963); A. F. G. Wyatt, D. Phil. Thesis, Oxford Univ., 1963, unpublished.

3. D. P. Landau, Proc. Int. Magnetism Conf., Grenoble, September, 1970, J. Physique, 32, Suppl. to No. 2-3, C1-1012 (1971).

4. D. H. Martin, Magnetism In Solids, (Iliffe Books, Ltd., London, 1967) p. 224.

5. I. S. Jacobs, J. Appl. Phys., 34, 1106 (1963).

6. I. S. Jacobs and P. E. Lawrence, J. Appl. Phys., 35, 996 (1964).

7. T. Nagamiya, K. Yosida, and R. Kubo, Adv. Phys., 4, 1 (1955).

8. F. B. Anderson and H. B. Callen, Phys. Rev., 136, A1068 (1964).

## PART 2

## SPIN ORDERING IN ZERO MAGNETIC FIELD

### I. CRITICAL POINT THEORIES

Thermodynamic transitions in solids involving discontinuous behavior of many thermodynamic functions have been the subject of much research interest for many years. First order transitions, characterized by a finite change in the entropy, can be understood in terms of the standard idea of independent phases in equilibrium. Particular interest has been attached to higher order transitions which do not involve a finite change in entropy.

The Curie point in a ferromagnet is a classic example of such a transition. There is an anomalous peak in the specific heat at the Curie point, the spontaneous magnetization disappears and the susceptibility becomes infinite.

Quantitative predictions, based on various models, are usually presented in terms of a set of critical indices, which describe the asymptotic behavior of the thermodynamic functions in the critical region near the ordering temperature. For instance, in the example above the spontaneous magnetization is predicted to disappear according to $M \sim (T_c - T)^\beta$, where $\beta$ is the critical index in this case.

A number of excellent reviews of the entire field of critical point phenomena are to be found in the literature. Kadanoff et al. [1] and Heller [2] presented a comprehensive account of the experimental situation up to 1967. Included in their account was a discussion of magnetic systems, critical points in classic liquids, quantum liquid-gas phase transitions, superfluids, and ferroelectrics. At about the same time Fisher [3] presented a theoretical review of the best estimates of the various critical indices. Domb [4] and Weilinga [5] have presented a more recent analysis of critical behavior in magnetic systems with some discussion of the various mathematical techniques used to obtain numerical estimates of the critical indices.

The theoretical approach is based on an analysis of systems which can be described by an interaction Hamiltonian of the form

$$\mathcal{H} = -2 \sum_{(i,j)} (J_{ij}^x s_i^x s_j^x + J_{ij}^y s_i^y s_j^y + J_{ij}^z s_i^z s_j^z) - g\mu_0 H \sum_i s_i^z \tag{12}$$

where the sum in the first term includes, once only, each pair of neighboring sites in the lattice. Higher neighbor interactions have been considered in a few cases (see e.g. Refs. [6, 7]). If $J_{ij}^x = J_{ij}^y = J_{ij}^z = J$, Eq. 12 describes the so-called Heisenberg interaction. If $J_{ij}^x = J_{ij}^y = 0$, Eq. 12 is known as the Ising interaction. The use of the Hamiltonian, Eq. 12, can be reasonably justified for ionic systems, but not for metals. Thus the results obtained should be valid only for ionic systems. It is observed, however, that certain metals exhibit some of the same properties as do ionic systems. Equation 12 describes both ferromagnetic (J>0) lattices and antiferromagnetic (J<0) lattices. In the case of antiferromagnetism, however, the second term is usually rewritten as a sum over each of the two interpenetrating sublattices.

The ultimate goal in the case of critical point theories is to determine the expected behavior of thermodynamic properties of the system, such as the specific heat, magnetization, and susceptibility. One first attempts to find the free energy F(H,T), given by

$$F(H,T) = -kT \ln Z \tag{13}$$

where

$$Z = \text{Tr} \exp(-\mathscr{H}/kT) \tag{14}$$

where k is Boltzman's constant and T is the absolute temperature. The specific heat C, the spontaneous magnetization M, and the susceptibility $\chi$, are then obtained from the standard statistical mechanics for a system of volume V as,

$$C = \frac{\partial}{\partial T}\left[ kT^2 \frac{\partial}{\partial T}(\ln Z)\right] \tag{15}$$

$$M = \frac{kT}{V} \frac{\partial}{\partial H}(\ln Z) \tag{16}$$

and

$$\chi = \frac{kT}{V} \frac{\partial^2}{\partial H^2}(\ln Z) \tag{17}$$

The partition function, Z(H,T), can be expanded into an exact double power series in H and 1/T at high temperatures. One can then derive series expansions for the various thermodynamic functions of interest. These series have generally been analyzed by either the Padé approximate method or the ratio method to determine the asymptotic behavior as the critical temperature is approached from above. Similar low temperature expansions can be analyzed to determine the behavior as one approaches the critical temperature from below.

Following the technique of Baker et al. [8], since F is an even function of H, it is convenient to write

$$N^{-1}\ln Z = F_0(x) + \sum_{n=1}^{\infty} [(2n)!]^{-1} y^{2n} F_n(x) \tag{18}$$

where $x = J/kT$, $y = \mu H/kT$, and N is the number of lattice sites. The zero field specific heat C(O) and the zero field susceptibility $\chi(0)$ can be expressed as

$$C(0) = Nkx^2(\partial^2/\partial x^2)F_0(x) \tag{19}$$

and

$$\chi(0) = (N\mu^2/kT)F_1(x) \tag{20}$$

where $F_0(x)$ and $F_1(x)$ are each power series in x. The analyses of the above series yield the following asymptotic behaviors for the specific heat and the susceptibility:

$$C(0) \simeq A(1-T_c/T)^{-\alpha} = A\epsilon^{-\alpha} \tag{21}$$

and

$$\chi(0) \simeq B(1-T_c/T)^{-\gamma} = B\epsilon^{-\gamma} \tag{21}$$

For $T<T_c$, with a similar approach using low temperature expansions, one obtains

$$C(0) \simeq A'\epsilon^{-\alpha'} \tag{23}$$

$$\chi(0) \simeq B'\epsilon^{-\gamma'} \tag{24}$$

and

$$M(0) \simeq C\epsilon^{-\beta} \tag{25}$$

where M(0) is the spontaneous magnetization. For antiferromagnets M(0) is interpreted as the sublattice magnetization.

Table 1 lists some of the critical indices expected for the molecular field model, two and three dimensional Ising models and the three dimensional Heisenberg model. In the cases of the molecular field and the two dimensional Ising models the indices are the result of exact solutions, whereas for the three dimensional models the techniques described are used to obtain approximate solutions. The specific heat series are extremely hard to work with due to very slow convergence, and this leads to much more uncertainty in the values of $\alpha$ and $\alpha'$. For the S=1/2 Heisenberg model, Baker et al. [8] (using the Padé approximate method of evaluation of the series) conclude that $\alpha = 0.4 \pm 0.1$ if one assumes that a singularity of the type indicated in Eq. 21 exists. On the other hand scaling law arguments [9] would suggest that $\alpha$ should have the value -0.20. Baker et al. [8], found that $\alpha = -0.20$ was consistent with the results of a numerical evaluation of the specific heat series in the range $0.70 \leq T_c/T \leq 0.96$. However, they point out that a value $\alpha = -0.25$ fits the results equally well. It should be pointed out, on the other hand, that a value of $T_c/T = 0.96$ corresponds to $\epsilon = 0.04$ and the asymptotic behavior indicated in Eq. 21 probably would not be expected for a value of $\epsilon$ much greater than about

TABLE 1

Predicted Critical Indices
for Several Models

|  | Molecular Field | | Ising 2D S=1/2 | | Ising 3D S=1/2 | | Heisenberg 3D S=1/2 | |
|---|---|---|---|---|---|---|---|---|
| $\alpha$ |  |  | 0 |  | 1/8 | (11) | $0.4 \pm 0.1$ | (8) |
|  | Discontinuity in C(0) | | $\log \infty$ in C(0) | |  |  | $-0.2 \pm 0.05$ | (8) |
| $\alpha'$ |  |  | 0 |  | 1/8 |  |  |  |
|  |  |  |  |  | $0.066^{+0.16}_{-0.04}$ | (10) |  |  |
| $\gamma$ | 1 | (1) | 7/4 | (12) | $1.25 \pm 0.001$ |  | $1.43 \pm 0.01$ | (8) |
| $\gamma'$ | 1 | (1) | 7/4 | (12) | $1.31 \pm^{0.04}_{0.05}$ | (10) |  |  |
| $\beta$ | 1/2 | (1) | 1/8 | (12) | $0.312^{+0.002}_{-0.005}$ | (10 | $0.357 \pm 0.003$ |  |

0.1. Hence the above test does not extend sufficiently far into the critical region to provide a true test.

Because of this very unsatisfactory situation in the evaluation of the specific heat series, it is difficult to say what the model predicts with any degree of certainty. With the negative value of $\alpha$ the specific heat is predicted to be of the form

$$C(0) = A - B\epsilon^{0.20} \tag{26}$$

which produces a cusp at the critical temperature instead of a singularity. It is tempting to interpret the experimental rounding of the data near the critical temperature in this way. It is known, however, that crystal inhomogenities can produce rounding of the data of the magnitude observed, due to the simple fact that there will be a distribution of critical temperatures throughout the crystal. This point will be further explored in the discussion of experimental results.

## II.  EXPERIMENTAL RESULTS

### A.  Analysis of Experimental Data

The determination of critical constants from experimental data is in general hampered by a variety of problems. The data seems to follow an asymptotic power law behavior only over a rather restricted temperature range. As the critical temperature is approached rounding occurs, for values of $\epsilon$ sometimes as large as $5\text{-}10 \times 10^{-3}$. The contribution to the thermodynamic properties from nonsingular, slowly varying terms becomes important as one gets farther from the critical temperature. In trying to extract the asymptotic behavior from the data, therefore, one is restricted to values of $\epsilon$ less than about $5 \times 10^{-2}$. If the rounding occurs over the range mentioned above, it is difficult to make a very precise fit to the data.

Another very important problem which results from the rounding of the data is the inability to determine $T_c$ with any accuracy. For the specific heat, for instance, one is thus faced with the problem of trying to fit the data to an equation of the form:

$$C = A(1 - T/T_c)^{-\alpha} + B$$

$$= A\epsilon^{-\alpha} + B \tag{27}$$

where none of the four constants A, B, $T_c$, or $\alpha$ is known. A number of techniques for determining the constants in this case have been used by the various experimenters. However, one fact is always clear. The value of $\alpha$ determined by fitting Eq. 27 to the experimental data is very sensitive to the value of $T_c$ chosen. In much of the earlier work one of the constants was usually assumed to be known, and the remainder determined by a fitting routine. In some cases $T_c$ was chosen as the temperature of the maximum specific heat. In other cases the specific heat for $T<T_c$ was assumed to follow a logarithmic behavior, and $T_c$ could then be determined from the best fit of the data for $T<T_c$ to a logarithmic function. Both of these techniques suffer from the fact that a definite bias is introduced in the first step.

One method which eliminates most of the above problems is to differentiate Eq. 27 with respect to T and plot the logarithm of dC/dT against log $\epsilon$. From Eq. 27 it is straightforward to show that

$$dC/dT = (A\alpha/T_c)\epsilon^{-(\alpha+1)} \tag{28}$$

This reduces the problem to a one parameter problem. $T_c$ is varied to produce the best linear fit. Once $T_c$ is determined the slope of the straight line gives the constant $\alpha$.

The nature of the rounding of the data near $T_c$ has been the subject of considerable discussion for some time. Since all of the model calculations have been done for samples of infinite size, it was suggested that perhaps the finite size of real samples might account for the rounding. Domb [13] considered the effect of a finite size on the specific heat and concluded that rounding would become significant only for

$$1-T_c/T \leq 1/M \tag{29}$$

where M is the number of lattice sites along one dimension of a cubic sample. Even for a normal sample with many grain boundaries, one expects $M \cong 10^6 - 10^7$, so that rounding would occur only for values of $\epsilon$ less than about $10^{-6}$ at worst.

It has also been suggested by Fisher [14] and others that perhaps the specific heat should not exhibit a singularity, but rather a cusp at the ordering temperature. Indeed, Baker et al. [8] found that a relation such as Eq. 26 was not inconsistent with the series expansions for the Heisenberg model. However, the possible validity of Eq. 26 was not tested close enough to the critical temperature to be conclusive.

Recently Fisher [15] considered the effect of impurities on the asymptotic behavior of the thermodynamic variables. Under rather general assumptions he has shown that the "ideal" critical exponents should be renormalized. This renormalization, in the case of the specific heat exponents ($\alpha$ and $\alpha'$), can actually change their numerical sign. Thus a singularity in a pure system might be changed to a cusplike behavior in an impure system. To date, however, this author knows of no experimental results which can be completely understood in terms of a positive power law cusp. The rounding does not appear to fit a simple power law behavior.

It is quite reasonable to assume that inhomogeneities and strains in the crystals are mainly responsible for the observed rounding. The ordering temperature is directly related to the magnitude of the exchange energy. The exchange interaction, in turn, is very sensitive to the separation and orientation of the magnetic ions in the crystal. Thus strains and defects in the crystal will cause a variation of the magnitude of the exchange interaction throughout the crystal. The ordering temperature will no longer be a constant for the entire crystal, but will be distributed over a range of values. A rounding of the thermodynamic properties will result. It is difficult to make any quantitative estimates of the magnitude of this effect, however observations on different crystals of the same material indicate that transition temperatures sometimes vary as much as two to three tenths of a percent. This is just the order of magnitude needed to explain the rounding observed in most cases.

### B. Susceptibility and Magnetization Data

A look at the experimentally determined critical exponents in a number of low temperature insulators reveals that there is a wide divergence in the values obtained in different materials. Table 2 gives some values of $\gamma$ and $\beta$ obtained for both ferromagnets and antiferromagnets. The copper double salts are all cubic, spin 1/2 ferromagnets with very little anisotropy. They would be expected to be very good examples of materials for which the Heisenberg hamiltonian is valid. Indeed, the values of $\gamma$ obtained, 1.36, 1.37, and 1.31, are in very good agreement with the earlier accepted value of 1.33 for the Heisenberg model. For the first two compounds the authors did not state the accuracy of their results, but considering the narrow range of $\epsilon$ over which they fit their data, it is possible that the results are not in serious disagreement with the presently accepted value of 1.43. It is known that next nearest neighbor interactions are quite important in these compounds and this could have a noticable effect on the values of the exponents.

The values of $\beta$ obtained for a number of salts seem to lie somewhat between the theoretical values of the Ising model and the Heisenberg model. Since most of these compounds have a sizable anisotropy, it is difficult to make a quantitative comparison. Again next nearest neighbor interactions will probably have some effect.

The compound $CoCl_2 \cdot 6H_2O$ appears to be an exception with a value of $\beta = 0.18$. This small value of $\beta$ is about what might be expected for a two dimensional magnetic structure. In the analysis of the specific heat data, Robinson and Friedberg [32] found that only 48 percent of the total magnetic entropy change occurred below the critical temperature. According to

TABLE 2

Experimental Critical Exponents Determined from
Magnetic Susceptibility and Magnetization Data

| Material | $T_c(K)$ | Spin | Range of $\epsilon$ for fit | $\gamma$ | $\beta$ | Ref. |
|---|---|---|---|---|---|---|
| Ferromagnets | | | | | | |
| $CuK_2Cl_4 \cdot 2H_2O$ | 0.88 | 1/2 | $3 \times 10^{-2} - 3 \times 10^{-1}$ | 1.36 | | 16 |
| $Cu(NH_4)_2Cl_4 \cdot 2H_2O$ | 0.70 | 1/2 | $4 \times 10^{-2} - 4 \times 10^{-1}$ | 1.37 | | 16 |
| $Cu(NH_4)_2Br_4 \cdot 2H_2O$ | 1.773 | 1/2 | $10^{-2} - 3 \times 10^{-1}$ | 1.31 | | 17 |
| EuS | 16.50 | 7/2 | $10^{-2} - 10^{-1}$ | | 0.38 | 18 |
| $CrBr_3$ | 32.56 | 3/2 | $7 \times 10^{-3} - 5 \times 10^{-2}$ | | 0.330 | 19 |
| | | | | | 0.365 | 20 |
| Antiferromagnets | | | | | | |
| $MnF_2$ | 67.336 | 5/2 | $6 \times 10^{-5} - 8 \times 10^{-2}$ | | 0.335 | 21 |
| $FeF_3$ | 363.11 | 5/2 | $5 \times 10^{-4} - 6 \times 10^{-2}$ | | 0.352 | 22 |
| $FeF_2$ | 78.11 | 2 | $5 \times 10^{-4} - 10^{-1}$ | | 0.325 | 23 |
| $CoCl_2 \cdot 6H_2O$ | 2.29 | 1/2 | $10^{-2} - 5 \times 10^{-1}$ | | 0.18 | 24 |
| $KMnF_3$ | 88.06 | 5/2 | $10^{-2} - 10^{-1}$ | | 0.33 | 25 |

Domb and Miedema [7] this is consistent with a two dimensional square lattice with coordination number 4, rather than with a three dimensional lattice with a higher coordination number.

## C.  Specific Heat Data

The specific heat results in Table 3 are not so easy to interpret.  In some of the earlier work the experimenters were perhaps influenced by the very fine results for liquid helium near the lambda point reported by Kellers, Fairbank and Buckingham [33].  The observed logarithmic singularity on both sides of the lambda point was in agreement with the exact solution to the two dimensional Ising model [34].  At that time the approximate solutions to the three dimensional Ising and Heisenberg models were not available.  Some of these results were plotted so as to exhibit logarithmic behavior over a narrow range of temperatures near the critical point.

A good example of this is demonstrated in the case of $CoCl_2 \cdot 6H_2O$.  Skalyo and Friedberg [29] reported that a logarithmic dependence was consistent with the data over a very small range in $\epsilon$.  They also reported a rounding of the data over about 12 mK near the critical temperature.  White, Song, Rives and Landau [30] have recently remeasured this material with higher temperature resolution.  By using Eq. 28 to analyze the data it was found that $\alpha = 0.35 \pm 0.05$ and $\alpha' = -0.19 \pm 0.04$.  In addition the data could be fit quite well over the entire temperature range, including the rounding region, with the above quoted values of $\alpha$ and $\alpha'$ by assuming a gaussian distribution of critical temperatures.  A gaussian width of 5 mK was found to give a very good fit to the data.  Using these same values for $\alpha$ and $\alpha'$ but with a gaussian width of 13 mK, a good fit to the earlier results of Skalyo and Friedberg was found.

With the exception of $MnF_2$ all of the high resolution results are consistent with the value of $\alpha = 0.4 \pm 0.1$ obtained by Baker et al. [8] from the series solutions of the Heisenberg model.

In reviewing the experimentally determined critical exponents it is clear that there is considerable discrepency between the measured critical behavior and the best theoretical estimates from model calculations.  Renormalization due to impurities, according to Fisher [15], does not appear to completely rectify the disagreement.  Rounding of the data near $T_c$ can be qualitatively understood in terms of a distribution of ordering temperatures.  Inhomogenities in the lattice due to strains, dislocations and the like might cause such a broadening, however it is not clear what effect they might have on the critical indices.

In a recent article Barmatz and Hohenberg [35] show that a gravitational field can produce an anomalously large value of $\alpha$ near the critical point in a liquid.  Domb [4] suggests that a similar effect might arise from a magnetic field.  The dipolar field is important in a number of magnetic

TABLE 3

Experimental Critical Exponents Determined from Specific Heat Data

| Material | $T_c$ (K) | Spin | Range of $\epsilon$ for fit | $\alpha$ | $\alpha'$ | Ref. |
|---|---|---|---|---|---|---|
| Antiferromagnets | | | | | | |
| $MnF_2$ | 67.336 | 5/2 | $2 \times 10^{-4} - 5 \times 10^{-2}$ | $\cong 0.16$ | ~0 (log) | 26 |
| DAG | 2.53 | 1/2 | $3 \times 10^{-4} - 3 \times 10^{-2}$ | $0.31 \pm .02$ | ~0 (log) | 27 |
| $MnCl_2 \cdot 4H_2O$ | 1.62 | 5/2 | $10^{-4} - 3 \times 10^{-2}$ | $0.35 \pm .02$ | ~0 (log) | 28 |
| $CoCl_2 \cdot 6H_2O$ | 2.289 | 1/2 | $10^{-2} - 10^{-1}$ | ~0 (log) | ~0 (log) | 29 |
| | | | $4 \times 10^{-3} - 7 \times 10^{-2}$ | 0.35 | -0.19 | 30 |
| Ferromagnets | | | | | | |
| $CuK_2Cl_4 \cdot 2H_2O$ | 0.88 | 1/2 | $10^{-3} - 10^{-1}$ | ~0 (log) | ~0 (log) | 31 |

compounds, but since it is due to internal interactions, it is not clear that one would expect an effect similar to that produced by an external field. The earth's magnetic field is present in most so-called zero field measurements, however, the energy associated with it is, in temperature units, about $10^{-4}$ K. This could effect the data over a noticeable temperature region only for those compounds which order around 1.0 K or below.

## REFERENCES

1. Leo P. Kadanoff, Wolfgang Gotze, David Hamblen, Robert Hecht, E. A. S. Lewis, V. V. Palciauskas, Martin Rayl and J. Swift, Rev. Mod. Phys. 39, 395 (1967).

2. P. Heller, Repts. Prog. Phys. 30, pt. II, 731 (1967).

3. M. E. Fisher, Repts. Prog. Phys. 30, p. II, 615 (1967).

4. C. Domb, Adv. Phys. 19, 339 (1970).

5. R. F. Weilinga, Prog. in Low Temperature Phys., Vol. VI, ed. C. J. Gorter. (North-Holland Publ. Co., Amsterdam, 1970).

6. N. W. Dalton and D. W. Wood, Phys. Rev. 138, A779 (1965).

7. C. Domb and A. R. Miedema, Prog. in Low Temperature Phys., Vol. IV, ed. C. J. Gorter (North-Holland Publ. Co., Amsterdam, 1964).

8. George A. Baker, Jr., H. E. Gilbert, J. Eve and G. S. Rushbrooke, Phys. Rev. 164, 800 (1967).

9. L. P. Kadanoff, Physics 2, 263 (1966); C. Domb and D. L. Hunter, Proc. Phys. Soc. (London) 86, 1147 (1965); M. E. Fisher, J. Appl. Phys. 38, 981 (1967); J. W. Essam and M. E. Fisher, J. Chem. Phys. 39, 842 (1963).

10. George A. Baker, Jr. and David S. Gaunt, Phys. Rev. 155, 545 (1967).

11. D. S. Gaunt and C. Domb, J. Phys. Chem. (Proc. Phys. Soc.) 1, 1038 (1968); M. F. Sykes, J. L. Martin and D. L. Hunter, Proc. Phys. Soc. (London) 91, 671 (1967).

12. M. E. Fisher, Proceedings of the International Conference on Phenomena near Critical Points, Washington (1965) NBS Misc. Publication No. 273.

13. C. Domb, Proc. Phys. Soc. (London) 86, 933 (1965).

14. M. E. Fisher, Bull. Am. Phys. Soc. 13, 57 (1968).

15. M. E. Fisher, Phys. Rev. 176, 257 (1968).

16. A. R. Miedema, H. van Kempen and W. J. Huiskamp, Physica 29, 1266 (1963).

17. L. J. DeJongh, A. R. Miedema and R. F. Weilinga, to be published.

18. R. F. Weilinga and W. J. Huiskamp, Physica 40, 602 (1969).

19. P. Heller and G. B. Benedek, Phys. Rev. Letters 14, 71 (1965).

20. S. D. Senturia and G. B. Benedek, Phys. Rev. Letters 17, 475 (1966).

21. P. Heller and G. B. Benedek, Phys. Rev. Letters, 8, 428 (1962).

22. G. K. Wertheim, H. J. Guggenheim, and D. N. E. Buchanan, Phys. Rev. 169, 465 (1968).

23. G. K. Wertheim and D. N. E. Buchanan, Phys. Rev. 161, 478 (1967).

24. E. Sawatzky and M. Bloom, Can. J. Phys. 42, 647 (1964).

25. M. J. Cooper and R. Nathans, J. Appl. Phys. 37, 1041 (1966).

26. D. T. Teaney, Phys. Rev. Letters 14, 898 (1965).

27. B. E. Keen, D. P. Landau, and W. P. Wolf, J. Appl. Phys. 38, 967 (1967).

28. George S. Dixon and John E. Rives, Phys. Rev. 177, 871 (1969).

29. J. Skalyo, Jr. and S. A. Friedberg, Phys. Rev. Letters 13, 133 (1964).

30. J. J. White, H. I. Song, J. E. Rives, and D. P. Landau, Phys. Rev. B 4, 4605 (1971).

31. A. R. Miedema, R. F. Wielinga, and W. J. Huiskamp, Phys. Letters 17, 87 (1965).

32. W. K. Robinson and S. A. Friedberg, Phys. Rev. 117, 402 (1960).

33. C. F. Kellers, thesis, Duke University (1960); M. J. Buckingham and W. M. Fairbank, Progress in Low Temperature Physics (North-Holland Publ. Co., Amsterdam, 1961), Vol. III.

34. L. Onsager, Phys. Rev. 65, 117 (1944).

35. M. Barmatz and P. C. Hohenberg, Phys. Rev. Letters 24, 1225 (1970).

PART 3

## ANTIFERROMAGNETISM IN EXTERNAL MAGNETIC FIELDS

Part I included a brief survey of some of the properties of antiferromagnets along with a very elementary description of the behavior of an antiferromagnet in an external field. In this part a more detailed description will be presented.

The theory of antiferromagnetism based on the Weiss molecular field approximation was developed by Néel [1], Bitter [2], and Van Vleck [3].

More recently improvements and extensions have been made by a large number of authors. Consider the simplest model consisting of two inter-penetrating sublattices, A and B, for which the nearest neighbors of an A ion are all on the B sublattice and vice-versa. The A and B sublattices are equivalent in every way, their lattice sites being equal in number and identical in environment.

An ion in the A sublattice behaves according to the molecular field hypothesis as though it were acted upon by an effective field, $H_A$, which has two sources. The A ions interact with the ions of the B sublattice and with other ions of the A sublattice. We have, therefore,

$$H_A = -AM_B - \Gamma M_A \tag{30}$$

and similarly for the molecular field acting on a B ion

$$H_B = -AM_A - \Gamma M_B \tag{31}$$

$M_A$ and $M_B$ are the overall magnetizations of the A and B sublattices, and the constants A and $\Gamma$ measure the interaction between the nearest neighbor ions and the next-nearest neighbor ions, respectively. If $A>0$ and $A>>\Gamma$ then at absolute zero the magnetically ordered state will be the alignment of all A ions in one direction and of all B ions in the opposite direction.

## I.  MAGNETIZATION AND SUSCEPTIBILITY

### A.  Molecular Field Theory

The magnetization of a system of noninteracting ions is given by the expression

$$M = M_o B_S (\mu H/kT) \tag{32}$$

where H is the external magnetic field, and $\mu$ is the maximum value for the aligned moment of an ion, $g\mu_B S$, where g is the g-factor for the ion, $\mu_B$ is the Bohr magneton, and S is the spin quantum number of the ion. $B_S$ is the Brillouin function defined in Eq. 6. In neglecting the orbital moment of the ion we are assuming that the crystal fields quench the orbital part of the total moment. This is approximately true for a large number of antiferromagnetic crystals, and in any event the treatment can be generalized for all cases by replacing the spin quantum number, S, by the total angular momentum quantum number, J.

The hypothesis of the molecular field method is that interactions between A and B ions produce the same magnetizations within each sublattice as would external fields of magnitudes given by Eqs. 30 and 31. Hence the spontaneous magnetizations $M_A$ and $M_B$ are given by Eq. 32, where H is replaced by $H_A$ and $H_B$, respectively. In zero external magnetic field $M_A$ should be equal in magnitude to $M_B$, therefore, with the above restrictions on the constants A and $\Gamma$ we have that $M_A = -M_B$. From Eqs. 30, 31, and 32 we get

$$M_A = M_o B_S [ \mu (A-\Gamma) M_A/kT] \tag{33}$$

and

$$M_B = M_o B_S [ \mu (A-\Gamma) M_B/kT] \tag{34}$$

Each of these equations is identical with the corresponding result obtained for the ferromagnetic case with $A-\Gamma$ serving as the resultant molecular field constant. At absolute zero $M_A$ and $M_B$ are equal to $M_O$ which has the magnitude $Ng\mu_B S/2$ where N is the total number of magnetic ions per unit volume. As the temperature is increased the spontaneous magnetization decreases and finally reaches zero at the ordering temperature, or Néel point. It follows from Eq. 33 or 34 that

$$T_N = C(A-\Gamma)/2 \tag{35}$$

where

$$C = (Ng^2\mu_B^2/3k)S(S+1) \tag{36}$$

is the Curie constant.

The magnetic internal energy is given by

$$\begin{aligned} E_m &= -\frac{1}{2}(-AM_A \cdot M_B - \Gamma M_A \cdot AM_B \cdot M_A - \Gamma M_B \cdot M_B) \\ &= -(A-\Gamma)M_A^2 = -(A-\Gamma)M_B^2 \end{aligned} \tag{37}$$

The magnetic contribution to the specific heat is thus

$$C_m = \frac{dE_m}{dT} = -2(A-\Gamma)\frac{dM_A}{dT} \tag{38}$$

This leads to a discontinuity at the Néel point, but gives zero contribution above the Néel point, in contradiction to the experimentally observed specific heat tail at high temperatures. Moreover, below the Néel point the specific heat varies more rapidly with temperature than the molecular field theory predicts. These deviations are closely associated with the neglect of the energy associated with short range order mentioned in Part I, the Introduction.

The foregoing treatment leads to the familiar Curie-Weiss behavior of the magnetic susceptibility at high temperatures. Above the Néel point the spontaneous magnetization is zero, but an external magnetic field will induce a small magnetic moment. Using the fact that for small argument

$$B_S(x) \rightarrow (S+1) \, x/3S \quad \text{as } x \rightarrow 0 \tag{39}$$

one obtains an isotropic volume susceptibility which obeys the Curie-Weiss law, as follows

$$\chi = (M_A + M_B)/H = C/(T+\theta) \tag{40}$$

where C is given by Eq. 36 and $\theta = C(A+\Gamma)/2$. The Curie-Weiss $\theta$ is not equal to the critical temperature, $T_N$, but,

$$\theta/T_N = (A+\Gamma)/(A-\Gamma) \tag{41}$$

The value of the Curie-Weiss $\theta$ can be determined from experimentally measured susceptibilities at high temperature. Since the magnitude of A can be determined independently, as follows, it is then possible to determine experimentally the magnitude of the next-nearest neighbor interaction constant, $\Gamma$.

The susceptibility below the Néel point can also be derived using the molecular field model. So far we have not included anisotropy forces which will tend to select out one direction as the preferred direction for the alignment of $M_A$ and $M_B$. We will defer this for now and just assume that there is some preferred direction along which the sublattice moments will align parallel and antiparallel. There are two main cases of interest: (a) the applied field parallel to the preferred axis, and (b) the applied field perpendicular to this axis. In the first case the applied field changes the magnitude of the sublattice moments only. Van Vleck [3] has shown that in small fields the parallel susceptibility is given by

$$\chi_{||} = (\delta M_A + \delta M_B)/H = \frac{Ng^2 \mu_B^2 S^2 B_S'(y)}{k[T+(3T_c SB_S'(y))/(S+1)]} \tag{42}$$

where

$$y = [(A-\Gamma)M_S Sg\mu_B]/kT$$

and $B_S'(y)$ is the derivative of the Brillouin function with respect to y. $M_S$ is the magnitude of $M_A$ or $M_B$ in zero applied field. The parallel suscepti-bility given by Eq. 42 drops rapidly from its value at the Néel point to zero. The behavior of $\chi_{||}$ is shown in Fig. 6.

With the field perpendicular to the preferred axis the sublattice moments will be turned out of the preferred direction by the field. The rotation of the moments is opposed by the forces associated with the molecular fields, as well as by the neglected anisotropy forces. The situation is depicted in Fig. 7, where $AM_B$ and $AM_A$ are the molecular fields due to the B and the A sublattices, respectively. At equilibrium there will be no resultant torque acting on the ions, and we, therefore, have

$$H \cos\theta = AM_A \sin 2\theta = 2AM_A \sin\theta\cos\theta \qquad (43)$$

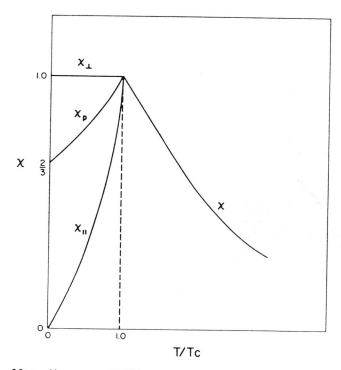

FIG. 6. Magnetic susceptibility against temperature for a simple anti-ferromagnetic substance.

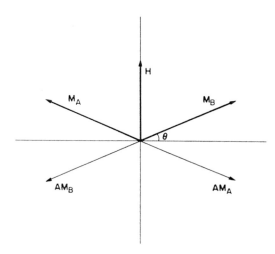

FIG. 7. The effect of a magnetic field, perpendicular to the preferred axis, on the sublattice moments.

and

$$H \cos\theta = AM_B \sin 2\theta = 2AM_B \sin\theta\cos\theta \qquad (44)$$

Now the perpendicular susceptibility will depend on the component of the magnetization parallel to the applied field, hence,

$$\chi_\perp = (M_A + M_B) \sin\theta / H = 1/A \qquad (45)$$

Thus the perpendicular susceptibility is independent of temperature as shown in Fig. 6. A measurement of $\chi_\perp$ gives a direct measure of the magnitude of the interaction constant A.

If the applied field makes an arbitrary angle $\phi$ with the preferred axis, then the susceptibility in the direction of the field $\chi_\phi$ is given by

$$\chi_\phi = \chi_{||} \cos^2\phi + \chi_\perp \sin^2\phi \qquad (46)$$

In a polycrystalline sample there will be crystals of all orientations relative to the field. By taking an average over all orientations one finds that $\chi_p = (1/3)\chi_{||} + (2/3)\chi_\perp$. Since $\chi_{||} = \chi_\perp$ at the Neel point and $\chi_{||} = 0$ at T = 0, this means that $\chi_p = (2/3) \chi_p$ at T = 0. This behavior of $\chi_p$ is also shown in Fig. 6.

A number of cubic antiferromagnetic compounds, such as MnO, MnS, $MnF_\perp$, and $FeF_\perp$ show a behavior somewhat similar to the molecular field predictions, but the temperature dependence of the susceptibilities above and below the Néel point are not in very good agreement with the results of the above treatment. Still the molecular field model is useful because of its simplicity and because it does contain, qualitatively at least, all of the essential features observed in antiferromagnetic crystals. Improvements in the model to bring it into better agreement with experiment have been introduced by many authors. A good discussion of the various effective field models is given by Smart [4].

Before proceeding with the introduction of the anisotropy energy and its effect on the susceptibility, it would be well at this point to discuss the relationship between the molecular field model presented so far and the microscopic quantum theory. Consider an isotropic system which can be approximately described by the Heisenberg Hamiltonian in which the spins are coupled with each other by exchange interactions. The exchange energy of a given ion i with its nearest neighbors is given by

$$V_{ex} = -2J \sum_j S_i \cdot S_j \tag{47}$$

where it is assumed that the exchange interaction between each pair of ions ($J_{ij}$) is the same. The essential assumption of the molecular field model is that the dipole moment of the z nearest neighbors,

$$m = \sum_j (g_j S_j \mu_B)/h \tag{48}$$

is constant and does not fluctuate from point to point in the crystal. The intensity of magnetization of the sample is

$$M = N\overline{m}/z \tag{49}$$

where $\overline{m}$ is the value of m averaged over all the groups of z ions in a unit volume and N is the number of magnetic ions per unit volume. Since the g-factor is the same for ions which differ only in their orientations we can write the exchange energy as

$$V_{ex} = -2J S_i \cdot \sum_j S_j = -2J S_i \cdot m/g\mu_B \tag{50}$$

Now since $m = \overline{m}$ we have

$$V_{ex} = \frac{-2Jz}{g^2 \mu_B^2 N} \mu_i \cdot M \tag{51}$$

where $\mu_i = g\mu_B S_i$ is the magnetic moment of the ith ion.

This expression can be compared to the energy of a dipole of moment $\mu_i$ in a hypothetical molecular field, $\gamma M$,

$$E_M = -\gamma \mu_i \cdot M \tag{52}$$

There is a direct correspondence with the molecular field constant $\gamma$ equal to

$$\gamma = (2Jz)/g^2 \mu_B^2 N \tag{53}$$

The above treatment can be extended to cover the case where ions outside the z nearest neighbor must be taken into account. If there are $z_1$ and $z_2$ neighbors in the nearest and next-nearest groups, respectively, with corresponding exchange constants $J_1$ and $J_2$, a similar treatment leads to

$$\gamma = \frac{2}{N\mu_B^2}\left[\frac{J_1 z_1}{g_1^2} + \frac{J_2 z_2}{g_2^2}\right] \tag{54}$$

with an obvious generalization for more distant interactions.

For a body centered cubic lattice with antiferromagnetic ordering of the type discussed above the two molecular field constants are given by

$$A = -32J_1/Ng^2\mu_B^2 \quad \text{and} \quad \Gamma = -24J_2/Ng^2\mu_B^2 \tag{55}$$

where we have used $N/2$ on each sublattice and $z_1$ and $z_2$ are 8 and 6, respectively. All of the relations previously expressed in terms of the molecular field constants could now be expressed in terms of the exchange constants $J_1$ and $J_2$.

The effects of anisotropy on the properties of antiferromagnets can be quite dramatic as was pointed out in Part 1. There are three kinds of anisotropic interaction which are of importance in magnetic crystals: magnetic dipolar interactions, anisotropic exchange, and crystalline field anisotropy. The relative importance in a given substance will vary and each will, in general, have a different effect on the properties of the system. We will leave this discussion open for the present and consider a phenomenological approach.

In order to simplify the notation below let us replace the sublattice notation (A and B) with the symbols + and -, so that $M_A$ and $M_B$ become $M^+$ and $M^-$. Further, let the direction cosines of $M^+$ be $\gamma_+$, $\beta_+$, and $\gamma_+$, with similar notation for the negative sign. The anisotropy energy is introduced in the simplest form, i.e., uniaxial case:

$$E_a = -K(\gamma_+^2 + \gamma_-^2)/2 \qquad (56)$$

and biaxial or orthorhombic case:

$$E_a = K_1(\beta_+^2 + \beta_-^2)/2 + K_2(\gamma_+^2 + \gamma_-^2)/2 \qquad (57)$$

More general forms could be used but in the case where $M^+$ and $M^-$ are almost antiparallel to each other, they reduce to the above forms. We further assume an isotropic g-factor and let the anisotropy energy be small compared to the exchange energy. This last criterion is not strictly fulfilled in a number of crystals which order at low temperatures, with the effect that highly anisotropic crystals will have to be treated as special cases. One would expect that the Ising model would be more appropriate to these cases.

Returning to the above case we now introduce an effective anisotropy magnetic field through the relation

$$E_a = -H_a^+ \cdot \delta M^+ - H_a^- \cdot \delta M^- \qquad (58)$$

where $\delta$ denotes the arbitrary variation of the quantity. It is assumed that the magnitudes of $M^+$ and $M^-$ are constant and only their orientations change. If we first consider the uniaxial case, we have

$$E_a = -K(\gamma_+ \delta\gamma_+ + \gamma_- \delta\gamma_-)$$

$$= -\Sigma_\pm (H_{ax}^+ \delta\alpha_+ + H_{ay}^+ \delta\beta_+ + H_{az}^+ \delta\gamma_+)M^+$$

and, hence,

$$H_{ax}^\pm = H_{ay}^\pm = 0, \quad H_{az}^\pm = (K/M^\pm)\gamma_\pm = (K/M_o)\gamma_\pm \qquad (59)$$

where $M_o$ is the magnitude of $M^\pm$ in zero external field.

In the biaxial case we obtain

$$H_{ax}^\pm = 0, \quad H_{ay}^\pm = -(K_1/M_o)\beta_\pm, \quad H_{az}^\pm = -(K_2/M_o)\gamma_\pm \qquad (60)$$

Let us consider the susceptibility first assuming uniaxial anisotropy. In this case the sublattice moments are in the direction of the axis of anisotropy, the z-axis, in zero field. For small fields applied along the z-axis the susceptibility measured along the z-axis is not effected by the anisotropy, so long as the field is less than the critical field mentioned in Part 1. However with H applied perpendicular to the z-axis there is a correction to the perpendicular susceptibility. The torque acting on, say, $M^+$ including the anisotropy field is $M^+ \times (H+H_e+H_a)$ which must vanish in equilibrium. This gives

$$\chi_\perp = 1/(A+K/2M_o^2) \tag{61}$$

which may be compared to Eq. 45 when the anisotropy is neglected. When H is applied along the z-axis and $H>H_c$ the sublattice moments are more or less perpendicular to the z-axis, if H is not much larger than $H_c$. In this case one gets

$$\chi_{sf} = 1/(A-K/2M_o^2) \tag{62}$$

where $\chi_{sf}$ is the susceptibility measured along the z-axis, but with the system in the spin-flop state.

In case of biaxial anisotropy the two perpendicular directions are not equivalent and each must be considered in turn. In this case we consider the x-axis to be the preferred axis, so that in zero field the sublattice moments are in the + and - x-directions. For small fields applied along the y-axis one obtains, using the same technique as above,

$$\chi_\perp = 1/(A+K_1/2M_o^2) \tag{63}$$

Similarily, for H applied along the z-axis one has

$$\chi_\perp = 1/(A+K_2/2M_o^2) \tag{64}$$

Now for H applied along the x-axis and $H>H_c$ there are two possibilities. Depending on the relative magnitudes of $K_1$ and $K_2$ the sublattice moments will either be in the xy plane or the xz plane. If the moments lie in the xy plane one has

$$\chi_{sf} = 1/(A-K_1/2M_o^2) \tag{65}$$

and if they lie in the xz plane one has

$$\chi_{sf} = 1/(A-K_2/2M_o^2) \tag{66}$$

The above predictions of the molecular field model of antiferromagnetism are in qualitative agreement with a number of simple antiferromagnetic substances. However, it falls far short of giving a complete description of antiferromagnetism. When the temperature dependence of the magnetization, susceptibility and specific heat, for instance, are compared to experiment rather serious discrepancies are at once apparent. Despite this fact molecular field models are helpful, in that they are rather simple to understand and to apply.

## B.  Spin Wave Theory

The success of the spin wave theories in predicting the correct low temperature behavior of magnetic materials has led to considerable effort in this area. The basic techniques and results of modern spin wave theories are well described in most recent books on magnetism and will not be discussed here. It is of interest, however, to compare some of the spin wave results to those obtained above using the molecular field theory.

The application of spin wave theory to the case of antiferromagnetism is described quite well in articles by Anderson [5], Kubo [6], and Oguchi [7]. Anderson and Callen [8], using a temperature-dependent Green function approach, with an appropriate choice of decoupling procedure, obtained results similar to those of Oguchi.

In small fields near absolute zero Anderson and Callen [8], using a random phase approximation (RPA), found for $S \geq 1/2$,

$$\chi_\perp = \mu^2 N/2zJ \tag{67}$$

where $\mu = g\mu_B$. A comparison with Eq. 45 shows that the RPA gives the same result as the molecular field theory when anisotropy is neglected. On the other hand when the so-called Callen decoupling (CD) scheme is employed [9] for $S \geq 1/2$ they obtain

$$\chi_\perp = \frac{\mu^2 N}{2zJ} [1 - a/2S - bT^2/S^3 + \dots] \tag{68}$$

The leading temperature dependent term agrees with the spin wave result of Oguchi except for the magnitude of the constant b. Experimental results on a number of compounds do show a slight temperature dependence of $\chi_\perp$ which agrees reasonably well with the above result.

The parallel susceptibility for small fields and low temperatures is given, for $S \geq 1/2$ by Oguchi as

$$\chi_{||} = (\mu^2 N/zJS) [c(1-d/S)T^2/S^2 + \dots] \tag{69}$$

For $S > 1/2$ Anderson and Callen obtain an identical result with CD, but not for $S = 1/2$. For RPA their result is incorrect for all S.

For a review of the more standard results of spin wave theory the reader is referred to any recent text on magnetism.

## II.  MAGNETIC PHASE BOUNDARIES

A typical H–T phase diagram for antiferromagnetism with the external field along the preferred axis was shown in Fig. 5. In Part 1 a brief description of the three distinct phases was given. The low temperature, low field phase has the sublattice moments aligned antiparallel to one another and along the preferred direction. At $H_{cl}$ the moments flip over almost perpendicular to the preferred direction, but with the sublattice moments still more or less antiparallel to one another. As the external field is increased further, the moments tilt more and more toward the preferred axis by the action of the field. When the field is large enough to completely overcome the exchange interaction a transition to a strongly paramagnetic state occurs.

### A.  Molecular Field Theory

Let us now review the details of such a phase diagram in terms of a molecular field approach and spin wave theory. In the former case the results quoted below follow primarily from the work of Nagamiya et al. [10], and Gorter and Peski-Tinbergen [11], while the spin wave theory has been considered in detail by Anderson and Callen [8], Anderson [5], Kubo [6], Wang and Callen [12], and Feder and Pytte [13]. Consider first the case of the external field along the preferred direction at T=0 for a simple two sublattice antiferromagnet with uniaxial anisotropy. As the external field is increased the anisotropy will keep the sublattice moments aligned along the preferred axis. However, above a critical field given by Eq. 11 the free energy is minimized if the moments are more or less perpendicular to the preferred axis. From Eq. 56, which defines the anisotropy constant K, and Eq. 53, 62, and 11 we get for the critical field

$$H_{cl} = [\, 2K/(\chi_{\perp} - \chi_{||})\,]^{\frac{1}{2}}$$

$$= [\, 2K(A + K/2M_o^2)\,]^{\frac{1}{2}} \text{ at } T = 0$$

$$= [\, 2K(2zJ/g^2\mu_B^2 N + K/2M_o^2)\,]^{\frac{1}{2}} \tag{70}$$

We can express this in a somewhat simpler form if we define an effective exchange field, $H_e$, as

$$H_e = 2zSJ/g\mu_B \tag{71}$$

Using Eq. 59 and the fact that $M_o = g\mu_B SN$, we can write

$$H_{cl}^2 = 2H_e H_a + H_a^2 \text{ at } T=0 \tag{72}$$

For T>0 if follows from Eq. 70 that

$$H_{cl}^2 = (2H_e H_a + H_a^2)/(1 - \chi_{//}/\chi_\perp) \tag{73}$$

At T=0, according to the molecular field approach the net magnetization in the direction of the field (preferred direction) is zero for all fields up to $H_{cl}$. Above $H_{cl}$ the sublattice moments are not quite antiperpendicular to the preferred direction. The applied field has partially overcome the exchange interaction leaving the moments tilted somewhat toward the preferred direction. There is, thus, a net magnetization in the direction of the field. In going from the antiferromagnetic to the spin flop state across the phase boundary $H_{cl}$ there has been an increase in the magnetization in the preferred direction and thus the transition is of first order.

In the biaxial case with $K_2>K_1$ Eq. 73 still gives the correct expression for the critical field if we replace K by $K_1$ in the expressions for $H_a$ and $\chi_\perp$. For a discussion of the more general case of the applied field in an arbitrary direction relative to the crystalline directions the reader is referred to Nagamiya et al. [10].

For $H>H_{cl}$ the sublattice moments tilt more and more toward the preferred direction as the field is increased. When the field is large enough to completely overcome the exchange energy the moments are all aligned along the preferred direction and a strongly paramagnetic state results. The transition form the spin flop state to the paramagnetic state is higher than first order, but a distinct, temperature dependent phase boundary separates the two states. At T=0 Nagamiya et al. [10] have shown that the transition from the spin flop state to the paramagnetic state will occur at a field given by

$$H_{c2} = 2H_e - H_a \tag{74}$$

where $H_a = K/M_0$ for uniaxial anisotropy, and $H_a = K_1/M_0$ in the biaxial case. According to their treatment of the molecular field theory the critical field, $H_{c2}$ decreases in proportion to $T^{\frac{1}{2}}$ as the temperature increases. In most cases $H_{cl}$ increases slightly with increasing temperature, hence for some temperature, $T_t$, less than $T_N$ we have $H_{cl}(T_t)=H_{c2}(T_t)$. For temperatures between $T_t$ and $T_N$ and for fields less than $H_{cl}(T_t)$ there is a second order phase boundary separating the antiferromagnetic state from the paramagnetic state. This boundary terminates at $T_N$. At $T_t$ there is a triple point where all three phases are in equilibrium.

When the field is applied perpendicular to the preferred direction, for all temperatures less than $T_N$ there is but a single transition from the antiferromagnetic state directly to the paramagnetic state. The sublattice moments are perpendicular to the applied field in the limit of zero field. As the field is increased the moments tilt toward the applied field. Using the approach of Nagamiya et al. [10], it follows that at $T=0$ the phase boundary will occur at a field, $H_c$, given by

$$H_{c_\perp} = 2H_e + H_a \qquad (75)$$

where $H_a = K/M_0$ in the uniaxial case, and for the biaxial case $H_a$ is either $K_1/M_0$ for the field along the y-axis, or $K_2/M_0$ for the field along the z-axis. Since it was assumed that $K_2 > K_1$, the z-axis is the least preferred axis.

If the values of all the critical fields mentioned above can be experimentally determined as T approaches zero, then by using Eqs. 72, 74, and 75 the exchange constant and the anisotropy constant(s) can be uniquely determined for $T=0$.

The molecular field theory described here has some fairly serious drawbacks. It is basically a phenomenological macroscopic theory which leads to a rather good qualitative description of magnetism. However, it has nothing to say about the truly quantum aspects of magnetism, hence experimental results are often found to be in quite serious disagreement with the predictions of the molecular field theory. On the other hand, some of the more recent treatments using the basic molecular field hypothesis give the experimentalist a very useful tool for the analysis of a wide variety of experimental data.

## B. Spin-Wave Theory

The quantum mechanical spin-wave theory is particularly applicable at very low temperatures where the lowest energy excitations in the magnetic system are simple quantized spin waves, or magnons. For instance, the experimentally determined behavior of such properties as specific heat, spontaneous magnetization and ferromagnetic and antiferromagnetic resonance are quite well understood in terms of spin-wave theory at low temperatures.

For the case in hand, the field induced phase transitions in antiferromagnets, we need to investigate the field dependence of the spin-wave dispersion relations. It is well known that for isotropic ferromagnets the spin wave energy depends quadratically on the wave vector for small wave vectors. Anisotropy can produce an energy gap in the spin wave energies. For isotropic antiferromagnets in zero field Anderson [5] showed that the spin-wave energy is

$$\hbar\omega(k) = 2zJS(1-\gamma_k^2)^{\frac{1}{2}} \tag{76}$$

where k is the wave vector and

$$\gamma_k = (1/z)\Sigma e^{ik\cdot\rho} = (1/z)\Sigma e^{i\lambda} \tag{77}$$

In the previous equation $\rho$ represents the vectors from a lattice point to its nearest neighbors, and the sum extents over all nearest neighbors. For crystal lattices of high symmetry, for instance the cubic lattices, Kubo [6] and others have shown that the above summation can be expressed as a simple D term summation of terms like $\cos(\lambda_j)$, where D is the dimensionality of the lattice and $k_j$ is the ith component of the wave vector k.

For small wave vector $\gamma_k \cong 1-\lambda^2$, and Eq. 76 reduces to

$$\hbar\omega(k) = 2zJS\lambda \tag{78}$$

Thus for isotropic antiferromagnets the dispersion is linear in k for small k in contrast to the quadratic behavior for ferromagnets. Anisotropy, however, will again lead to an energy gap at k=0 in the dispersion.

In the case of uniaxial single ion anisotropy the spin wave energy can be expressed by the relation

$$\hbar\omega(k) = 2zJS(\eta^2-\gamma_k^2)^{\frac{1}{2}} \tag{79}$$

with

$$\eta = 1+L\xi^2/zJ \tag{80}$$

In the last equation $\xi^2 = 1-1/2S$ and L is the anisotropy constant defined by the anisotropy term in the Hamiltonian

$$\mathcal{H}_a = -L[\Sigma_\alpha S_z^2(\alpha) + \Sigma_\beta S_z^2(\beta)] \tag{81}$$

where the sums are extended over each sublattice.

Feder and Pytte [13] have extended the work of earlier workers by treating spin-wave interactions to higher order and find that results such as Eq. 79 should include higher order corrections. However, the main features of the theory are not altered significantly, so for the sake of clarity the corrections will not be included in what follows.

Due to the identical sublattices the dispersion relation (Eq. 79) is doubly degenerate and is defined only over half of the standard k space. An external field splits this degeneracy since

$$\hbar\omega(H) = \hbar\omega(k) \pm \mu H \tag{82}$$

As the field is increased the k=0 mode of the lower branch will reach zero energy at a field such that $\mu H = \hbar\omega(k=0)$. The antiferromagnetic state becomes unstable at this point. Using an argument similar to that used earlier it can be shown that the spin flop state is the stable state for fields greater than this value but smaller than the exchange field.

For k=0 Eq. 79 reduces to

$$\hbar\omega(0) = \mu H_{cl}^{a} = 2S[L\xi^{2}(2zJ+L\xi^{2})]^{\frac{1}{2}} \tag{83}$$

This value of the critical field which locates the phase boundary between the antiferromagnetic and the spin-flop states is identical to Eq. 72 obtained from molecular field theory, provided we define an effective anisotropy field $H_a = 2SL\xi^2/\mu$.

Within the spin-flop state there are still two sublattices and the two dispersion curves, obtained by Wang and Callen [12], as well as by Feder and Pytte [13], are given by

$$\hbar\omega(k) = 2zJS[1\bar{+}\gamma_k + \xi(1-\xi)u^2L/2zJ]^{\frac{1}{2}}$$

$$[1\bar{+}\gamma_k(1-2u^2) - \xi(1-\xi)u^2L/2zJ]^{\frac{1}{2}} \tag{84}$$

where u=sin$\theta$. The angle $\theta$ is the angle between the sublattice moments and the external field H and is given by

$$\cos\theta = \mu H/2S(2zJ-\xi^2L) \tag{85}$$

Within the spin-flop state if the field is decreased the phase boundary between the spin flop and the antiferromagnetic state is reached when the k=0 mode of the lower branch of the dispersion curves reaches zero energy. Wang and Callen [12] found that this occurs at a critical field given by

$$\mu H_{cl}^{f} = 2S(2zJ-\xi^2L) \left[ \frac{(1+\xi)L}{4zJ+\xi(1+\xi)L} \right]^{\frac{1}{2}} \tag{86}$$

Since $H_{cl}^{f} < H_{cl}^{a}$ there is a predicted hysteresis effect, similar to the familiar supercooling and superheating effect associated with the liquid–solid first-order transition.

Feder and Pytte [13] also found an identical result for the case of single-ion anisotropy. However, they also treated the case of anisotropic exchange. They consider an anisotropic exchange hamiltonian of the form

$$\mathcal{H}_a = K \sum_{(\alpha,\beta)} S_z(\alpha) S_z(\beta) \tag{87}$$

where the sum is over all nearest neighbor pairs, and K is the anisotropy constant. The dispersion relations reflect the different form of the anisotropy hamiltonian and lead to no hysteresis at the antiferromagnetic spin-flop phase boundary. They obtain

$$\mu H_{cl}^a = \mu H_{cl}^f = 2Sz[K(J+K/4)]^{\frac{1}{2}} \tag{88}$$

Experimentally it might be quite difficult to observe any hysteresis, because of impurities and inhomogenities. In one case of note, $MnCl_2 \cdot 4H_2O$ [14], differential susceptibility measurements were taken going through the transition in both directions with no evidence of hysteresis.

In the paramagnetic phase there is no distinction between the two sublattices and a single dispersion relation over the entire Brillouin zone results. According to Feder and Pytte [13] this can be expressed as

$$\hbar\omega(k) = \mu H - 2zSJ(1-2L\xi^2/2zJ-\gamma_k) \tag{89}$$

Typical dispersion curves for the three phases are shown in Figs. 8, 9, and 10, for the k-vector in the (111) direction of a cubic lattice.

From Fig. 10 it is observed that if the spin flop phase is approached from the paramagnetic phase the transition occurs when the large k portion of the curve first reaches zero energy ($\gamma_k = -1$). Thus at T-0 the upper critical field is given by

$$\mu H_{c2} = 2S(2zJ-\xi^2 L) \tag{90}$$

If the transition is approached from the spin-flop side the critical field is determined by calculating the field for which the angle $\theta \rightarrow 0$, where $\theta$ was defined by Eq. 85. By inspection of Eq. 85 it is seen that one gets an identical result to Eq. 90. The transition is higher than first order with no hysteresis. It is further seen that Eq. 90 is identical to Eq. 74 obtained from the molecular field theory if we let $H_a = 2S\xi^2 L/\mu$.

For T>0 Anderson and Callen [8] find that the upper critical field, $H_{c2}(T)$ decreases with increasing temperature as $T^{3/2}$. This comes about because the spin-wave energies are renormalized by the magnetization at finite temperatures. This is in contrast with the well known Dyson [15] result for ferromagnets, where the spin wave energies are renormalized by energy, leading to a $1-cT^{5/2}$ dependence at low temperatures for the spontaneous magnetization. The $T^{3/2}$ dependence also does not agree with the molecular field results of Nagamiya et al. [10], which predicts $T^{1/2}$ dependence.

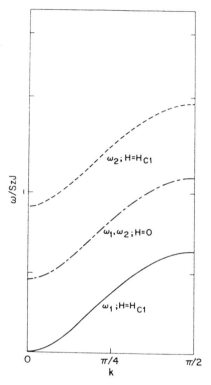

FIG. 8. Typical spin-wave spectra for the antiferromagnetic phase at T=0.

## III. EXPERIMENTALLY DETERMINED PHASE BOUNDARIES

The experimental verification of the predicted phase transitions in antiferromagnetic materials is hampered, in many cases, by the large exchange energy. In order for the external field to completely overcome the antiferromagnetic exchange one needs a field of the order of $H=kT_N/\mu$. For $S=1/2$ this means $H \sim 7 \times 10^3 T_N$ Oe. For reasonable laboratory fields this limits $T_N$ to the liquid helium or liquid hydrogen temperature range. The spin flop critical field is approximately proportional to $(H_e H_a)^{1/2}$ so that if $H_a$ is considerably less than $H_e$ this effect can be observed in crystals with somewhat higher Néel temperatures.

The first reported observation of the spin-flop transition was in $CuCl_2 \cdot 2H_2O$ by Van den Handel et al. [16]. They found that, for the field along the a-axis at a temperature of about 1.5 K, there was an almost discontinuous increase in the magnetization at a field of about 6.5 kOe. Further studies revealed

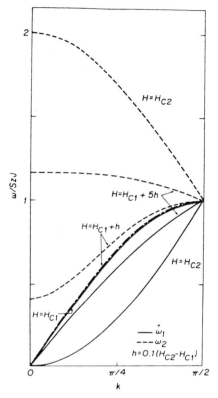

FIG. 9. Typical spin-wave spectra for the spin-flop phase at T=0.

that the magnetic susceptibility above 6.5 kOe had almost the same value as the perpendicular susceptibility at zero field, as predicted. The slight difference is due to the anisotropy as can be seen by comparing Eq. 61 and 62.

CuCl$_2$·2H$_2$O was also the first substance in which antiferromagnetic resonance was observed. Spin waves of infinite wavelength (k=0) can be excited by electromagnetic radiation of frequency equal to the spin-wave frequency. In ferromagnetic substances the k=0 spin-wave energies are usually quite low, since the energy gap is due primarily to anisotropy. Ferromagnetic resonance has long been a useful tool for the investigation of the dynamic behavior of ferromagnetic materials. In antiferromagnetic materials, on the other hand the energy gap for k=0 is roughly proportional to $(H_e H_a)^{1/2}$ as seen from Eq. 79 and 83. Thus the zero field resonant frequencies for materials with relatively high Néel temperatures are found to be in the extreme microwave region or even in the far infrared. In order

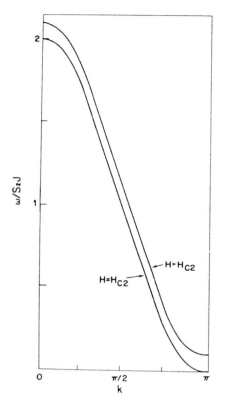

FIG. 10.  Typical spin-wave spectra for the paramagnetic phase at $T=0$.

to reduce the resonant frequencies to reasonable microwave frequencies very large magnetic fields would be required for these substances. However, for materials like $CuCl_2 \cdot 2H_2O$ resonant frequencies are found to be in the millimeter or submillimeter range. Ubbink [17] was the first to report such measurements in $CuCl_2 \cdot 2H_2O$. Since that time measurements have been made on $MnF_2$ (18), $Cr_2O_3$ (19), as well as many other materials which order antiferromagnetically at low temperatures. Foner [20] presented an excellent review of the subject together with the experimental results up to 1963. In a recent book by White [21] the subject is treated from a more modern quantum approach.

## A.  $CoCl_2 \cdot 6H_2O$

In recent years a number of antiferromagnetic compounds with ordering temperatures in or near liquid helium temperatures have been studied in external fields. Monoclinic $CoCl_2 \cdot 6H_2O$ was investigated by Van der Lugt and Poulis [22] using magnetic resonance techniques, and by Date [23] using electron resonance. Earlier susceptibility measurements of Flippen and Friedberg [24] and specific heat results of Robinson and Friedberg [25] verified that the zero-field ordering temperature, $T_N$, is 2.29 K. The c-axis is the preferred axis in this crystal.

The transition from the paramagnetic to the antiferromagnetic state is characterized by a sudden disappearance of the proton magnetic resonance lines that are typical of the paramagnetic state. With fields varying from 0-10 kOe directed along the c-axis, Van der Lugt and Poulis were able to determine the paramagnetic to antiferromagnetic phase boundary, which terminates at about 7.8 kOe and 2.05 K. Above 7.8 kOe they were able to map out part of the spin flop to paramagnetic boundary.

At a field $H=H_{c1}$ it is possible to observe electron magnetic resonance at (or near) zero frequency. Thus the spin flop critical field can be determined by measuring the field at which zero frequency electron resonance occurs. The rf frequencies used in ordinary proton magnetic resonance are ideal for these observations, since they are very close to zero frequency compared to the microwave frequencies required to observe electron magnetic resonance at zero field. Using the same apparatus as for the above proton work Van der Lugt and Poulis [22] were able to locate the spin flop critical field from 1.0 K up to the triple point of about 2.05 K. The critical field increased with temperature almost linearly from 7.14 kOe to 7.8 kOe. Schmidt and Friedberg [26] confirmed the above results by direct isothermal magnetization measurements over essentially the same temperature interval as the resonance work.

The antiferromagnetic to spin-flop phase boundary has also been investigated using an adiabatic method for observing spin flop. The variation of the temperature of an adiabatically isolated substance with magnetic field is related to the substance's magnetic susceptibility by [27]

$$\left(\frac{\partial T}{\partial H}\right)_S = -\frac{TH}{C_H} \left(\frac{\partial \chi}{\partial T}\right)_H \tag{91}$$

where S is the entropy and $C_H$ is the specific heat at constant field. For an antiferromagnet the right hand side of Eq. 91 is negative for $H<H_{c1}$ if H is along the preferred direction, since then $\chi_{||}$ must be used in the equation. On the other hand, for $H>H_{c1}$ the appropriate susceptibility is $\chi_\perp$ (or more precisely $\chi_{SF}$), and the right hand side of Eq. 91 is very close to zero. Thus as the field is increased adiabatically from zero the temperature will at first decrease, then as the spin-flop phase boundary is crossed it becomes independent of temperature. Measurements of this sort should make it reasonably easy to pick out the critical field. Actually in most cases the situation is not quite that simple. The phase transition is of first order, hence there is a latent heat associated with the transition.

In every case observed so far $dH_{c1}/dT$ is positive so that there should be a discontinuous increase in temperature at the phase boundary.

This discontinuity, however, occurs as a function of internal field, not external field, and sample shape becomes important. For a homogeneous ellipsoidal sample one has

$$H_i = H_o - NM \qquad (92)$$

where $H_i$ is the internal field, $H_o$ the external field and N is the classical demagnetizing factor. For an infinitly long needle magnetized along its length N=0, and only in this case is $H_i = H_o$. For other shapes the internal field must be calculated from Eq. 92. One of the effects of this shape dependence is to spread out the first order transition over a range of external fields. The extent of this spread depends on the magnitude of N and the magnetization of the substance. A more detailed discussion of the importance of the shape on the observed thermodynamic behavior of anti-ferromagnetic substances in external fields is given in a recent paper by Landau et al. [28] in which experimental results for the Ising-like antiferro-magnet, Dysprosium Aluminum Garnet (DAG), are presented. Additional papers on the subject of a more theoretical nature are in preparation by the same authors.

In the region of the spin flop transition the magnetization is increasing rapidly with external field. According to Eq. 92 it is entirely possible for the internal field to remain constant over a range of external fields. In the case of DAG the internal field is found to be constant over about 3.5 kOe at a temperature of 1.14 K. In most of the hydrated transition element halides, on the other hand, the systems are less magnetically dense and the effect will be limited to external field changes of the order of 50-500 Oe. Indeed, in all cases where thermal techniques have been used to observe the spin flop transition in these compounds the transition does appear to be spread out over external fields of the above range.

McElearney et al. [29] used this adiabatic technique to study spin flop in $CoCl_2 \cdot 6H_2O$. Their determination of the spin flop critical field in the range 1.2 to 1.7 agreed quite well with the previous work of Van Der Lugt and Poulis [22] and Schmidt and Friedberg [26]. A plot of temperature against external field did not reveal a discontinuous increase in temperature at the spin-flop transition, but a more gradual increase. No attempt was made to plot the data against internal field.

## B. $CoBr_2 \cdot 6H_2O$

$CoBr_2 \cdot 6H_2O$ is isomorphic to $CoCl_2 \cdot 6H_2O$ with an antiferromagnetic ordering temperature of 3.2 K. Using the adiabatic technique McElearney et al., [29] mapped out $H_{c1}$ from 1.2 K to the triple point which was found to be 2.91 K. They found that $H_{c1}$ could be fit very well to the equation

$H_{c1}$ = 7523 - 230.7T + 292.5T$^2$ Oe. The antiferromagnetic to paramagnetic phase transition was also determined by specific heat measurements in fields up to 10 kOe.

## C.  $MnCl_2 \cdot 4H_2O$

Both $H_{c1}$ and $H_{c2}$ were determined, down to 0.25 K, in monoclinic $MnCl_2 \cdot 4H_2O$ by Rives [30] using differential susceptibility measurements. Since the differential susceptibility $\chi = dM/dH$, is the slope of the magnetization as a function of field, it is a particularity sensitive method for locating phase boundaries. At the first order spin flop transition there is a discontinuity in the magnetization as a function of internal field. According to Landau et al. [28], as a function of external field the susceptibility should rise sharply to a value of $1/N$, then remain constant over a range of field which depends on the magnetization and on the value of N.

The original susceptibility measurements on $MnCl_2 \cdot 4H_2O$ were taken on unshaped flat platelets for which no exact demagnetizing corrections were possible. Fig. 11 shows the data taken at a temperature of 0.47 K. The spin flop transition is located by the sharp peak in the susceptibility. No evidence of a flattened portion at the top of the peak is obtained with this crystal. Above the spin flop transition the value of the susceptibility reflects the fact that the sublattice moments are aligned almost perpendicular to the field direction. The susceptibility remains almost constant up to about 19.5 kOe, where there is very sudden drop to zero.

The field at which the slope of the susceptibility changes discontinuously is taken as $H_{c2}$, the transition from the spin flop state to the paramagnetic state. Above about 21 kOe complete saturation of the spin system occurs giving zero differential susceptibility. Actually with higher resolution there is evidence of a small peak in the susceptibility at $H_{c2}$.

A plot of the critical fields, $H_{c1}$ and $H_{c2}$, obtained in this manner as a function of temperature is given in Fig. 12. Gijsman et al. [31], determined $H_{c1}$ down to 1.0 K using zero frequency electron resonance, and obtained identical results, within experimental error, to those shown in Fig. 12. McElearney et al. [29], observed the onset of spin flop by their adiabatic magnetization technique down to 0.85 K. The adiabatic cooling stopped at fields consistent with those obtained for $H_{c1}$ by Rives [30], but no sharp transition was observed. The relatively low Néel temperature of $MnCl_2 \cdot 4H_2O$ (1.62 K) makes the observation of the spin flop transition above 1.0 K rather difficult. Better adiabatic results should be possible at lower temperatures.

The temperature dependence of the spin flop critical field could be fit quite well by the relation $H_{c1}$ = 7550+434T Oe. Up to about 0.7 K the data of Rives [30] for the spin flop to paramagnetic critical field could be fit to the relation $H_{c2}$ = 20.6(1-0.288T$^n$), with n = 1.82. According to the theory

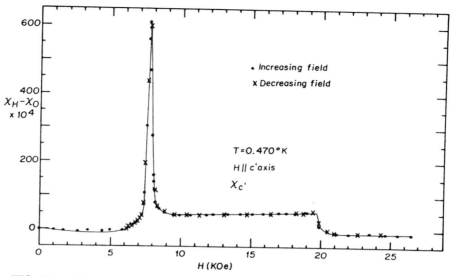

FIG. 11. Differential susceptibility along the preferred direction against field for $MnCl_2 \cdot 4H_2O$ at T=0.47 K. (Reprinted from Ref. 30, p. 492, by courtesy of Am. Phys. Soc.)

of Anderson and Callen [8] the value of n should be 1.5. It is not known whether higher order spin wave interactions would change this value sufficiently to be consistent with experiment.

Forstat et al. [32], studied the effect of deuteration on the exchange interaction by looking at the magnetic phase diagram of $MnCl_2 \cdot 4D_2O$. The Néel temperature was found to be about 2% lower than for the hydrated form. The triple point was 1.19 K compared to 1.3 K for $MnCl_2 \cdot 4H_2O$, and the spin flop critical field was found, by adiabatic measurements, to follow the relation $H_{cl} = 6196 + 114.8T$ Oe.

### D. $MnBr_2 \cdot 4H_2O$

$MnBr_2 \cdot 4H_2O$ undergoes an antiferromagnetic transition in zero field at 2.13 K and should be expected to behave similiarily to $MnCl_2 \cdot 4H_2O$. A number of studies have been made in fields including, specific heat [26,33, 34,35], susceptibility [26, 31], optical absorption [36] and antiferromagnetic resonance [37] experiments. The latter two experiments indicated spin flopping near 9 kOe between 1.2 and 1.4 K, but specific heat and susceptibility measurements do not substantuate that result. Adiabatic measurements [29] also show no evidence of spin flopping down to 0.9 K. Differential susceptibility measurements [38] down to 0.3 K indicate that the

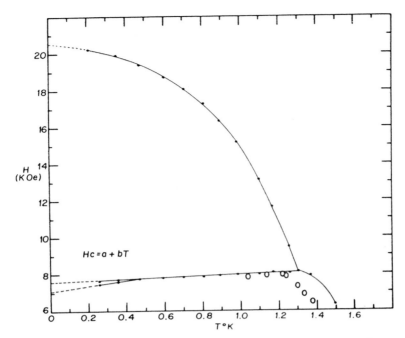

FIG. 12.  H–T phase diagram for $MnCl_2 \cdot 4H_2O$; Hc = a+bT; · = Present work and o = Gijsman et al.; a = 7550 ± 10 Oe and  b = 434 ± 10 Oe. (Reprinted from Ref. [30], p. 493, by courtesy of Am. Phys. Soc.)

behavior of $MnBr_2 \cdot 4H_2O$ is not as simple as that of the chloride.  It is quite possible that the magnetic structure of the bromide is different from that of the chloride and that measurements have been taken along the wrong direction.

There has long been a question in the case of the chloride as to just exactly what the preferred direction is.  $MnCl_2 \cdot 4H_2O$ is monoclinic with the angle $\beta$ = 99°25'.  Thus the direction perpendicular to the ab-plane (c') is a little over 9° from the crystalline c-axis.  A number of researchers have quoted the preferred direction as the c-axis, whereas others have claimed the c' direction is preferred.  Recent neutron diffraction measurements by Altman et al. [39] indicate that the preferred direction is between the c and the c' directions, about 2° from the c' direction. Neutron diffraction studies on the bromide might clear up the question of the unusual behavior of that compound.

### E.  $CuCl_2 \cdot 2H_2O$

Poulis and Hardeman [40] located the spin flop critical field in $CuCl_2 \cdot 2H_2O$ with NMR measurements.  Butterworth and Zidell [41] verified their results using the adiabatic magnetization technique mentioned earlier. $CuCl_2 \cdot 2H_2O$ is orthorhombic with a Néel temperature of 4.357 K. Using specific heat measurements Butterworth and Zidell were also able to locate the antiferromagnetic to paramagnetic phase boundary and the spin flop to paramagnetic boundary up to 20 kOe. They find the triple point to be 4.31 K at 8.50 kOe. At the lowest temperature measured, 1.3 K they find $H_{c1} =$ 6.5 kOe. Below about 2.0 K the adiabatic measurements on this crystal did produce an almost discontinuous increase in temperature as expected for a first order transition.

## F.  $FeCl_2 \cdot 4H_2O$

$FeCl_2 \cdot 4H_2O$ was also investigated using the adiabatic magnetization technique [29]. It has a monoclinic structure and orders antiferromagnetically at 1.1 K in zero field. Because of a limiting starting temperature of about 1.0 K the results only extend slightly below the triple point which was found to be about 0.76 K at a field of about 5500 Oe. Specific heat measurements verified that the antiferromagnetic to paramagnetic phase boundary extends from 1.1 K in zero field to 0.76 K at 5500 Oe. Up to 10 kOe the spin flop to paramagnetic phase boundary appears to be independent of temperature. It is of interest to note that the ratio of the triple point temperature to the zero field Néel temperature, $T_t/T_N$, is considerably greater than in the previously discussed crystals.

A number of anhydrous compounds have also been investigated recently. In general these compounds are more magnetically dense than the hydrated compounds and therefore have a larger exchange interaction which leads to a higher Néel temperature. This makes a complete investigation of the phase boundaries difficult with reasonable laboratory magnets. Fields of 100 - 150 kOe are fairly common today with the advent of high field superconducting magnets. Conventional Bitter magnets and pulsed magnets are also capable of achieving fields in this range and higher.

## G.  $MnF_2$

$MnF_2$ is probably the most thoroughly investigated antiferromagnet of all. It appears to be a classic example of a simple uniaxial antiferromagnet. $MnF_2$ has a tetragonal lattice and orders antiferromagnetically at 67.4 K with the preferred direction being the c-axis. The spin flop transition has been investigated by Shipira et al. [42] using ultrasonic and differential magnetization measurements, and by Shipira and Zak [43] using ultrasonic techniques. The latter authors observed a sharp peak in the ultrasonic attenuation at 93 kOe at a temperature of 4.2 K. They attributed this to spin flopping. Shipira et al. [42], measured dM/dH with pulsed fields along the c-axis up to 200 kOe. Their measurements extend from 4.2 K up to the triple point which was found to be 65.1 K at a field of 120±4 kOe. At 4.2 K and 20.3 K ultrasonic attenuation measurements gave values of $H_{c1}$ slightly higher than those obtained from the magnetization data, but no explanation was given for the discrepency. The antiferromagnetic to paramagnetic phase boundary was also located using both techniques. The phase boundary was found to fit the relation $T_N - T = DH^2$, with $D = (1.65\pm0.15) \times 10^{-10}$ deg/Oe$^2$. Standard molecular field theory predicts such a $H^2$ dependence, but with a considerably smaller constant D. Shipira et al. [42] found much better agreement with their data by starting with Fisher's [44] relation which connects the magnetic contribution to the specific heat at zero field, and the temperature dependence of the susceptibility $\chi_{||}$ for the field along the preferred axis, viz.,

$$C = Ad(\chi_{||}T)/dT \tag{93}$$

where A is a slowly varying function of T near $T_N$. Skalyo, et al. [45] have shown that near $T_N$ the antiferromagnetic to paramagnetic phase boundary satisfies the relation

$$d^2T/dH^2 = -A^{-1} \tag{94}$$

Using the specific heat data of Teaney [46] and the susceptibility data of Foner [47], Shipira et al. obtained $D = (1.56\pm0.16) \times 10^{-10}$deg/Oe$^2$ in good agreement with their experimental value.

### H.  BaMnF$_4$

Spin flopping has also been observed in BaMnF$_4$ by Holmes et al. [48] at temperatures up to 20 K. BaMnF$_4$ has an orthorhombic structure and undergoes antiferromagnetic ordering at about 25 K. The magnetization was measured from 1.4 to 20 K in fields up to 20 kOe along the preferred b-axis. The characteristic sharp increase in magnetization was observed at a field of 10.4 kOe at 1.4 K. The triple point was not precisely determined, however at 20 K there was still evidence of a spin flop transition at a field of 14.5 kOe, so it is presumed that the triple point is only slightly higher.

### I.  GdAlO$_3$

Blazey and Rohrer [49] observed antiferromagnetic ordering in GdAlO$_3$ below 3.89 K by susceptibility measurements. GdAlO$_3$ appears to behave very much like a simple two sublattice uniaxial antiferromagnet. The crystal structure is orthorhombic with the b-axis being the preferred direction for spin alignment. Cashion et al. [50] reported a specific heat peak at 3.69 K and assign that value to $T_N$. In addition they measured the magnetization at 0.6 K in fields up to 70 kOe along all three crystallographic axis. A typical spin flop transition was observed at 11.2 kOe along the b-axis, and the spin flop to paramagnetic transition was observed at 36.8 kOe. Along the b-axis(a-axis) a single transition at 43.0(42.5) kOe was observed, thus establishing the essential features of a simple two sublattice uniaxial antiferromagnet. Using Eq. 72, 74, and 75 they could calculate both $H_e$ = 19.9 kOe and $H_a$ = 3.0 kOe. These values were somewhat different than those calculated from the susceptibilities. This is probably a shortcoming of the molecular field theory, since a similar discrepency was observed in the case of $MnCl_2 \cdot 4H_2O$(30).

Blazey and Rohrer [49] measured the differential susceptibility in fields up to 200 kOe between 1.25 K and 40 K. Their determination of the critical fields are in general agreement with those of Cashion et al. but by taking data at many temperatures they were able to plot out the entire phase diagram. The behavior of the susceptibility near the upper critical field $H_{c2}$ was very similar to that observed in $MnCl_2 \cdot 4H_2O$ by Rives [30], a small peak followed by a rapid drop to zero(saturation).

The differential susceptibility results in the neighborhood of the antiferromagnetic to spin flop transition are of particular interest. Below 3.0 K they observed that the height of the susceptibility peak was independent of temperature, but that as the temperature was lowered the sharp peak became broadened. At 1.25 K the width was of the order of 200-300 Oe wide, with the magnitude of the susceptibility almost constant over this range. This is the same behavior as that observed in DAG and discussed by Landau et al. [28]. It points out the importance of the demagnetizing factor, and the fact that the internal field is really the thermodynamic variable which is to be considered in the calculations.

If the anisotropy energy is larger than, or comparable to, the exchange energy the spin flop phase disappears and there will be a transition directly from the antiferromagnetic state to the paramagnetic state. This has been observed in a number of compounds including $FeCo_3$[51], $FeBr_2$[52], $FeCl_2$ [53], and Dysprosium Aluminum Garnet(DAG) [28, 54].

## J. $FeCl_2$

In the case of $FeCl_2$ which has an ordering temperature of 23.5 K, Jacobs and Lawrence [53] found from magnetization isotherms that a first order transition from the antiferromagnetic to the paramagnetic state occured at 10.6 kOe at 4.2 K. Above about 20 K the first order transition apparently goes over into one of higher order. This is similar to the case of DAG (28) where the first order transition goes over to one of higher order above 1.66 K.

## K. $Dy_3Al_5O_{12}$ (DAG)

Wolf and his co-workers [28,54] have made a very comprehensive study of the Ising antiferromagnetic DAG. Although the large anisotropy makes DAG behave like an almost ideal Ising system, many of the thermodynamic arguments discussed in Ref. [28] are completely general and hold for any magnetic system. DAG becomes antiferromagnetic in zero field at a temperature of 2.544 K. The crystal structure is cubic and at liquid helium temperatures the system can be represented by an effective spin of 1/2. The g-value is very anisotropic ($g_x \cong g_y \cong 0$, $g_z = 18.2$), and it is this large value of $g_z$ which allows the entire phase diagram to be investigated in quite small fields.

As mentioned before below 1.66 K there is a range of external fields over which the internal field remains almost constant. This produced a constant magnitude for the differential susceptibility over this range of fields. As a function of internal field the transition from the antiferromagnetic state to the paramagnetic state is found to be first order below 1.66 K. As a function of external field, on the other hand, a mixed state exists over the range of fields for which the susceptibility is constant. Above 1.66 K the transition is of higher order and there is a direct transition, as a function of external field, from the antiferromagnetic to the paramagnetic state. At 1.14 K the transition takes place at an internal field of 3.74 kOe.

The magnetic measurements on DAG were taken on a spherical sample with a demagnetizing factor of 4.19. Below about 1.0 K the mixed state extends over about 3.5 kOe. In a recent experiment [55] on a spherical sample of $MnCl_2 \cdot 4H_2O$ aligned along the preferred direction the differential susceptibility was measured at 0.30 K with more field resolution than in earlier measurements [30]. At the spin flop transition the top of the large peak in the susceptibility was found to be flat over about 20 Oe. When proper account is taken for the relatively small magnetization of $MnCl_2 \cdot 4H_2O$ compared to that of DAG, one would expect the existance of a mixed state over about that range of field.

It is clear from the work of Landau et al. [28] and the recent measurements on a spherical sample of $MnCl_2 \cdot 4H_2O$ [55] that careful measurements on samples of spherical or ellipsoidal shape are necessary in order to gain precise thermodynamic information on the phase diagrams of antiferromagnetic compounds. Proper account must be taken of the internal field and this requires samples of high enough symmetry to allow demagnetizing factors to be calculated accurately. In most cases experiments need to be extended to lower temperatures and higher fields in order to derive enough information to completely describe the nature of the various transitions which take place.

## REFERENCES

1. L. Néel, Ann. Phys. (Paris) (10) 18, 5 (1932); Ann. Phys. (11) 5, 232 (1936); Ann. Phys. (12) 3, 137 (1948).

2. F. Bitter, Phys. Rev. 54, 79 (1937).

3. J. H. Van Vleck, J. Chem. Phys. 9, 85 (1941).

4. J. Samuel Smart, <u>Effective Field Theories Of Magnetism</u>, (W. B. Saunders Co., Philadelphia, 1966).

5. P. W. Anderson, <u>Phys. Rev.</u> <u>83</u>, 694 (1952).

6. R. Kubo, <u>Phys. Rev.</u> <u>87</u>, 568 (1952).

7. T. Oguchi, <u>Phys. Rev.</u> <u>117</u>, 117 (1960).

8. F. Burr Anderson and Herbert B. Callen, <u>Phys. Rev.</u> <u>136</u>, A1068 (1964).

9. H. B. Callen, <u>Phys. Rev.</u> <u>130</u>, 890 (1963).

10. T. Nagamiya, K. Yosida, and R. Kubo, <u>Advances in Phys.</u> <u>4</u>, 1 (1955).

11. C. J. Gorter and Tineke Van Peski-Tinbergen, <u>Physica</u> 22, 237 (1956). Tineke Van Peski-Tinbergen and C. J. Gorter, <u>Physica</u> <u>20</u>, 592 (1954).

12. Yung-Li Wang and Herbert B. Callen, <u>J. Phys. Chem. Solids</u> 25, 1459 (1964).

13. J. Feder and E. Pytte, <u>Phys. Rev.</u> <u>168</u>, 640 (1968).

14. John E. Rives, <u>Phys. Rev.</u> <u>162</u>, 491 (1967).

15. F. J. Dyson, <u>Phys. Rev.</u> <u>102</u>, 1217 (1956); <u>Phys. Rev.</u> <u>102</u>, 1230 (1956).

16. J. Van den Handel, H. M. Gijsman, and N. J. Poulis, <u>Physica</u> <u>18</u>, 862 (1952).

17. J. Ubbink, <u>Physica</u> <u>19</u>, 919 (1953).

18. F. M. Johnson and A. H. Nethercot, Jr., <u>Phys. Rev.</u> <u>114</u>, 705 (1959).

19. S. Foner, <u>J. Phys. Radium</u> <u>20</u>, 336 (1959); <u>Phys. Rev.</u> <u>130</u>, 183 (1963).

20. S. Foner, <u>Magnetism</u>, Vol. I, ed. George T. Rado and Harry Suhl, (Academic Press, New York, 1963).

21. Robert M. White, <u>Quantum Theory Of Magnetism</u>, McGraw-Hill Book Co., New York, 1970.

22. W. Van der Lugt and N. J. Poulis, <u>Physica</u> <u>26</u>, 917 (1960).

23. M. Date, <u>J. Phys. Soc. Japan</u> <u>14</u>, 1244 (1959).

24. M. B. Flippen and S. A. Friedberg, <u>J. Appl. Phys.</u> <u>31</u>, 338S (1960).

25. W. K. Robinson and S. A. Friedberg, <u>Phys. Rev.</u> <u>117</u>, 402 (1960).

26. V. A. Schmidt and S. A. Friedberg, <u>J. Appl. Phys.</u> 38, 5319 (1967).

27. F. Reif, <u>Fundamentals of Statistical and Thermal Physics</u>, McGraw-Hill Book Co., New York, 1965, p. 450.

28. D. P. Landau, B. E. Keen, B. Schneider and W. P. Wolf, <u>Phys. Rev.</u> April 1, 1971.

29. J. M. McElearney, H. Forstat and P. T. Bailey, Phys. Rev. 181, 887 (1969).

30. John E. Rives, Phys. Rev. 162, 491 (1967).

31. H. M. Gijsman, N. J. Poulis and J. Van den Handel, Physica 25, 954 (1959).

32. H. Forstat, P. T. Bailey and J. R. Ricks, Phys. Letters 30A, 52 (1969).

33. S. A. Friedberg and J. H. Schelleng, Proceedings of the International Conference on Magnetism, Nottingham, 1964 (The Institute of Physics and The Physical Society, London, 1965).

34. D. G. Kapadnis and R. Hartsman, Physica 22, 181 (1956).

35. J. H. Schelleng and S. A. Friedberg, J. Appl. Phys. 34, 1017 (1963).

36. I. Tsujikawa and J. Kanda, J. Phys. Radium 20, 352 (1959).

37. B. Bolger, Conference de Physique des Basses Temperatures, Paris, 1955, Centre National de la Recherche Scientifique, and UNESCO, Paris, 1956.

38. T. Barber and J. E. Rives, unpublished.

39. R. F. Altman, S. Spooner, J. E. Rives, and D. Landau Winter Meeting Am. Cryst. Soc., Columbia, S. C., 1971.

40. N. J. Poulis and G. E. G. Hardeman, Physica 20, 7 (1954).

41. G. J. Butterworth and V. S. Zidell, J. Appl. Phys. 40, 1033 (1969).

42. Y. Shipira, S. Foner, and A. Misetich, Phys. Rev. Letters 23, 98 (1969).

43. Y. Shipira and J. Zak, Phys. Rev. 170, 503 (1968).

44. M. E. Fisher, Phil. Mag. 7. 1731 (1962).

45. J. Skalyo, Jr., A. F. Cohen, S. A. Friedberg and R. B. Griffiths, Phys. Rev. 164, 705 (1967).

46. D. T. Teaney, Phys. Rev. Letters 14, 898 (1965).

47. S. Foner, in Magnetism, edited by G. T. Rado and H. Suhl, Academic Press, Inc., New York, 1963, Vol. I, p. 383.

48. L. Holmes, M. Eibschutz, and H. J. Guggenheim, Solid State Comm. 7, 973 (1969).

49. K. W. Blazey and H. Rohrer, Phys. Rev. 173, 574 (1968).

50. J. D. Cashion, A. H. Cooke, J. F. B. Hawkes, M. J. M. Leask, T. L. Thorp, and M. R. Wells, J. Appl. Phys. 39, 1360 (1968).

51. I. S. Jacobs, J. Appl. Phys. 34, 1106 (1963).

52. I. S. Jacobs and P. E. Lawrence, J. Appl. Phys. 35, 996 (1964).

53. I. S. Jacobs and P. E. Lawrence, Phys. Rev. 164, 866 (1967).

54. M. Ball, W. P. Wolf, and A. F. G. Wyatt, Phys. Letters 10, 7 (1964).
B. E. Keen, D. Landau, B. Schneider and W. P. Wolf, J. Appl. Phys.
37, 1120 (1966). B. E. Keen, D. P. Landau and W. P. Wolf, Phys.
Letters 23, 202 (1966).

55. J. E. Rives and V. Benedict, unpublished.

## PART 4

## EXPERIMENTAL TECHNIQUES

The literature is quite plentiful on the topic of methods of thermal and magnetic measurements. A recent book by Schieber [1] gives an excellent review of modern experimental techniques for studying most of the interesting properties of insulating magnetic materials. It also includes a discussion of crystal growth and reviews some of the properties of a large group of magnetic compounds. Several other books [2] discuss various solid state experimental methods which have some interest in magnetism. Calorimetry is treated quite extensively in Volume I of Experimental Thermodynamics [3].

The discussion below on high resolution specific heat and differential susceptibility measurements are intended only to expand on the above mentioned works.

### I. HIGH RESOLUTION SPECIFIC HEAT MEASUREMENTS

There are a large number of techniques available for carrying out precise specific heat measurements. There are two main points which must be considered regardless of which technique is used. One must be able to precisely determine the amount of heat introduced into the sample to be measured, and one must have an accurate means of measuring the temperature. In practice one measures the total heat capacity of the sample, and for small temperature intervals it is possible to write

$$\Delta Q = C \Delta T \tag{95}$$

where $\Delta Q$ is the heat added to the sample, $\Delta T$ is the resulting temperature increase, and C is the average value of the heat capacity over the temperature range $\Delta T$. The accuracy of the calculation of the heat capacity depends on how well one can determine $\Delta Q$ and $\Delta T$. In order to obtain specific heat results with high temperature resolution, it is also important to make $\Delta T$ as small as possible.

The method one chooses usually depends on just what type of measurement one is attempting to perform. If it is of interest to measure the heat capacity over large ranges of temperature then an adiabatic technique is best. This method is well covered in Ref. [3]. At very low temperatures it is usually not convenient to do an adiabatic measurement. In an adiabatic method a thermal shield surrounds the sample and its temperature is controlled to remain as close as possible to the temperature of the sample. At very low temperatures it is usually more convenient to hold the shield at a constant temperature. Due to the difference in temperature between the sample and the shield, heat will either be conducted to the sample from the shield, or vice versa, during the experiment. It is necessary to know the magnitude of this so-called heat leak in order to make accurate heat capacity calculations.

There are two commonly used techniques which allow for the simultaneous determination of the heat leak and the heat capacity of the sample. The most convenient method of supplying heat to the sample is to supply electrical energy to a resistor in good thermal contact with the sample. Energy can either be supplied to the resistor continuously or in the form of pulses. In the pulse technique one supplies an electric current of known magnitude and duration and notes the increase in temperature. Between pulses one monitors the rate of change of temperature due to the residual heat leak. A typical temperature versus time plot is known in Fig. 13. The heat pulse is turned on at $t_1$ and turned off at $t_2$. During the time the electrical heat is on ($\Delta t$), the total heat input to the sample is given by $I^2 R \Delta t$, the Joule heat, pulse the integrated residual heat leak $Q_r$, thus

$$\Delta Q_t = \int_{t_1}^{t_2} (I^2 R + \dot{Q}_r) \, dt \qquad (96)$$

With the electrical heat off the rate of change of temperature is given by $\dot{Q}_r = C\dot{T}$, where the dots represent derivatives with respect to time. If the heat capacity is fairly constant over the range of temperatures covered by the pulse then $\dot{Q}_r$ can be replaced in Eq. 96 by the above relation, and from Eq. 95 one obtains

$$\Delta Q_t = C \Delta T = (I^2 R + C\dot{T}) \, \Delta t$$

from which it is trivial to solve for the value of the heat capacity obtaining

$$C = I^2 R \Delta t / (\Delta T - \dot{T} \Delta t) \qquad (97)$$

In practice one observes the drift in temperature before and after the heat pulse and takes account of any change in the residual heat leak by

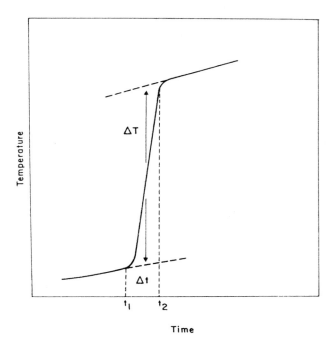

FIG. 13.  Temperature against time profile for a single heat pulse.

averaging the results.  If the heat leak is sufficiently small one can use
small pulses and obtain reasonably high resolution data.  Best results are
obtained if $I^2R$ is several times the magnitude of $\dot{Q}_r$.  The temperature
resolution of the results depend primarily on the duration of the pulse
and on how long after the pulse it takes for the system to reach equilibrium
so that the drifting rate can be determined.  In the liquid helium range of
temperatures one can usually determine the specific heat of magnetic
crystals by this technique with a resolution of the order of 0.1 mK to 10
mK.  This is generally acceptable unless one is interested in investigating
the behavior of the system very near a phase transition such as the zero
field ordering temperature.  In this case it is usually necessary to measure
the specific heat with a temperature resolution of from 1.0 to 10 $\mu$K.  The
continuous heating method is the most convenient way to achieve such
resolution.

In the continuous heating method one supplies a constant rate of heating
and observes the rate of change in the temperature with the entire system
in a steady state.  The residual heat leak can be determined in a manner
similar to the pulse method by periodically switching off the electrical
heat and observing the drift rate.  If the electrical heating rate is denoted

by $\dot{Q}_e$, the time derivative of the temperature with the heat on and heat off by $\dot{T}_t$ and $\dot{T}_r$, respectively, one obtains with the heat on

$$\dot{Q}_t = \dot{Q}_e + \dot{Q}_r = C \, \dot{T}_t \qquad (98)$$

and with the heat off

$$\dot{Q}_r = C \, \dot{T}_r \qquad (99)$$

From the known values of $\dot{Q}_e$, $\dot{T}_t$, and $\dot{T}_r$ one can eliminate $\dot{Q}_r$ and calculate the heat capacity.

In practice this technique is particularly advantageous in the liquid helium range of temperatures where the temperature of the thermal shield can be controlled quite well, and thus the value of $\dot{Q}_r$ can be controlled.

In order to get a clearer picture of the advantages of the continuous heating method let us consider a specific example. In the neighborhood of the Néel temperature in $MnCl_2 \cdot 4H_2O$, for instance, the specific heat was found to reach a peak value of $C/R = 7$ [4], where R is the universal gas constant. A number of other compounds have been found to have specific heat peaks near their ordering temperatures of this order of magnitude. For convenience let us consider a case where $C/R = 5$. A typical sample might have a mass which represents about 0.01 moles of the compound. In that case the heat capacity would be approximately 0.4 Joules/deg or 4 ergs/$\mu$deg. If the total heating rate in this case was 4 ergs/sec, then the temperature would increase at a rate of 1.0 $\mu$deg/sec. It is generally possible to keep the residual heat leak smaller than 0.1 erg/sec in the liquid helium temperature range. Under these conditions reasonably accurate specific heat measurements with $10^{-6}$ deg resolution are possible.

In the typical example given it was assumed that one could achieve good temperature equilibrium over the entire sample during the measuring period. It is essential to know how well this assumption is achieved in practice. Due to the finite value of the thermal conductivity of the material, and the fact that heat is continuously added to the sample, there will always be some temperature gradient across the sample. The temperature resolution of the specific heat results will be limited by the difference in temperature between different points in the sample.

For the example given above let us make an estimate of the temperature gradient in the sample under the conditions considered above. For convenience consider a rectangular crystal of cross sectional area A and length L. Let the heater be attached to one end and the other end be thermally connected to a constant temperature bath. In this case one can express the heat flow as

$$\dot{Q} = \kappa A \cdot \nabla T = \kappa A \Delta T / L \qquad (100)$$

The thermal conductivity of $MnCl_2 \cdot 4H_2O$ has been measured by Rives et al. [5], from 0.3 K to 2.6 K. In the neighborhood of the Néel temperature the magnitude of the conductivity is approximately 600 mW/cm-°K. A typical sample might have A = 0.5 cm$^2$ and L = 1.0 cm. Using the 4 ergs/ sec figure above for $\dot{Q}$ we obtain, upon substitution in Eq. 100, a value of $\Delta T = 1.0 \times 10^{-6}$ K. Thus the maximum temperature variation over the sample under these circumstances should be 1.0 $\mu$K. In actual practice the sample is not thermally connected to a constant temperature bath, but is isolated thermally from its surroundings, so that the actual temperature gradients should be somewhat smaller. For the example considered, therefore, it should be possible to achieve a temperature resolution of almost 1.0 $\mu$K.

In a number of magnetic compounds the thermal conductivity is considerably smaller in magnitude than for $MnCl_2 \cdot 4H_2O$. In these cases the temperature resolution will be primarily limited by the resulting temperature gradients, unless the heating rates can be reduced proportionately.

From the discussion above it is clear that it is possible to achieve, in certain practical cases, temperature resolution of the order of 1.0 to 10 $\mu$K in the liquid helium temperature range. To complete the calculation of the specific heat it is also necessary to be able to measure temperature changes of the order of magnitude mentioned. In the liquid helium temperature range carbon resistors are the most commonly used secondary thermometers. The electrical resistance R of a carbon resistor is roughly proportional to $T^{-1}$, with the constants of proportionality dependent on the type and room temperature value of resistor.

Several factors influence the selection of the proper resistor for the particular experiment to be performed. In order to resolve small temperature differences one wants to select a resistor with a large dR/dT in the temperature range of interest. This factor alone is relatively easy to obtain due to the temperature dependence of the carbon resistors. However, because of limitations in most measuring circuits it is usually convenient to limit the value of the resistor in the working temperature range to a value between about $10^4$ - $10^5$ ohms. As a typical example a nominally 70 ohm Allen-Bradley 1/2 watt carbon resistor will have a resistance of about $6 \times 10^4$ ohms in the neighborhood of 1.5 K, dR/dT should be of the order of $10^5$ ohms/deg. A change in resistance of 0.1 ohm thus represents a change in temperature of 1.0 $\mu$K. The measuring instrument must be capable of resolving 0.1 ohm changes out of a total resistance of $5 \times 10^4$, or a relative change $\Delta R/R = 2 \times 10^{-6}$.

In order to insure that the resistance thermometer itself remains in good temperature equilibrium with the sample the power dissipated in the

resistor must be quite small. In a standard 1/2 watt resistor operating
in the liquid helium temperature range experience shows that a power
level much greater than $10^{-9}$ watts will cause significant self heating in
the resistor. In addition it is necessary to reduce the power level as the
temperature is decreased, although this is not very important if the range
of temperature covered in the experiment is not too great.

Both dc and ac techniques may be used for resistance measurement, but
the ac method generally offers more sensitivity at lower power levels
through the use of phase sensitive detection. A standard ac Wheatstone
bridge is commonly used in such applications. In Fig. 14 a modification
of the Wheatstone bridge is presented. It can be thought of as a two arm
comparator, or as a bridge in which the secondary of a step-down trans-
former takes the place of two arms of the bridge.

The circuit shown in Fig. 14 has several advantages over the standard
Wheatstone bridge. For a particular setting of the source level a constant

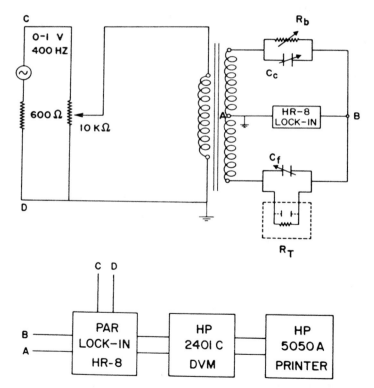

FIG. 14. AC resistance bridge for high resolution specific heat measurements.

voltage is applied across the resistance thermometer, independent of the value of the resistance. This means that the power level is automatically decreased as the resistance increases. The capacitors $C_c$ and $C_f$ are used to balance the capacitance between the leads which extend into the cryostat. This capacitance can be reduced considerably by using individually shielded leads from the bridge to the cryostat. In order to eliminate the effects of the lead wires to the resistance thermometer, $R_T$, a pair of identical leads, shorted inside the cryostat, may be placed in the circuit between the point B and the balancing resistor, $R_b$. This insures that the balancing resistor gives the value of the thermometer alone, and corrections for the temperature dependence of the leads are unnecessary. Within the cryostat one normally uses wire of high thermal resistance which also has a very small temperature coefficient of resistance, but even these small changes can seriously effect the accuracy of high resolution specific heat results.

The point B in the circuit is connected to the imput of a phase sensitive detector. In the circuit shown a Princeton Applied Research Model HR-8 Lock-In Dectector is used. For maximum sensitivity a high input impedence preamplifier preceeds the HR-8. The dc output of the phase sensitive detector is proportional to the unbalance voltage of the bridge. Within the linear region of the circuit this is also proportional to the unbalance in the resistance. The dc output of the phase sensitive detector can be monitored in a number of ways. A strip chart recorder can be used to give a continuous record of the data. The specific heat can then be determined by measuring the slope of the recorded data and then introducing this into Eq. 98 and 99 in the appropriate manner.

An alternative method involves measuring the output of the phase sensitive detector with a digital voltmeter and printing a permanent record of the output of the voltmeter. One can then use a high speed digital computer to calculate the specific heat results. With high resolution specific heat experiments, where the amount of data is quite overwhelming, the latter method is preferred.

To complete the calculation of the specific heat the resistance thermometer must be calibrated against a standard thermometer throughout the range of interest. In the liquid helium range of temperatures the saturated vapor pressure of liquid $^4$He or $^3$He is usually taken as the standard. A number of functional relationships for the temperature dependence of carbon resistors have been used in the literature [6-8]. The absolute calibration is limited by measuring errors to about 1.0 mK, however the relative calibration can be several orders of magnitude better with the proper choice of the functional relation between R and T.

Recently, several commercial firms have offered specially doped germanium resistance thermometers for use at low temperatures. They offer several advantages over carbon resistors. They are offered in a variety of dopings, each have a particular usefulness within a restricted temperature range. It is possible to select a given thermometer which has a reasonable value of R and dR/dT in any range of interest between 0.2 to 100 K. In addition the calibration of the germanium resistors does not seem to drift with time or temperature cycling nearly as much as do carbon resistors.

The experimental cryostats for measuring specific heats in the liquid helium temperature range are generally of fairly standard low temperature design. The one important feature for specific heat measurements is the heat switch. Some provision must be made for thermally connecting the sample to a refrigerator in order to cool the sample initially to the temperature of interest. Thereafter the thermal link must be broken in order to thermally isolate the sample from its surroundings as well as possible. This is usually done mechanically in a variety of ways.

Figure 15 shows a typical cryostat designed for operation between 1.0 and 4.2 K. The cryostat is immersed in a 4.2 K liquid helium bath, which in turn is surrounded by a liquid air bath. Inside the outer vacuum can is located a $^4$He refrigerator which is filled from the main helium bath through the valve I. By pumping on the vapor of the helium in the refrigerator its temperature can be controlled down to about 1.0 K. The sample D is located inside the inner vacuum can which is sealed to the bottom of the $^4$He refrigerator, and remains in thermal equilibrium with it. The sample is suspended by a stainless steel wire, which in turn is connected to the upper end of a metal bellows J in the main helium bath. A stainless steel rod extends out the top of the cryostat and is connected to a screw arrangement which is used to raise and lower the sample. A copper disc is connected to the top of the sample and as the sample is raised this disc comes into contact with three copper rods which are in turn thermally connected to the refrigerator. Thus thermal contact between the sample and the refrigerator is made by raising the sample by means of the bellows. To achieve isolation the sample is lowered. The electrical heater and the resistance thermometer are connected to the sample by the use of phosphur bronze clamps. The thermometer can be calibrated against the vapor pressure of the liquid helium by introducing helium gas inside the inner vacuum can.

A more detailed outline of the low temperature portion of the calorimeter is shown in Fig. 16. The mechanical heat switch and the mounting procedure for the sample are shown in somewhat more detail than in Fig. 15.

A mechanical heat switch of a rather different design is described by Landau et al., [9]. In this case the sample is rigidly supported by an insulating rod to the inner vacuum can. This has a decided advantage over the freely hanging method shown in Fig. 16, since vibrations of the sample can produce a significant heat leak at liquid helium temperatures. A gold plated strip extends from the sample into the jaws of a gold plated clamp which is in thermal equilibrium with the refrigerator. The clamp is designed so that its jaws can be opened and closed by a screw arrangement

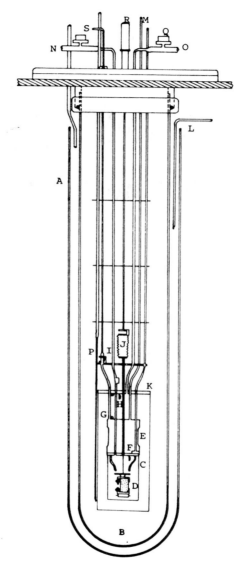

FIG. 15. Simple calorimeter for specific heat measurements in the liquid helium temperature range. A = air dewar; B = helium dewar; C = calorimeter; D = sample; E = He$^4$ refrigerator; F = manometer bulb; G = resistance thermometer; H = thermal ground; I = refr. filling valve; J = sample elevator bellows; K = radiation trap; L = liquid air control; M = refr. pumping line; N = IVC pumping line; O = OVC pumping line; Q = electrical connectors; P = transfer line; R = sample elevator screw; and S = refr. valve handle.

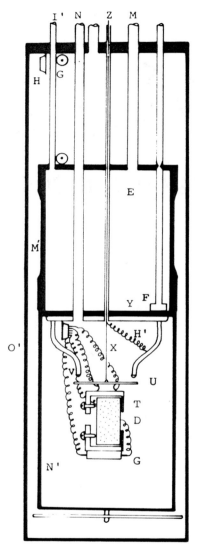

FIG. 16. Detailed view of specific heat calorimeter showing mechanical heat switch and mounting procedure. D = sample; E = helium refrigerator; F = manometer bulb; G = resistance thermometer; H' = thermal ground; I' = refr. filling line; M = refr. pumping line; M' = refrigerator heater; N = IVC pumping line; N' = IVC can; O' = OVC can; T = sample heater; U = copper hat; V = indium tips; W = heat switch; X = nylon thread; Y = liquid helium; and Z = sample elevator tube.

at the top of the cryostat. Other variations of these basic mechanical heat switches are also in use in many laboratories. For more detailed description of the overall techniques in low temperature calorimetry the reader is referred to Refs. [1-3].

## II.  DIFFERENTIAL SUSCEPTIBILITY

The magnetization of a magnetic material is generally measured by either the force method or the induction method. The most common force methods are based on either the Faraday or the Gouy technique [1,2] and involve a measurement of the force acting on the sample which is placed in an inhomogenious magnetic field. This method is usually quite sensitive, but suffers from the fact that zero, or almost zero, field measurements are not possible. In most systems fields of the order of 1 kOe are usually required in order to achieve maximum sensitivity.

The susceptibility is calculated by taking the ratio of the measured magnetization to the average value of the applied field. Since there is a field gradient across the sample, the calculated susceptibility is actually the average value over the range of fields used in the measurement. This is usually not serious, but very near a magnetic transition considerable error is possible.

The ballistic induction method allows for the use of small applied fields, but is also a measure of the magnetization of the sample [1,2]. In this method the sample is located within the coils of a mutual inductance. When the current in the primary coil is changed, the output of the secondary coil depends partially on the magnetization of the sample. It is common to use a sensitive ballistic galvanometer as the detector in such a circuit. If the period of the galvanometer is long compared to the time over which the primary current is changed, then the deflection of the galvanometer is proportional to the total charge which flows through the coil of the galvano-meter. If the galvanometer is connected directly across the secondary coil, then a simple analysis of the circuit gives

$$Q = (N/R) \, \Delta\phi \qquad (101)$$

where Q is the total charge induced, R is the total resistance in the circuit, N the number of turns in the secondary coil, and $\Delta\phi$ is the change in the magnetic flux due to the change in the primary current. Now

$$\phi = \alpha H + \beta M \qquad (102)$$

where H is the applied field due to the current in the primary, and M is the magnetization of the sample. If the applied field is changed by an amount $\Delta H$, then

$$\Delta\phi = \alpha \Delta H + \beta \Delta M \qquad (103)$$

and the deflection of the galvanometer is dependent on the change in magnetization.

In order to determine the constants in the above equations the system has to be calibrated using a sample with known susceptibility. This method can then be used to calculate $\chi = \Delta M/\Delta H$, which is just the average slope of the magnetization versus field over the interval $\Delta H$. In practice $\Delta H$ is usually of the order of 1.0 Oe or less, so that the susceptibility can be measured in much smaller fields than with the force method.

Ballistic measurements are commonly made with a circuit similar to the one shown in Fig. 17 which is a modification of the familiar Hartshorn bridge. $M_1$ is a concentrically wound set of coils with the sample S located symmetrically inside. The secondary generally consists of two oppositely wound sections so that the total mutual inductance, with the sample absent, is zero, or close to zero. This essentially eliminates the first term on the right hand side of Eq. 103 and greatly increases the accuracy with which $\Delta M$ can be determined. It is sometimes advantageous, from symmetry considerations, to wind the compensating secondary, a, in two symmetrically placed sections a and a' as shown in Fig. 17. For maximum sensitivity it is desired to have $M_1$ as close to the sample as possible. At low temperatures this usually means that the coils are located inside the helium cryostat. This has the additional advantage that the resistance

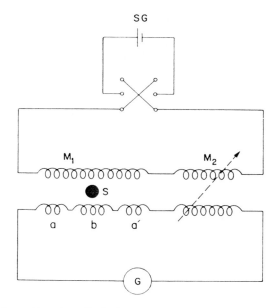

FIG. 17. Schematic circuit of Hartshorn bridge used for ballistic susceptibility measurements.

of the secondary is considerably reduced, and it is possible to operate the galvanometer in a critically damped condition.

The variable standard mutual inductor $M_2$ is connected in the circuit so as to oppose the effect of $M_1$ and allows for the bridge to be operated in two different ways. Even though $M_1$ is almost zero without the sample in place, the presence of the sample can produce quite large galvanometer deflections, particularly at low temperatures. $M_2$ can be adjusted to give sufficiently small deflections at some desired temperature. The change in deflection, as the temperature is changed, can then be measured, and the combination of the value of $M_2$ and the galvanometer deflection gives a measure of $\Delta M$. It is also possible to adjust $M_2$, at each temperature, to give zero deflection with the value of $\Delta M$ read directly from the settings of $M_2$.

Capacitive effects between the two coils can complicate the measurement, so it is desirable to place an electrically grounded shield between the primary and the secondary coils. A grounded metal can between the coils and the sample is also desirable to eliminate electrostatic charge effects. It is worth pointing out one caution, however, if the coils are located at liquid helium temperatures. A copper can inside the coils will produce undesirable eddy currents which will interfere with the measurements. A metal with lower conductivity, such as brass, is preferred. The eddy current effect is much more serious when the bridge is operated as an ac bridge, but can cause some problems in a ballistic measurement.

The ac induction method offers several advantages over the ballistic method. Tuned amplifiers can be used to increase the sensitivity of the bridge. The ac susceptibility can be described in complex notation by $\chi = \chi' - i\chi''$, where $\chi'$ is the inductive component, and $\chi''$ the out of phase, or resistive, component. Both the in phase, as well as the out of phase, components must be balanced in order to obtain a proper ac balance. A schematic circuit of an ac Hartshorn bridge is shown in Fig. 18 $M_2$ is used to balance the inductive component, and the variable resistance R is used to balance the resistive component. The resistive component is proportional to the losses in the sample, which are in turn proportional to the product of the frequency, $H_{ac}^2$ and $\chi''$. These can be minimized by using low frequencies and small applied fields.

The output voltage of the ac bridge is proportional to $d\phi/dt$, and from Eq. 102 one has

$$\frac{d\phi}{dt} = \alpha \frac{dH}{dt} + \beta \frac{dM}{dt}$$

$$= (\alpha + \beta \frac{dM}{dH}) \cdot \frac{dH}{dt}$$

$$= (\alpha + \beta \chi') \cdot \frac{dH}{dt} \tag{104}$$

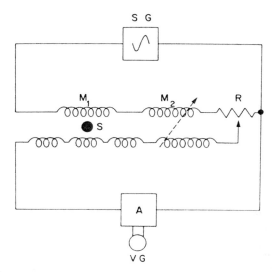

FIG. 18. Schematic circuit af ac Hartshorn bridge used for differential susceptibility measurements.

If $H = H_0 \sin\omega t$ then, neglecting the term in $\alpha$ which can be canceled out as in the ballistic case, the output voltage is proportional to $\beta\chi'\omega H_0 \cos\omega t$. It is thus seen that the magnitude of the output voltage is directly proportional to the differential susceptibility. The value of $\chi$ determined in this way is actually the average value of the differential susceptibility, averaged over the field $2H_0$, the total variation of the applied field.

In most ac bridges, in order to keep the losses a minimum, $H_0$ is generally limited to a value of 0.1 - 0.2 Oe. If $\chi$ is a slowly varying function of H then its variation over $2H_0$ will be negligable. If, in addition to the ac field, a static field H is applied to the sample then the value of the susceptibility calculated as outlined above can be considered as the value of the susceptibility at the field H with very little error.

If, on the other hand, the susceptibility is being measured near a magnetic phase transition where $\chi$ is changing rapidly with field it is important to keep $H_0$ as small as possible, and to consider its effect on the calculated value of the susceptibility.

A mutual inductance bridge with considerably more sensitivity than the simple Hartshorn bridge shown in Fig. 18 was first proposed by Pillinger, Jastram, and Daunt [10] for use in low temperature thermometry. A commercial version of this bridge is now available from Cryotronics, Inc. [11]. The sensitivity of this bridge is roughly $4 \times 10^{-3}$ microhenries which, with properly designed coils, makes it extremely convenient for measuring the susceptibility of small quantities of dilute magnetic samples. Abel, Anderson, and Wheatley [12] have given an excellent discussion of the operation of this bridge, as well as information on the design of mutual inductance coils for use in very accurate susceptibility measurements using the Crytronics, Inc. bridge.

The Cryotronics, Inc. bridge was used by Rives to measure the differential susceptibility of $MnCl_2 \cdot 4H_2O$ [13] in external fields up to 27 kOe in the liquid helium temperature range. A compensated mutual inductance coil, similar to that described by Abel et al. [12], was wound on a brass can which was kept at 4.2 K. The two sections of the secondary coil consisted of 1000 turns each of No. 40 AWG insulated copper wire. The primary consisted of a single layer of No. 36 AWG insulated copper wire. The sample was thermally connected to a liquid $He^3$ refrigerator and was located in the center of one section of the secondary coil. A superconducting solenoid was used to supply the static external field. The presence of the superconducting solenoid altered somewhat the mutual inductance of the measuring coils, but sufficient compensation was achieved to enable susceptibility measurements with a relative accuracy of $\pm 0.01\%$.

When such a system is used in conjunction with a superconducting solenoid, one can minimize the effects of the solenoid by winding the secondary coil in three sections as shown in Fig. 17. Since the sample is generally placed in the center of the solenoid, the measuring coils can then be symmetrically placed within the solenoid. This will reduce the mutual inductance between the secondary coil and the solenoid windings.

If the compensating section of the secondary coil is wound in two sections as in Fig. 17 there is one further comment of note. It is usually not convenient to extend the primary windings too far beyond the secondary windings on each end of the coil. Since the field due to the current in the primary is somewhat less near the ends of the primary coil, than near the center, exact compensation will not be possible unless additional turns are added to each of the compensating coil sections. For a given design configuration it is possible to calculate the necessary number of additional turns by calculating the field profile within the primary coil. In a two section secondary this is unnecessary since each section is symmetrically placed within the primary coil.

## REFERENCES

1.  Michael M. Schieber, Experimental Magnetochemistry, North-Holland Publishing Co., Amsterdam, 1967.

2.  Methods Of Experimental Physics, Vol. 6, Part B, ed. K. Lark-Horovitz and Vivian A. Johnson, Academic Press, New York, 1959.

    Guy Kendall White, Experimental Techniques In Low-Temperature Physics, Oxford University Press, London, 1959.

Handbuch Der Physik, Vols. XIV and XV, ed. S. Flugge, Springer-Verlag Co., Berlin, 1956.

3. Experimental Thermodynamics, Vol. I, ed. John P. McCullough and Donald W. Scott, Plenum Press, New York, 1968.

4. George S. Dixon and John E. Rives, Phys. Rev. 177, 871 (1969).

5. J. E. Rives, D. Walton and G. S. Dixon, J. Appl. Phys. 41, 1435 (1970).

6. J. R. Clement and E. H. Quinnell, Rev. Sci. Instr. 23, 213 (1952).

7. A. Brown, M. W. Zemansky, and H. A. Boorse, Phys. Rev. 84, 1050 (1952).

8. O. V. Lounasmaa, Phil. Mag. (8) 3, 652 (1958).

9. D. P. Landau, B. E. Keen, B. Schneider and W. P. Wolf, Phys. Rev. B 3, 2310 (1971).

10. W. L. Pillinger, P. S. Jastram, and J. G. Daunt, Rev. Sci. Instr. 29, 159 (1958).

11. Cryotronics, Inc., West Main Street, High Bridge, New Jersey.

12. W. R. Abel, A. C. Anderson, and J. C. Wheatley, Rev. Sci. Instr. 35, 444 (1964).

13. John E. Rives, Phys. Rev. 162, 491 (1967).

PART 5

## PARTIAL LIST OF LOW TEMPERATURE ANTIFERROMAGNETIC AND FERROMAGNETIC COMPOUNDS AND THEIR PROPERTIES

In this concluding part some of the more thoroughly investigated magnetic compounds are listed, along with a number of the important magnetic parameters. The list includes a number of examples of different classes of chemical compounds which order at relatively low temperatures. For a rather exhaustive list of all the known inorganic magnetic compounds the reader is referred to the book by Schieber [1]. In addition Wolf [2] has discussed the properties of a number of highly anisotropic antiferromagnetic compounds.

### I. ANTIFERROMAGNETIC COMPOUNDS

In Table 4 the data for each compound is catagorized under the following headings: the Néel temperature, $T_N$, in degrees Kelvin, the crystalline structure, the effective magnetic moment in Bohr magnetons, the preferred direction for magnetic alignment, the exchange energy in degrees Kelvin, the paramagnetic Curie temperature, $\theta$, in degrees Kelvin, the antiferro-magnetic to spin flop critical field, $H_{sf}$, the spin flop to paramagnetic critical field, $H_c$, the triple point temperature, the effective exchange field, $H_E$, and the effective anisotropy field, $H_A$. For the exchange energy, $J_{1,2,3}$ refer to the nearest neighbor, next-nearest neighbor, and next-next-nearest neighbor interactions, respectively. $J_T$ refers to the intralayer exchange, $J_L$ to the interlayer exchange, and z to the number of neighbors. For the critical fields $H_{sf}$ and $H_c$ the numbers in parentheses refer to the temperatures associated with the quoted values of the critical fields. The triple point is defined in Fig. 5.

A number of the compounds listed do not order in the simple two-sub-lattice antiferromagnetic structure assumed earlier. For instance, $CoCl_2$, $FeCl_2$, and $FeBr_2$ consist of ferromagnetically ordered layers with alter-native layers aligned antiferromagnetically. At the critical fields listed in Table 4 the antiferromagnetic interaction is broken, resulting in a transition to a ferromagnetic state. This transition is of first order for temperatures sufficiently below the Néel temperature. $CrCl_3$ appears to behave in a similar manner.

The dihydrates $CoCl_2 \cdot 2H_2O$ and $FeCl_2 \cdot 2H_2O$ appear to order in linear chains. At low temperatures two transitions are found as the field in increased. The upper transition again results in a ferromagnetic state. Narath [101] found that the lower transition could be understood in terms of a transition from linear chains to a six-sublattice structure. On the other hand Kobayashi and Haseda [97] suggested the possibility of a spin flop transition at the lower critical field with the spins canted toward the b axis.

$Dy_3Al_5O_{12}$(DAG) is another example of a highly anisotropic antiferromag-net with a direct transition to a ferromagnetic state at an internal field of 3.74 kOe. This transition is of first order up to T=1.66 K.


## II.  FERROMAGNETIC COMPOUNDS

A short list of some insulating ferromagnetic compounds are included in Table 5 for reference. The hydrated copper double salts are of partic-ular interest, since they should be well understood in terms of the Heisen-berg model. The distortion from cubic symmetry is slight, and they are almost completely isotropic. Because of this, and the fact that their ordering temperatures are in the convenient liquid helium range, they have been of particular interest.

TABLE 4

Some Low Temperature Antiferromagnetic Compounds
and Their Properties

| Compound | $T_N$(K) | Structure | $n_{eff}$ | Pre-ferred axis | Exchange energy $J/k$(K) |
|---|---|---|---|---|---|
| 1. $CoF_2$ | 37.7 | Tetragonal | 5.15 | c | $J_1/k=-9.57$ $(J_2+3J_3)/k=-15.8$ |
| 2. $MnF_2$ | 67.4 | Tetragonal | 5.71 | c | $J_1/k=+0.32$ $J_2/k=-1.76$ |
| 3. $FeF_2$ | 78.4 | Tetragonal | 5.56 | c | |
| 4. $NiF_2$ | 73.2 | Tetragonal | 3.50 | c | |
| 5. $CoCl_2$ | 25 | Hexagonal | 3.15 | (ab) | |
| 6. $FeCl_2$ | 23.5 | Hexagonal | 4.3 | c | $z_T J_T/k=2.5$ $z_L J_L/k=-0.5$ |
| 7. $FeBr_2$ | 11 | Hexagonal | 4.5 | c | $z_T J_T/k=0.3$ $z_L J_L/k-1.1$ |
| 8. $MnBr_2$ | 2.16 | Hexagonal | 5.84 | b | |
| 9. $CrF_3$ | 80 | Rhombohedral | 3.85 | | |
| 10. $CrCl_3$ | 16.8 | Hexagonal | 4.7 | c | $z_T J_T/k=6.7$ $z_L J_L/k=-0.067$ |
| 11. $MnCl_2 \cdot 4H_2O$ | 1.6 | Monoclinic | 5.94 | c' | $-0.06$ |
| 12. $FeCl_2 \cdot 4H_2O$ | 1.1 | Monoclinic | 3.40 | b | $J_1/k=-0.007$ $J_2/k=0.047$ $J_3/k=0.103$ |
| 13. $CoCl_2 \cdot 6H_2O$ | 4.61 | Monoclinic | 4.61 | c | $zJ/k=-4.6$ |
| 14. $NiCl_2 \cdot 6H_2O$ | 3.11 | Monoclinic | 3.11 | a | $zJ/k=-7.4$ |

Table 4 (Cont'd)

| $\theta$ (K) | $H_{sf}$(kOe) | $H_c$(kOe) | $T_t$(K) | $H_E$(kOe) | $H_A$(kOe) | Refs. |
|---|---|---|---|---|---|---|
| -20 | | | | | | 3,4,5,6,7,21 |
| -113 | 93(4.2) | | 65.1 | | 0.003 | 4,6,8,9,10,11 12,13,14,15, 16,17,21,35, 232,233,234,235 |
| -117 | | | | | | 8,12,6,14,21 |
| -116 | | | | | | 3,4,5,6,18,19 20,21,22,35 |
| +38 | | 34[a] | | | | 23,24,25 |
| +48 | | 10.6[a] | | | | 24,25,26,27,28 29,23,30,31,32 231 |
| +6 | | 23.8[a] a (31.5) b | | | | 23,29,27,33,34 |
| -2 | | | | | | 35,36,37,38,231 |
| -124 | | | | | | 39,40,41,42,43 |
| -31 | | $H_i$=1.6[a] | | 1.45 | 1.50 | 39,40,44,45,46 47,48,203,204 205,25 |
| -1.9 | 7.5(0) | 20.6(0) | 1.3 | | | 54,55,56,57,58 59,60,61,62,63 64,65,66,67,68 229 |
| -2.0 | 5.5(.4) | | 0.76 | | | 69,71,72,73,74 63 |
| -2.0 | 7.2(1.0) | | 2.06 | | | 76,77,78,79,80 81,82,63,83 |
| -7.0 | | | | | | 70,77,81,84 |

| Compound | $T_N(K)$ | Structure | $n_{eff}$ | Preferred axis | Exchange energy $J/k(K)$ |
|---|---|---|---|---|---|
| 15. $MnBr_2 \cdot 4H_2O$ | 2.2 | Monoclinic | 5.93 | c(c') | -0.09 |
| 16. $CoBr_2 \cdot 6H_2O$ | 3.2 | Monoclinic | | c | |
| 17. $CuCl_2 \cdot 2H_2O$ | 4.3 | Orthorhombic | 1.9 | a | |
| 18. $CoCl_2 \cdot 2H_2O$ | 17.5 | Orthorhombic | 3.2 | b | $z_1J_1/k=-17.4$ $z_2J_2/k=-1.9$ |
| 19. $FeCl_2 \cdot 2H_2O$ | 23 | Monoclinic | 4.25 | (ac) | $z_1J_1/k=-24.8$ $z_2J_2/k=-1.2$ |
| 20. $NiCl_2 \cdot 4H_2O$ | 2.99 | Monoclinic | | c | $zJ/k=-10.53$ |
| 21. $Cu_3(CO_3)(OH)_2$ (azurite) | 1.86 | Monoclinic | | (ac) | |
| 22. $RbMnF_3$ | 83 | Cubic | | | |
| 23. $CsMnF_3$ | 53.5 | Hexagonal | | c | |
| 24. $KMnF_3$ | 88.3 | Cubic | 6.15 | | |
| 25. $TlMnF_3$ | 76 | Cubic | | | $J_1/k=-3.42$ $J_2/k=-0.246$ |
| 26. $BaMnF_4$ | 25 | Orthorhombic | | b | |
| 27. $K_2IrCl_6$ | 3.08 | Cubic | | | $J_1/k=-5.4$ (-11.5) $J_2/k\cong0.1$ |
| 28. $K_2ReCl_6$ | 11.9 | Cubic | 3.63 | | |
| 29. $K_2ReBr_6$ | 15.2 | Cubic | 3.53 | | |
| 30. $(NH_4)_2IrCl_6$ | 2.16 | Cubic | | | $J_1/k=-7.5$ $J_2/k\cong0.1$ |
| 31. $GdAlO_3$ | 3.89 | Orthorhombic | 7.0 | b | -0.065 |

| $\theta$ (K) | $H_{sf}$(kOe) | $H_c$(kOe) | $T_t$(K) | $H_E$(kOe) | $H_A$(kOe) | Refs. |
|---|---|---|---|---|---|---|
| -2.5 | 9(1.2) | | | | | 54,56,61,63,65 66,82,85,86 |
| | 7.5 | | 2.91 | | | 63,81,87,88,89 |
| -5.0 | 6.5(1.0) | | 4.31 | | | 90,91,92,93,94 95,96 |
| $\theta_c=\theta_a=$ -29 $\theta_b=-1.0$ | $31.3^b$ | $44.9^b$ | | | | 29,27,97,98,99 100 |
| $\theta_1=-14$ $\theta_2=-5$ | $39.2^b$ | $45.6^b$ | | | | 100, 29,101,102 103 |
| | | | | | | 104 |
| | $19.5(T_t)$ | | 1.6 | | | 105,106,107, 108,118 |
| -136 | 2.3 | | | 816 | 0.004 | 109,110,111,112, 113,114,115,116, 117,119,120,121, 122,123 |
| | 0.9 | | | 350 | $H_{A1}=10$ $H_{A2}=.001$ | 113,109,124,125, 126 |
| -158 | 5.0 | | | | | 13,113,127,128, 129,130,131 |
| -138 | 3.0 | | | 630 | 0.0068 | 132,133 |
| | 10.4(1.4) | | 20 | 500 | 0.10 | 134,135 |
| -32 | | | | | | 143,144,145 |
| -55 | | | | | | 136,138,139,141 |
| -80 | | | | | | 136,138,139,140, 141 |
| -20 | | | | | | 136.138,143 |
| -4.6 (-14.0) | 11.5 | 35.5(b) 42.0(a,c) | 3.7 | 20 | 3.0 | 146,147,148,149 150 |

| Compound | $T_N(K)$ | Structure | $n_{eff}$ | Pre-ferred axis | Exchange energy $J/k(K)$ |
|---|---|---|---|---|---|
| 32. $Dy_3Al_5O_{12}$ | 2.53 | Cubic | | $(111)^c$ | c |
| 33. $CsMnCl_3 \cdot 2H_2O$ | 4.89 | Orthorhombic | | a | -3.0 |
| 34. $Cs_2MnCl_4 \cdot 2H_2O$ | 1.8 | Triclinic | | | $zJ/k=-0.54$ |
| 35. $Rb_2MnCl_4 \cdot 2H_2O$ | 2.24 | Triclinic | | | |
| 36. $Cu(NH_4)_4SO_4 \cdot H_2O$ | 0.37 | | | (ab) | $J_1/k=-3.15$ $J_2/J_1=0.01$ |
| 37. $Cu(NO_3)_2 \cdot 2.5H_2O$ | 3.2 | Monoclinic | | b | -5.14 (-2.56) |
| 38. $Mn(HCOO)_2 \cdot 2H_2O$ | 3.7 1.72 | Monoclinic | | (ac) b | $J_1/k=-0.65$ $J_2/k=+0.10$ |

[a] These compounds order in layer-type magnetic structure. (See text).
[b] These compounds order in linear chains. (See text).
[c] In $Dy_3Al_5O_{12}$ (DAG), Schneider et al. [157] indicate that the total interaction is a sum of Ising-like terms with the interaction constants, $K^{ij}$, being the sum of dipolar plus non-dipolar terms: $K^{ij} = K_j^{dip} + K_j^{non-dip}$

| $\theta$ (K) | $H_{sf}$(kOe) | $H_c$(kOe) | $T_t$(K) | $H_E$(kOe) | $H_A$(kOe) | Refs. |
|---|---|---|---|---|---|---|
| | | $H_i =$ $3.74^c$ (1.0) | $1.66^c$ | 6 | | 151,152,153,154, 155,156,157,158 |
| | | | | | | 159,160,161,163 165,178 |
| -3.0 | | | | 20 | 4.0 | 164,165,166,167, 168,178 |
| | | | | | | 165,167,168,166 162 |
| | | | | | | 169,170,171,172, 173,174,175,176 |
| | | $35^d$ | | | | 176,177,75,179 |
| $\theta_1$=-.05 $\theta_2$=-3.0 | | | | | | 180,181,182,183 184,185,186,187 188,189,190,191 192 |

$= (1 + \alpha_j)K_j^{dip}$, for j = 1,2,3. They conclude that $\alpha_1 = + 0.52\pm0.11$, $\alpha_2 = -0.13\pm0.18$, $\alpha_3 = -0.14\pm0.16$, $|K_1|/k = 0.770\pm0.052K$, $|K_2|/k = 0.159\pm0.032K$, and $K_3/k = -0.402\pm0.073K$. For details of the phase boundaries and transitions see Ref. [158].

[d] This compound orders in binary pairs (See Ref. [177]).

TABLE 5

Some Low Temperature Insulating Ferromagnetic Compounds
and their Properties

| Compound | $T_c$ (K) | Structure | $n_{eff}$ | Exchange energy $J/k$ (K) | $\theta$ (K) | Refs. |
|---|---|---|---|---|---|---|
| 1. $CrBr_3$ | 32.5 | Hexagonal | 3.85 | $J_T/k=8.25$<br>$J_L/K=0.5$ | +47 | 49, 50, 51, 52, 53, 193, 194, 195, 196, 197, 39 |
| 2. $EuS$ | 16.3 | Cubic | 6.8 | $J_1/k=0.20$<br>$J_2/k=-0.08$ | +19 | 198, 199, 200, 201, 202, 206, 208 |
| 3. $EuO$ | 69 (77) | Cubic | 6.9 | $J_1/k=0.76$<br>$J_2/k=-0.0974$ | +77 | 207, 208, 209, 210, 211, 198, 200, 213 |
| 4. $GdCl_3$ | 2.2 | Hexagonal | 7.0 | $J_1/k=-0.039$<br>$J_2/k=0.048$ | +2.9<br>(2.1) | 214, 215, 216, 217, 218, 219, 220, 221 |
| 5. $CuK_2Cl_4 \cdot 2H_2O$ | 0.88 | Tetragonal |  | $J_1/k=0.28$<br>$J_2/k=0.056$ | +1.2 | 222, 223, 224, 225, 226, 227, 228, 229 |
| 6. $Cu(NH_4)_2Cl_4 \cdot 2H_2O$ | 0.70 | Tetragonal |  | $J_1/k=0.20$<br>$J_2/k=0.060$ | +0.95 | 222, 223, 224, 225 |
| 7. $Cu(NH_4)_2Br_4 \cdot 2H_2O$ | 1.73 | Tetragonal |  |  |  | 224, 230, 229 |
| 8. $CuRb_2Cl_4 \cdot 2H_2O$ | 1.02 | Tetragonal |  |  |  | 224, 229 |

## REFERENCES

1. Michael M. Schieber, Experimental Magnetochemistry, North-Holland Publishing Co., Amsterdam, 1967.

2. W. P. Wolf, "Proc. Int. Magnetism Conf.", Grenoble, September, 1970, J. Physique 32, Suppl. to No. 2-3, Cl-26 (1971).

3. H. Bizette, Ann. Phys. Paris 1, 316 (1946).

4. J. W. Stout and E. Catalano, Phys. Rev. 92, 1575 (1953).

5. P. L. Richards, J. Appl. Phys. 34, 1237 (1963); J. Appl. Phys. 35, 850 (1964).

6. R. A. Erickson, Phys. Rev. 90, 779 (1953).

7. H. Kamimura, J. Appl. Phys. 35, 844 (1964).

8. H. Bizette, J. Phys. Radium 12, 161 (1961).

9. I. S. Jacobs, J. Appl. Phys. 32, 61S (1961).

10. J. C. Burgiel and M. W. P. Strundberg, J. Appl. Phys. 35, 852 (1964).

11. Y. Shipira, S. Foner, and A. Misetich, Phys. Rev. Letters 23, 98 (1969).

12. J. W. Stout and L. M. Matarrese, Rev. Mod. Phys. 25, 338 (1953).

13. R. G. Leisure and R. W. Moss, Phys. Rev. 188, 840 (1969).

14. Ojiro Nagai and Toshijiro Tanaka, Phys. Rev. 188, 821 (1969).

15. Dale T. Teaney, Phys. Rev. Letters 14, 898 (1965).

16. S. Foner, Magnetism, ed. G. T. Rado and H. Suhl, Academic Press, Inc., New York, 1963, Vol I, p. 383.

17. G. G. Low, A. Okazaki, R. W. H. Stevenson, and K. C. Turberfield, J. Appl. Phys. 35, 998 (1964).

18. M. Balkanski, P. Moch, and R. G. Shulman, J. Chem. Phys. 40, 1897 (1964).

19. E. Catalano and J. W. Stout, J. Chem. Phys. 23, 1284 (1955).

20. R. G. Shulman, Phys. Rev. 121, 125 (1961).

21. J. W. Stout and S. A. Reed, J. Am. Chem. Soc. 76, 5279 (1954).

22. W. H. Baur, Naturwiss, 44, 349 (1957).

23. M. K. Wilkinson, J. W. Cable, E. O. Wollan, and W. C. Koehler, Phys. Rev. 113, 1679 (1959).

24. I. S. Jacobs, S. Roberts, and P. E. Lawrence, J. Appl. Phys. 36, 1197 (1965).

25. C. Starr, F. Bitter, and A. R. Kauffman, Phys. Rev. 58, 977 (1940).

26. A. Ito and K. Ono, J. Phys. Soc. Japan 20, 784 (1965).

27. I. S. Jacobs and P. E. Lawrence, Phys. Rev. 164, 866 (1967).

28. O. N. Trapeznikova and L. V. Shubnikov, Physik z Sowjet Union 7, 66 (1935).

29. Albert Narath and J. E. Schirber, J. Appl. Phys. 37, 1124 (1966).

30. R. P. Kenan, R. E. Mills, and C. E. Campbell, J. Appl. Phys. 40, 1027 (1969).

31. P. Carrara, J. de Gunzbourg, and Y. Allain, J. Appl. Phys. 40, 1035 (1969).

32. H. Bizette, C. Terrier, and B. Tsai, Compt. Rend. 243, 895 (1956).

33. H. Bizette, C. Terrier, and B. Tasi, Compt. Rend. 245, 507 (1957).

34. I. S. Jacobs and P. E. Lawrence, J. Appl. Phys. 35, 996 (1964).

35. D. W. J. DeHaas, B. H. Schultz, and J. Koolhaas, Physica 7, 57 (1940).

36. E. O. Wollan, W. C. Koehler, and M. K. Wilkinson, Phys. Rev. 110, 630 (1958).

37. Wayne. B. Hadley and J. W. Stout, J. Chem. Phys. 39, 2205 (1963).

38. Alan J. Friedman and Robert A. Kromhout, Bull. Am. Phys. Soc., Ser. II, 16, 83 (1971).

39. W. N. Hansen, J. Appl. Phys. 30, 304S (1959).

40. W. N. Hansen and M. Griffel, J. Chem. Phys. 30, 913 (1959).

41. R. M. Bozorth and V. Kramer, J. Phys. Radium 20, 393 (1959).

42. K. Knox, Acta Cryst. 13, 507 (1960).

43. E. O. Wollan, H. R. Child, W. C. Koehler, and M. K. Wilkinson, Phys. Rev. 112, 1132 (1958).

44. H. Bizette and C. Terrier, J. Phys. Radium 23, 486 (1962).

45. J. W. Cable, M. K. Wilkinson, and E. O. Wollan, J. Phys. Chem. Solids 19, 29 (1961).

46. Albert Narath, J. Appl. Phys. 35, 838 (1964); Phys. Rev. 131, 1929 (1963).

47. J. W. Cable, M. K. Wilkinson, E. O. Wollan, and W. C. Koehler, Phys. Rev. 127, 714 (1962).

48. H. Bizette, C. Terrier, and A. Adams, Compt. Rend. 252, 1571 (1961).

49. I. Tsubokawa, J. Phys. Soc. Japan 15, 1664 (1960).

50. E. Legrand and R. Plumier, Physica Status Solidi 2, K112 (1962).

51. J. F. Dillon, J. Appl. Phys. 33, 1191 (1962); J. Phys. Soc. Japan 19, 1662 (1964).

52. M. Yamada, J. Phys. Soc. Japan 18, 1696 (1963).

53. A. C. Gossard, V. Jaccarino, E. D. Jones, and J. P. Remeika, Phys. Rev. 135, A1051 (1964).

54. H. M. Gijsman, N. J. Poulis, and J. Van den Handel, Physica 25, 954 (1959).

55. John E. Rives, Phys. Rev. 162, 491 (1967).

56. Warren E. Henry, Phys. Rev. 94, 1146 (1954).

57. M. Abkowitz and A. Honig, Phys. Rev. 136, A1003 (1964).

58. T. A. Reichert, R. A. Butera, and E. J. Schiller, Phys. Rev. B, 1, 4446 (1970).

59. George S. Dixon and John E. Rives, Phys. Rev. 177, 871 (1969).

60. S. A. Friedberg and J. D. Wasscher, Physica 19, 1072 (1953).

61. A. R. Miedema, R. F. Weilinga, and W. J. Huiskamp, Physica 31, 835 (1965).

62. R. D. Spence and V. Nagarajan, Phys. Rev. 149, 191 (1966).

63. J. N. McElearney, H. Forstat, and P. T. Bailey, Phys. Rev. 181, 887 (1969).

64. R. F. Altman, S. Spooner, J. E. Rives, and D. P. Landau, Winter Meeting Am. Cryst. Soc., Columbia, S. C., 1971.

65. Ikuji Tsujikawa and Eize Kanda, J. Phys. Radium 20, 352 (1959).

66. B. Bolger, Conference de Physique des Basses Temperatures, Paris, 1955, Centre National de la Recherche Scientifique and UNESCO, Paris, 1956. p. 244.

67. J. E. Rives and D. Walton, Phys. Letters 27A, 609 (1968).

68. J. E. Rives, D. Walton, and G. S. Dixon, J. Appl. Phys. 41, 1435 (1970).

69. R. Pierce and S. A. Friedberg, J. Appl. Phys. 32, 66S (1961).

70. T. Haseda and M. Date, J. Phys. Soc. Japan 13, 175 (1958).

71. J. T. Schriempf and S. A. Friedberg, Phys. Rev. 136, A518 (1964).

72. Norikiyo Uryu, Phys. Rev 136, A527 (1964).

73. R. D. Spence, R. Au, and P. A. Van Dalen, Physica 30, 1612 (1964).

74. S. A. Friedberg, A. F. Cohen, and J. H. Schelleng, J. Phys. Soc. Japan 17, Suppl. B1, 515 (1962).

75. S. A. Friedberg and C. A. Raquet, J. Appl. Phys. 39, 1132 (1968).

76. T. Haseda and E. Kanda, J. Phys. Soc. Japan 12, 1051 (1957).

77. W. K. Robinson and S. A. Friedberg, Phys. Rev. 117, 402 (1960).

78. W. Van der Lugt and N. J. Poulis, Physica 26, 917 (1960).

79. J. Skalyo, Jr., and S. A. Friedberg, Phys. Rev. Letters 13, 133 (1964).

80. J. Skalyo, Jr., A. F. Cohen, S. A. Friedberg, and Robert B. Griffiths, Phys. Rev. 164, 705 (1967).

81. R. D. Spence, Paul Middents, Z. ElSaffer, and R. Klienberg, J. Appl. Phys. 35, 854 (1964).

82. V. A. Schmidt and S. A. Friedberg, J. Appl. Phys. 38, 5319 (1967).

83. E. Sawatzky and M. Bloom, Can. J. Phys. 42, 657 (1964).

84. Robert Klienberg, J. Appl. Phys. 38, 1453 (1967).

85. D. G. Kapadnis and R. Hartmans, Physica 22, 181 (1956).

86. J. H. Schelleng and S. A. Friedberg, J. Appl. Phys. 34, 1087 (1963).

87. H. Forstat, G. Taylor, and R. D. Spence, Phys. Rev. 116, 897 (1959).

88. M. Garber, J. Phys. Soc. Japan 15, 734 (1960).

89. H. Forstat, J. N. McElearney, and P. T. Bailey, Phys. Letters 27A, 70 (1968).

90. N. J. Poulis and G. Hardeman, Physica 18, 201 (1952).

91. J. Van den Handel, H. M. Gijsman, and N. J. Poulis, Physica 18, 862 (1952).

92. G. J. Butterworth and V. S. Zidell, J. Appl. Phys. 40, 1033 (1969).

93. W. J. O'Sullivan, W. W. Simmons, and W. A. Robinson, Phys. Rev. 140, A1759 (1965).

94. D. V. G. L. Narasimha Rao, and A. Narasimhamurty, Phys. Rev. 132, 961 (1963).

95. N. J. Poulis and G. E. G. Hardeman, Physica 18, 429 (1952).

96. S. A. Friedberg, Physica 18, 714 (1952).

97. Hanako Kobayashi and Taiichiro Haseda, J. Phys. Soc. Japan 19, 765 (1964).

98. A. Narath, Phys. Rev. 136, A766 (1964); J. Phys. Soc. Japan 19, 2244 (1964).

99. B. Morosin and J. Graeber, Acta Cryst. 16, 1176 (1963).

100. K. A. Hay and J. B. Torrance, Jr., J. Appl. Phys. 40, 999 (1969).

101. A. Narath, Phys. Rev. 139, A1221 (1965).

102. B. Morosin and E. J. Greaber, J. Chem. Phys. 42, 898 (1965).

103. K. Inomata and T. Oguchi, J. Phys. Soc. Japan 23, 765 (1967).

104. J. N. McElearney, D. B. Losee, S. Merchant, and R. L. Carlin, Bull. Am. Phys. Soc., Ser. II, 16, 50 (1971).

105. R. D. Spence and R. D. Ewing, Phys. Rev. 112, 1544 (1958).

106. W. Van der Lugt and N. J. Poulis, Physica 25, 1313 (1959).

107. H. Forstat, G. Taylor, and B. R. King, J. Chem. Phys. 31, 929 (1959).

108. H. Forstat, N. D. Love, T. K. Duncan and P. T. Bailey, Bull. Am. Phys. Soc., Ser. II, 16, 50 (1971).

109. Yu. P. Simanov, L. P. Batsanova, and L. M. Kovba, Russ. J. Inorg. Chem. (English transl.) 2, (10) 207 (1957).

110. D. T. Teaney, V. L. Moruzzi, and B. E. Argyle, J. Appl. Phys. 37, 1122 (1966).

111. R. L. Melcher and D. I. Bolef, Phys. Rev. 178, 864 (1969); Phys. Rev. 186, 491 (1969).

112. R. L. Melcher, D. I. Bolef, and R. W. H. Stevenson, Solid State Comm. 5, 735 (1967).

113. S. J. Pickart, H. A. Alperin, and R. Nathans, J. Phys. (France) 25, 565 (1964).

114. A. J. Heeger and D. T. Teaney, J. Appl. Phys. 35, 846 (1964).

115. J. J. Frieser, R. J. Joenk, P. E. Seiden, and D. T. Teaney, Proc. Ints. Conf. Magnetism, Nottingham, 1964 (The Inst. of Physics and the Phys. Soc. London, 1965) p. 432.

116. H. -Y. Lau, L. M. Corliss, A. Delapalme, J. M. Hastings, R. Nathans, and A. Tucciarone, J. Appl. Phys. 41, 1384 (1970).

117. William J. Ince, J. Appl. Phys. 40, 1595 (1969).

118. J. Van den Handel and H. C. Meijer, Low Temperature Physics LT9, Plenum Press, New York, 1965, p. 873.

119. D. E. Eastman, Phys. Rev. 156, 645 (1967).

120. W. J. Ince and F. R. Morgenthaler, Phys. Letters 29A, 106 (1969).

121. J. B. Merry and D. I. Bolef, Phys. Rev. Letters 23, 126 (1969).

122. D. T. Teaney, M. J. Freiser, and R. W. H. Stevenson, Phys. Rev. Letters 9, 212 (1962).

123. M. J. Freiser, P. E. Seiden, and D. T. Teaney, Phys. Rev. Letters 10, 293 (1963).

124. Kenneth Lee, A. M. Portis, and G. L. Witt, Phys. Rev. 132, 144 (1963).

125. Marden H. Seavey, J. Appl. Phys. 40, 1597 (1969).

126. A. Zalkin, K. Lee, and D. H. Templeton, J. Chem. Phys. 37, 697 (1962).

127. K. Hirakawa and T. Hashimoto, J. Phys. Soc. Japan 15, 2063 (1960).

128. A. J. Heeger, O. Beckman, and A. M. Portis, Phys. Rev. 123, 1652 (1961).

129. D. Beckman and K. Knox, Phys. Rev. 121, 376 (1962).

130. A. Nakamura, V. Minkiewicz, and A. M. Portis, J. Appl. Phys. 35, 842 (1964).

131. R. G. Shulman and K. Knox, Phys. Rev. 119, 94 (1960).

132. D. E. Eastman and M. W. Shafer, J. Appl. Phys. 38, 1274 (1967).

133. S. A. Kizhaev, A. G. Tutov, and V. B. Bokov, Soviet Phys., Solid State 7, 2325 (1966).

134. L. Holmes, M. Eibschutz, and H. J. Guggenheim, Solid State Comm. 7, 973 (1969).

135. M. Eibschutz and H. J. Guggenheim, Solid State Comm. 6, 737 (1968).

136. R. H. Busey and E. Sonder, J. Chem. Phys. 36, 93 (1962).

138. C. M. Nelson, G. E. Boyd, and W. T. Smith, Jr., J. Am. Chem. Soc. 76, 348 (1954).

139. B. Aminoff, Z. Krist. 94, 246 (1936).

140. D. H. Templeton and Carol H. Dauben, J. Am. Chem. Soc. 73, 4492 (1951).

141. J. Dalziel, N. S. Gill, R. S. Nyholm, and P. D. Peacock, J. Chem. Soc. (London) 1958, 4012 (1958).

143. J. H. E. Griffiths, J. Owen, J. G. Park, and M. F. Partridge, Proc. Roy. Soc. (London) A250, 84 (1959).

144. A. H. Cooke, R. Lazenby, F. R. McKin, J. Owen, and W. P. Wolf, Proc. Roy. Soc. (London) A250, 97 (1959).

145. C. A. Bailey and P. L. Smith, Phys. Rev. 114, 1010 (1959).

146. J. D. Cashion, A. H. Cooke, J. F. B. Hawkes, M. J. M. Leask, T. L. Thorp, and M. R. Wells, J. Appl. Phys. 39, 1360 (1968).

147. K. W. Blazey and H. Rohrer, Phys. Rev. 173, 574 (1968).

148. S. Geller and V. B. Bala, Acta Cryst. 9, 1019 (1956).

149. K. W. Blazey and H. Roher, Helv. Phys. Acta 40, 370 (1967).

150. K. W. Blazey and G. Burns, Proc. Phys. Soc. (London) 91, 640 (1967).

151. B. E. Keen, D. P. Landau, and W. P. Wolf, Phys. Letters 23, 202 (1966).

152. M. Ball, W. P. Wolf, and A. F. G. Wyatt, Phys. Letters 10, 7 (1964).

153. M. Ball, M. J. M. Leask, W. P. Wolf, and A. F. G. Wyatt, J. Appl. Phys. 34, 1104 (1963).

154. B. E. Keen, D. Landau, B. Schneider, and W. P. Wolf, J. Appl. Phys. 37, 1120 (1966).

155. W. P. Wolf and A. F. G. Wyatt, Phys. Rev. Letters 13, 368 (1964).

156. M. Ball, M. T. Hutchings, M. J. M. Leask, and W. P. Wolf, Proceedings of the Eighth International Conference on Low Temperature Physics, London, 1962, ed R. O. Davies, Butterworth Scientific Publications, Ltd., London, 1962 p. 248.

157. B. Schneider, D. P. Landau, B. E. Keen, and W. P. Wolf, Phys. Letters 23, 210 (1966).

158. D. P. Landau, B. E. Keen, B. Schneider, and W. P. Wolf, Phys. Rev. B 3, 2310 (1971).

159. T. Smith and S. A. Friedberg, Phys. Rev. 176, 660 (1968).

160. R. D. Spence, W. J. M. deJonge, and K. V. S. Ramo Rao, J. Chem. Phys. 51, 4694 (1969).

161. S. J. Jensen, P. Anderson, and S. E. Rasmussen, Acta Chem. Scand. 16, 1890 (1962).

162. H. Forstat, N. D. Love, and J. N. McElearney, Phys. Letters 25A, 253 (1967).

163. J. Skaylo, Jr., G. Shirane, S. A. Friedberg, and H. Kobayashi, Phys. Rev. B, 2, 4632 (1970).

164. T. Smith and S. A. Friedberg, Phys. Rev. 177, 1012 (1969).

165. R. D. Spence, J. A. Casey, and V. Nagarajan, J. Appl. Phys. 39, 1011 (1968).

166. S. J. Jensen, Acta Chem. Scand. 18, 2085 (1964).

167. N. D. Love, J. N. McElearney, and H. Forstat, Bull. Am. Phys. Soc., Ser. II, 12, 285 (1967).

168. R. D. Spence, J. A. Casey, and V. Nagarajan, Phys. Rev. 181, 488 (1969).

169. W. T. Duffy, Jr., J. Lubbers, H. Van Kempen, T. Haseda, and A. R. Miedema, Proceedings of the Eighth International Conference on Low Temperature Physics, London, 1962, ed. R. O. Davies, Butterworth Scientific Publications, Ltd., London, 1962 p. 245.

170. J. J. Fritz and H. L. Pinch, J. Am. Chem. Soc. 79, 3644 (1959).

171. T. Haseda and A. R. Miedema, Physica 27, 1102 (1961).

172. T. Watanabe and T. Haseda, J. Chem. Phys. 29, 1429 (1958).

173. F. Mazzi, Acta Cryst. 8, 137 (1955).

174. Takehiko Oguchi, Phys. Rev. 133, A1098 (1964).

175. Robert B. Griffiths, Phys. Rev. 135, A659 (1964).

176. L. Berger, S. A. Friedberg, and J. T. Schriempf, Phys. Rev. 132, 1057 (1963).

177. B. E. Myers, L. Berger, and S. A. Friedberg, J. Appl. Phys. 40, 1149 (1969).

178. T. Smith and S. A. Friedberg. Proc. Eleventh Int. Conf. on Low Temperature Physics, St. Andrews, 1968, Univ. of St. Andrews Printing Department., 1968 p. 1345.

179. S. Wittekock and N. J. Poulis, J. Appl. Phys. 39, 1017 (1968).

180. Hidetaro Abe, Hiroko Morigaki, Motohiro Matsuura, Kiyoshi Torii, and Kazuo Yamagata, J. Phys. Soc. Japan 19, 775 (1964).

181. A. Foner Cohen, S. A. Friedberg, and G. R. Wagner, Phys. Letters 11, 198 (1964).

182. G. R. Wagner and S. A. Friedberg, Phys. Letters 9, 11 (1964).

183. S. A. Friedberg and J. T. Schriempf, J. Appl. Phys. 35, 1000 (1964).

184. R. D. Pierce and S. A. Friedberg, Phys. Rev. 165, 680 (1968).

185. J. Skalyo, Jr., G. Shirane, and S. A. Friedberg, Phys. Rev. 188, 1037 (1969).

186. K. Osaki, K. Nakai, and T. Watanabe, J. Phys. Soc. Japan 19, 717 (1964).

187. R. B. Flippen and S. A. Friedberg, J. Chem. Phys. 38, 2652 (1963).

188. K. Yamagata and H. Abe, J. Phys. Soc. Japan 20, 906 (1965).

189. H. Abe and M. Matsurra, J. Phys. Soc. Japan 19, 1867 (1964).

190. K. Yamagata and H. Abe, J. Phys. Soc. Japan 21, 408 (1966).

191. R. D. Pierce and S. A. Friedberg, J. Appl. Phys. 38, 1462 (1967).

192. J. T. Schriempf and S. A. Friedberg, J. Chem. Phys. 40, 296 (1964).

193. H. Bizette and C. Terrier, J. Phys. Radium 23, 486 (1962).

194. S. D. Senturin and G. B. Benedek, Phys. Rev. Letters 17, 475 (1966).

195. John T. Ho and J. D. Litster, Phys. Rev. B 2, 4523 (1970); J. Appl. Phys. 40, 1270 (1969).

196. Richard Silberglitt, C. H. Cobb, and V. Jaccarino, J. Appl. Phys. 41, 952 (1970).

197. H. L. Davis and A. Narath, Phys. Rev. 134, A433 (1964).

198. U. Enz, J. F. Fast, S. van Houten, and J. Smit, Philips Res. Rept. 17, 451 (1962).

199. C. Domb and R. B. Bowers, J. Phys. C (Solid State Phys.) Ser. 2, 2, 755 (1969).

200. D. C. McCollum, R. L. Wild, and J. Callaway, Phys. Rev. 136, A426 (1964).

201. B. J. C. Van der Hoeven, Jr., D. T. Teaney, and V. L. Moruzzi, Phys. Rev. Letters 20, 719 (1968).

202. S. Methfessel, M. J. Freiser, G. D. Pettit, and J. C. Suits, J. Appl. Phys. 38, 1500 (1967).

203. Albert Narrath and A. T. Fromhold, Jr., Phys. Rev. Letters 17, 354 (1966).

204. A. Narath and H. L. Davis, Phys. Rev. 137, A163 (1965).

205. W. N. Hansen and M. Griffel, J. Chem. Phys. 28, 902 (1958).

206. S. H. Charap and E. L. Boyd, Phys. Rev. 133, A811 (1964).

207. B. T. Matthias, R. M. Bozorth, and J. H. Van Vleck, Phys. Rev. Letters 7, 160 (1961).

208. T. R. McGuire and M. W. Shafer, J. Appl. Phys. 35, 984 (1964).

209. K. Y. Ahn and M. W. Shafer, J. Appl. Phys. 38, 1197 (1967).

210. Y. Shipira and T. B. Reed, J. Appl. Phys. 40, 1197 (1969).

211. J. C. Dimmock, C. E. Hurwitz, and T. B. Reed, J. Appl. Phys. 40, 1336 (1969).

212. A. J. Henderson, Jr., G. R. Brown, T. B. Reed, and H. Meyer, J. Appl. Phys. 41, 946 (1970).

213. E. L. Boyd, Phys. Rev. 145, 174 (1966).

214. W. P. Wolf, M. J. M. Leask, P. Mangum, and A. F. G. Wyatt, J. Phys. Soc. Japan 17, Suppl. B-1, 487 (1962).

215. M. J. M. Leask, W. P. Wolf, and A. F. G. Wyatt, Proceedings of the Eighth International Conference on Low Temperatures, London, 1962, Butterworth Scientific Publications, Ltd., London, 1962.

216. G. Garton, M. J. M. Leask, W. P. Wolf, and A. F. G. Wyatt, J. Appl. Phys. 34, 1083 (1963).

217. R. B. Clover and W. P. Wolf, Solid State Comm. 6, 331 (1968).

218. C. D. Marquard, Proc. Phys. Soc. (London) 92, 650 (1967).

219. C. D. Marquard and R. B. Stinchcombe, Proc. Phys. Soc. (London) 92, 665 (1967).

220. E. L. Boyd and W. P. Wolf, J. Appl. Phys. 36, 1027 (1965).

221. R. J. Birgeneau, M. T. Hutchings, and W. P. Wolf, J. Appl. Phys. 38, 957 (1967); Phys. Rev. Letters 17, 308 (1966).

222. L. Chrobak, Z. Krist. 88, 35 (1935).

223. A. R. Miedema, H. Van Kempen, and W. J. Huiskamp, Physica 29, 1266 (1963); Proceedings of the International Conference on Low Temperatures, London, 1962, Butterworth Scientific Publications, Ltd., London, 1962 p. 228.

224. A. R. Miedema, R. F. Wielinga, and W. J. Huiskamp, Physica 31, 1585 (1965).

225. D. W. Wood and N. W. Dalton, Proc. Phys. Soc. (London) 87, 755 (1966).

226. A. Narasimhamurty, Indian J. Pure Appl. Phys. 2, 37 (1964).

227. N. C. Ford, Jr., and C. D. Jeffries, Phys. Rev. 141, 381 (1966).

228. G. S. Dixon and D. Walton, Phys. Rev. 185, 735 (1969).

229. G. S. Dixon, J. E. Rives, and D. Walton, Proc. Int. Magnetism Conf., Grenoble, September, 1970, J. Physique 32, Suppl. to No. 2-3, C1-528 (1971).

230. J. N. Haasbroek and A. S. M. Gieske, Phys. Letters 31A, 351 (1970).

231. W. C. Koehler, M. K. Wilkinson, J. W. Cable, and E. O. Wollan, J. Phys. Radium 20, 180 (1959).

232. E. D. Jones and K. B. Jefferts, Phys. Rev. 135, A1277 (1964).

233. Y. Shipira and J. Zak, Phys. Rev. 170, 503 (1968).

234. A. Okazaki, K. T. Tuberfield, and R. W. H. Stevenson, Phys. Letters 8, 9 (1964).

235. J. P. Timbie and R. M. White, Solid State Comm. 8, 513 (1970).

Chapter 2

# PREPARATION AND PROPERTIES OF HIGH VALENT FIRST ROW TRANSITION METAL OXIDES AND HALIDES

C. Rosenblum and S. L. Holt
Chemistry Department, University of Wyoming
Laramie, Wyoming

## I.  INTRODUCTION

It is the purpose of this chapter to survey the physical and chemical properties of transition metal compounds which possess an unusually high formal oxidation state. In this group we include Cr(VI,V,IV), Mn (VII,VI,V,IV), Fe(VI,V,IV), Co(V,IV), Ni(IV,III), and Cu(III). While compounds containing Cr(VI) and Mn(VII) are relatively common by comparison, they are included in order to present a more complete picture of the tetrahedral oxyanions $[MnO_4]^{x-}$. In addition, we have chosen not to discuss second and third row transition metal ions, not because they do not possess important physical and chemical properties, but because it is less easy to delineate an "unusually high oxidation state" for these elements. In an attempt to give a uniformity to the discussion of the physical properties of an individual ion we have omitted mention of compounds which involve ligands which have extensive $\pi$-systems and in the main have limited ourselves to halides and oxides of the first row transition metal complexes.

## II.  CHROMIUM

### A.  Chromium(VI)

The most common higher oxidation state of chromium is Cr(VI). Salts (e.g., the soluble K, Na, Mg, and insoluble Ba, Pb and Ag) of the chromate ion $[CrO_4]^{2-}$, are easily obtained from basic solutions of chromium(VI) oxide. The powerful oxidizing agent dichromate $[Cr_2O_7]^{2-}$ is formed upon lowering the pH with $HNO_3$ or $HClO_4$. Other common Cr(VI) species include $[CrO_3X]^-$ where X = halogen, $(OSO_3)^-$, or $OH^-$, and $[CrO_2X_2]$ where X = F or Cl. Structures have been determined for many Cr(VI) compounds. Details of bond distances and angles are given in Table 1.

The dichromate anion has been shown to exist as oxobridged chromium dimers as shown in Fig. 1 [1-5]. For example, the structure of $Sr_2Cr_2O_7$ has been shown to consist of two distinct $[Cr_2O_7]^{2-}$ units [5]. In each of these units two distorted $[CrO_4]^-$ tetrahedra are linked together by sharing of a common corner of the tetrahedra. The difference in the two $[Cr_2O_7]^{2-}$ units is largely due to a variation in the Cr-O-Cr angle. In the first case this angle is 133°, the bond distances being Cr-O (a), 1.68 Å; Cr-O (b), 1.59 Å; and Cr-O (c), 1.71 Å while in the second case the equivalent angle is 140° with bond distances of Cr-O (a), 1.60 Å; Cr-O (b), 1.58 Å; and Cr-O (c), 1.75 Å. Both anionic types possess <u>mm</u> symmetry.

The potassium [3,6], ammonium [1,2], and three forms of rubidium [5] dichromate have structures similar to the strontium salt, but with some variation in bond angles and distances. In the case of $(NH_4)_2Cr_2O_7$ [1,2,7] and the C2/c structural modification of the rubidium salt [5], the symmetry of the anion is only $\underline{m}$ due to a rotation of one $CrO_3$ unit with respect to the Cr-O (c) axis.

The temperature dependence of the infrared spectra of $(NH_4)_2Cr_2O_7$ and $(ND_4)_2Cr_2O_7$ has been studied in the temperature range 298 - 17°K [8]. This work confirmed an earlier report [9] of three phase transitions occurring for the $[NH_4]^+$ salt. The presence of two phase transitions for the $[ND_4]^+$ salt was also observed. It was further concluded that the Cr-O-Cr distances remain unchanged at lower temperatures, while large changes occur in the terminal Cr-O distances. (This report contains numerous references to earlier infrared measurements.)

Neutron diffraction studies [10,11] indicate a barrier to rotation of the $[NH_4]^+$ ion of ~4 kcal/mole for $(NH_4)_2Cr_2O_7$, about one half that observed for $(NH_4)_2CrO_4$. It has been suggested that this result indicates weaker hydrogen bonding interactions in the case of the dichromate compound.

Wilhelmi has determined the structure of a number of high valent oxides. Among these are $MCr_3O_8$ [M = K, Cs, Li] [12,14] and $Cr_5O_{12}$ [12]. The $MCr_3O_8$ structures consist of Cr(III)$O_6$ octahedra and Cr(VI)$O_4$ tetrahedra which are linked to one another by sharing corners and edges. The sodium [13,16], rubidium [13,16], and thallium [16] salts have the $KCr_3O_8$ structure [12]. The $Cr^{VI}$-O distances (~1.65 Å) agree well with values reported for $Cr^{VI}$-$O_4$ groups in $[CrO_4]^{2-}$ salts and not with those in $CrO_3$ and $(NH_4)_2Cr_2O_7$. The bond angles $Cr^{III}$-O-$Cr^{VI}$, on the other hand, do not reflect the valence asymmetry of this bridge (cf. $CrO_3$). The $Cr^{VI}$-O (tetrahedral) and $Cr^{III}$-O octahedral bond distances are consistent throughout the series. The $Cr_5O_{12}$ oxide can be formulated as $Cr_2^{III}$, $Cr_3^{VI}O_{12}$, or $Cr_2(CrO_4)_3$ [15], Fig. 2.

The crystal structure of $CrO_3$ has recently been redetermined [17]. This redetermination has confirmed the structure of $CrO_3$ as originally reported [18,19], but has provided an improvement in the molecular parameters. The structure of $CrO_3$ consists of infinite chains of corner-sharing $CrO_4$ tetrahedra with the major axis of the chain running parallel to the crystallographic $\underline{c}$ axis. The bridging Cr-O bond lengths are 1.748 Å, somewhat less than those distances found in various $[Cr_2O_7]^{2-}$ anions, while the terminal Cr-O bond lengths are 1.599Å. When prepared as large crystals heavy striations are often noted parallel to the chain direction and a crystal may be easily fractured parallel to the streaks. This is to be expected as the $CrO_3$ chains are held together in the $\underline{a}$ and $\underline{b}$ directions by van der Waals' forces only.

Ammonium chlorochromate, $NH_4CrO_3Cl$, has the same structure as the $KCrO_3Cl$ [20] and has almost identical bond lengths, Cr-O = 1.52 Å and

TABLE 1

Crystal Structures of Cr(VI) Compounds

| Compound | Space group | Bond dist Å | Bond angle° | Comment | Ref. |
|---|---|---|---|---|---|
| $(NH_4)_2Cr_2O_7$ | C2/c | Cr-O=1.55, 1.57, 1.78 [trans to bridging O], and 1.91 [bridging O] | Cr-O-Cr=115° | The dichromate ions consist of two tetrahedra having a corner in common. The symmetry of the $[Cr_2O_7]^{2-}$ anion in the ammonium salt is m. | 1,2,7 |
| $SrCr_2O_7$ | $P4_2/nmc$ | Cr(1)-O=1.59, 1.68 [trans to bridging O], and 1.71 [bridging O]. Cr(2)-O=1.58, 1.60 [trans to bridging O], and 1.75 [bridging O]. | Cr(1)-O-Cr(1)=133. Cr(2)-O-Cr(2)=140. | Two distinct anion configurations exist. Both have mm symmetry. The standard deviations do not rule out identical Cr-O distances for equivalent bonds in each conformer. | 5 |

| Compound | Space group | Cr–O distances | Cr–O–Cr angles | Comments | Ref. |
|---|---|---|---|---|---|
| $Rb_2Cr_2O_7$ | C2/c | Cr-O=1.58-1.63 [peripheral O] and 1.82 [bridging O] | Cr-O-Cr=116 | Preliminary data. Two other modifications belonging to space group P2 1/n. | 5 |
| $K_2Cr_2O_7$ | P$\bar{1}$ | Cr-O=1.786 (av) [bridging O] Cr-O=1.629 (av) [terminal O] | Cr-O-Cr=125 (av) | Two independent dichromate anions with somewhat different Cr-O-Cr angles. Distances are the same within experimental error. | 6 |
| $KCr_3O_8$ | C2/m | $Cr^{III}$-O=1.97 (av) $Cr^{VI}$-O=1.64 (av) | $Cr^{III}_{(1)}$-$O_{(2)}$-$Cr^{VI}$=138 $Cr^{III}_{(1)}$-$O_{(1)}$-$Cr^{VI}$=146 | The $MCr_3O_8$ structures are a three-dimensional network of $Cr^{III}O_6$ octahedra and $Cr^{IV}O_4$ tetrahedra, each $O_{(2)}$ forms a | 12 |

(continued)

Table 1 (continued)

| | | | | | |
|---|---|---|---|---|---|
| $CsCr_3O_8$ | Pnma | $Cr^{III}$-O=1.96 (av) $Cr^{VI}$-O=1.67 (av) | $Cr^{III}$-O-$Cr^{VI}$ = 126–140 and one = 166 | corner for two octahedra and one tetrahedron, Each $O_{(1)}$ forms a corner for one octahedron and one tetrahedron. The polyhedra lie in sheets between which lie K ions, surrounded by ten oxygen neighbors or Cs, in contact with fourteen oxygen atoms. | 14 |
| $LiCr_3O_8$ | Cmcm | $Cr^{III}$-O=2.05 (av) $Cr^{VI}$-O=1.65 (av) | $Cr^{III}$-O-$Cr^{VI}$ = 127 and 132 | The average of the $Cr^{III}$-O distances in the octahedra of the lithium salt reflects the increase of effective metal atom radius | 13 |

$Cr_5O_{12}$  Pbcn

$Cr^{III}$-O 1.97 (av)  $Cr$-O-$Cr$=101.5

$Cr^{VI}$-O 1.67 (av)  -133.5

$Cr^{VI}$-$O_{(2)}$=1.81  O-$Cr^{III}$-O=

$Cr^{VI}$-$O_{(6)}$=1.54  99.3-77.6

O-$Cr^{VI}$-O=113.1

-105.0

(Li,Cr) over Cr alone in an $O_h$ environment. The lithium atom maximum coordination number is six, i.e., they are not in between the layers as K and Cs are.

Pairs of octahedra are joined by sharing edges of $O_{(2)}$ atoms. Each $O_{(1)}$, $O_{(3)}$, $O_{(4)}$ and $O_{(6)}$ atom forms a common corner of one octahedron and one tetrahedron. The $O_{(6)}$ belongs only to one Cr tetrahedron.

15

(continued)

TABLE 1 (continued)

| Compound | Space group | Distances | Description | Ref. |
|---|---|---|---|---|
| $CrO_3$ | C2cm | $Cr-2O_{(1)} = 1.79$ $Cr-O_{(2)(3)} = 1.81$ | $Cr^{VI}-O-Cr^{VI} = 136°$ The oxygen atoms form distorted tetrahedra around the Cr-atoms, sharing corners in the $\underline{c}$-direction, forming parallel chains. | [18] |
| $KCrO_3Cl$ | $P2_1C$ | $Cr-O = 1.53$ $Cr-Cl = 2.16$ | The structure is composed of $K^+$ and $CrO_3Cl^-$ ions in distribution analogous to Sheelite. | [20] |
| $Na_2CrO_4$ | Cmcm | $Cr-O \simeq 1.60$ (av) | In all $M_2CrO_4$ cases each Cr is tetrahedrally surrounded by four O atoms and the M is surrounded by nine or ten O atoms. | [23] |
| $K_2CrO_4$ | Pnma | $Cr-O \simeq 1.60$ (av) | | [24] |

| $(NH_4)_2CrO_4$ | Pm | $Cr-O=1.60$ and 1.98 <br> $NH_4-O=2.60+$ <br> $O-O$ in Td $\simeq 2.75$ <br> $Cr-NH_4=2.97$ | Two parallel rows of distorted $CrO_4$ tetrahedra are held together only through electrostatic forces, $NH_4^+-O^{--}$. | 28 |

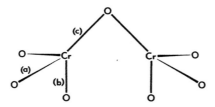

FIG. 1.  The structure of the $[Cr_2O_7]^{2-}$ anion.

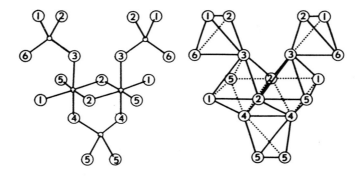

FIG. 2.  The structure of $Cr_5O_{12}$ [15].

Cr-Cl = 2.15 Å [21].  The ammonium fluorochromate on the other hand has a "barite" structure [21] while the potassium and cesium fluorochromates have the "sheelite" type structure [22], the $[CrO_3F]^-$ being an almost regular tetrahedron in $KCrO_3F$, with a Cr-O distance of 1.58 Å.

The structures of many salts of $[CrO_4]^{2-}$ have been determined.  Within the group $M_2CrO_4$, where M = Na, K, Rb, Cs, Tl, or $NH_4$, $Na_2CrO_4$ resembles $Na_2SO_4$ (III) [23] while the $K_2CrO_4$ [24], $Rb_2CrO_4$ [25], $Cs_2CrO_4$ [26], and $Tl_2CrO_4$ [27] salts are isomorphous with the $K_2SO_4$ structure.  The average Cr-O distance is 1.60 A in each case.  In $MCrO_4$ salts, where M = Ca [29], Ba [30], Mg [31], Pb [30], or Sr [30], the Cr-O distance within the tetrahedron of oxygens is a bit longer, 1.64 - 1.65 Å.  Chromates of Pb, Sr, and Ba belong to the "barite" class of structures, with coordination of twelve oxygens around the alkali ion, while $CaCrO_4$ is isomorphous with zircon, with an alkali ion coordination sphere of eight oxygens.  The magnesium compound is a heptahydrate, built of $[Mg(H_2O)_6]^{2+}$ octahedra and $[CrO_4]^{2-}$ tetrahedra, the extra $H_2O$ filling what would otherwise be a void in the structure [31].

Chromates (VI) of divalent transition metals, V, Ni, Co, Cu, Zn, and Cd are isomorphous, belonging to the space group Cmcm [32]. Each Cr-atom is surrounded by a distorted octahedron of oxygen atoms which share two edges with one another to form chains running parallel to the c axis. The other metal is tetrahedrally surrounded by four oxygen atoms belonging to three different Cr-O octahedral chains. The arrangement of octahedra and tetrahedra in $VCrO_4$ is the same as that found in $LiCr_3O_8$, formulated for example $(LiCr^{III})(Cr^{VI}O_4)_2$. ($CuCrO_4$ may be an exception, with the chromium tetrahedrally coordinated.) In $VCrO_4$: Cr-O = 1.95 or 2.03 Å, V-O = 1.72 or 1.80 Å; while in $CuCrO_4$: Cr-O = 1.66 or 1.68 Å, Cu-O = 2.05 or 2.15 Å.

Since Cr(VI) is a d° system, it exhibits neither a paramagnetic moment nor an electron spin resonance spectrum. Considerable effort has been put forward, however, in an attempt to understand the electronic structure of several Cr(VI) species. The most strenuous attempts have involved the tetrahedral oxyanion $[CrO_4]^{2-}$.

Only two reports of the low temperature polarized electronic spectrum (between 44,000 and 7,000 $cm^{-1}$) of a chromate containing crystal have appeared. The earliest work is that of Teltow [33]. Measurements were made at 20°K on the mixed crystal $K_2(S,Cr)O_4$. Recently Duinker and Ballhausen [34] have repeated the 20°K measurements and have extended them to 4°K. The overall features of both spectra are the same, two broad bands, one vibrationally structured band spanning 26,000 - 32,000 $cm^{-1}$ [34], centered at 26,810 $cm^{-1}$ (3.32 eV) [33], the other band, occurring in the range 34,000 - 44,000 $cm^{-1}$ $cm^{-1}$ [34], centered at 36,630 $cm^{-1}$ (4.54 eV) [33], Fig. 3.

Attempts to calculate an electronic structure for the $[CrO_4]^{2-}$ and $[MnO_4]^-$ with a view to providing an interpretation of the d° electronic spectrum in terms of transitions between particular molecular orbitals began in 1952 [35]. This initial work of Wolfsberg and Helmholz suggested an ordering of the metal $t_2$ and e orbitals ($t_2 < e$), Fig. 4. This result was considered to be of questionable accuracy. A molecular orbital scheme with the lowest unoccupied orbital, an e instead of a $t_2$, was subsequently suggested [36]. This is consistent with crystal field theory for tetrahedral ions and agrees with subsequent electron spin resonance findings pertaining to $[MnO_4]^{2-}$ [37]. A variety of molecular orbital theories appeared in the next ten years, for a summary see Ref. 38. These include specific calculations for the d° case, using a modification of the Wolfberg-Helmholz semiempirical approach [39] and have included electron repulsion or "ligand field" correction terms in diagonal and off-diagonal elements [40], variation of atomic orbital basis set and bond distances [41] and consideration of configuration interaction, i.e., eigenfunctions

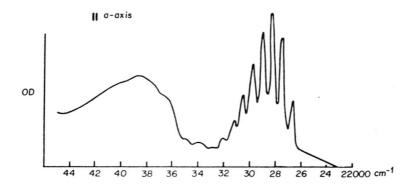

FIG. 3. The electronic spectrum of $[CrO_4]^{2-}$ at 20°K [34].

being linear combinations of spectroscopic configuration functions [42, 43].

These do not differ so much from one another in the ordering of highest-filled $(3t_2, t_1)$ and lowest-vacant $(2e, 4t_2)$ orbitals, but rather in assignment of transitions to corresponding absorption bands. (See for example, Refs. 38, 41, 42, and references cited therein.) The lack of significance of the absolute values of calculated molecular orbital energies was experimentally demonstrated for the d° tetrahedral oxyanions in 1966 [44]. Recently, the difficulties of correlating differences in the one-electron energy levels with electronic transitions in the visible and ultraviolet region of the spectrum have been appreciated (cf. Refs. 38, 40) and assignments of mixtures of different one-electron transitions to each absorption band have been made [41,42,43,45].

In all assignments of the d° spectrum of $[MnO_4]^-$ or $[CrO_4]^{2-}$, except one [41], the first, lower energy band (3.3 eV) is attributed predominantly to a $t_1 \rightarrow 2e(^1A_1 \rightarrow {}^1T_2)$ transition [34,36,42,43,47,49]. In the assignment of the second band, however, there is no consensus. Duinker and Ballhausen [34] cite similarity of magnetic circular dichroism between this band (4.5 eV), the third band of $MnO_4^-$ (3.99 eV), and the results of the calculation by Dahl and Johansen, on $[MnO_4]^-$ [48], in support of their $3t_2 \rightarrow 2e$ $(^1A_1 \rightarrow {}^1T_2)$ assignment. According to the calculation by Brown et al. [43], however, which includes interaction among excited configurations and reports closer correspondence between all calculated spectroscopic states and actual band positions than does Ref. 38 the third band is assigned predominantly to $t_1 \rightarrow 4t_2$ ($3t_2$ in their nomenclature). This latter assignment for the $[CrO_4]^{2-}$ 4.5 eV band implies a crystal field splitting, $\Delta$, for $[CrO_4]2-$ of 10,000 cm$^{-1}$. The Dahl and Johansen calculation on $[CrO_4]2-$ implies a $\Delta$ value of 9000 cm$^{-1}$ [41]. It is interesting to note that the

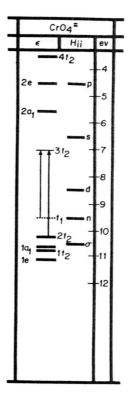

FIG. 4.  The one-electron energy levels for $[CrO_4]^{2-}$.  After Wolfsberg and Helmholz [35].

original ligand field calculation of Ballhausen and Liehr, assigning $t_1 \rightarrow e$ and $t_1 \rightarrow t_2$ to the first and second absorption bands also leads to a value of ~10,000 cm$^{-1}$ [36].  The $\Delta$ values of the tetrahedral oxyanions reflect in some measure the effective oxidation state of the metal ion and are, therefore, of particular interest.  These have been recently discussed [49].  Most recently, the first $\underline{ab}$ $\underline{initio}$ calculations of the one-electron energy levels of $[MnO_4]^-$ and $[CrO_4]^{2-}$ were reported [50].  The order of vacant, predominantly metal 3d orbitals is in accord with the electrostatic considerations of crystal field theory for the $[MnO_4]^-$, i.e., 2e below $7t_2$, but is reversed for $[CrO_4]^{2-}$, where the lowest unfilled orbital is $7a_1$ followed by $7t_2$, then 2e.

The electronic spectrum of crystalline $CrO_2Cl_2$, immersed in liquid helium was recently reported [51].  Because of the similarity of the band commencing at 5796 Å (17248 cm$^{-1}$, ~2.14 eV) in the chromyl chloride and fluoride, it was tentatively assigned as a transition involving predominantly oxygen atoms, rather than halogen atoms and was suggested to originate from the $t_1^6 e^0 \rightarrow t_1^5 e^1$ (Td approximation) excitation. As the symmetry of $CrO_2Cl_2$ is $C_{2v}$ and the sample was a frozen liquid (m.p. -96.5°C, b.p. 117°C), not a single crystal, it was not possible to obtain greater resolution of this band and perhaps discover the relative order of the $a_2$, $b_1$, and $b_2$ levels within the $t_1$ (Td) manifold.

The spectra of thirty-one chromates, seven dichromates, and $CrO_3$ measured in pellet form and at room temperature in the regions 250-4000 cm$^{-1}$ and 13,000-50,000 cm$^{-1}$ have been compared and discussed by Campbell [52]. Darrie et al. have measured the electronic spectrum of the hydrated and anhydrous chromates of Mg, La, Nd, and Sm [53]. Particular vibrational modes in the crystal for seven chromate salts [54], potassium dichromate [55], and $CrO_3$ and $K_2Cr_4O_{13}$ [56], have also been recently reported.

A review of the preparation and properties of the dark red liquid $CrO_2Cl_2$ appeared in 1958 [57]. The dark violet-red solid $CrO_2F_2$ (m.p. 31.6°C, yielding an orange liquid and red-brown vapor) can be prepared in a variety of ways; one fairly recent and straightforward preparation reacts $CrO_3$ with $SF_4$ at 5°C [58]. Six fundamental bands have been assigned from the infrared spectrum of gaseous $CrO_2F_2$ [59]. The mixed compound $CrO_2ClF$ [60] and the dark red $CrOF_4$ (m.p. 55°C) [61] have been reported.

The hexaflouride, a lemon-yellow solid, is formed by heating Cr(metal) to 400°C in the presence of 200 atmospheres of fluorine, and is unstable above -100°C [62]. An intense Cr-F stretching frequency of 785 cm$^{-1}$, at -180°C has been reported [63].

## B. Chromium(V)

The uncommon $3d^1$ chromium(V) species has been characterized in three different types of compounds; the dodecahedral $K_2CrO_8$, the tetrahedral $[CrO_4]^{3-}$ complexes, and the oxohalogen complexes, $[CrOX_4]^-$ and $[CrOX_5]^{2-}$. The esr and optical absorption spectral features of these species differ markedly. Details of structural and spin resonance properties are given in Tables 2 and 3.

The crystal structure of red-brown $K_3CrO_8$ was first determined in 1960 [64] and rerefined in 1962 [65]. The compound was observed to possess a distorted dodecahedral arrangement with the oxygen coordination occurring as pairs, Fig. 5. The average O-O distances were found to be trivially different from those normally observed for peroxo bonds ($O_2^{2-}$ = 1.49 Å) and it was observed that one oxygen atom of each pair lay 0.1 Å nearer the chromium than the other oxygen atom. Since the symmetry of the $[CrO_8]^{2-}$ anion is $D_2d$, the e and $t_2$ groups of d-orbitals are split.

TABLE 2

Crystal Structures of Cr(V) Compounds

| Compound | Space group | Bond dist. Å | Bond angle° | Comment | Ref. |
|---|---|---|---|---|---|
| $K_3CrO_8$ | I $\bar{4}$2m | $O_{(1)}$-$O_{(2)}$ pairs= 1.489<br><br>Cr-O=1.895 and 1.944 | $O_{(1)}$-Cr-$O_{(2)}$ = 45.6 | A dodecahedral configuration with Cr(V) atom tetrahedrally surrounded by four equidistant $O_2^{2-}$ ion pairs. | 64 |
| | | $O_a$-$O_b$ pairs = 1.405 ±0.039<br><br>Cr-$O_a$ =1.846 ±0.022<br><br>Cr-$O_b$=1.944 ±0.024 | $O_a$-Cr-$O_a$ = 86.8 ±1.4<br><br>$O_b$-Cr-$O_b$ = 173.7 ±2.5 | The point symmetry is $D_2$d. Refinement of data in Ref. 64 | 65 |
| $Ca_2CrO_4Cl$ | Pbcm | Cr-O=1.685 and 1.713 | O-Cr-O=119.1, 119.1, 104.6 and 105.1 | The $[CrO_4]^{3-}$ tetrahedra are held together along the $\underline{c}$ | 76 |

(continued)

TABLE 2 (continued)

| | | | | | |
|---|---|---|---|---|---|
| $Ca_5(CrO_4)_3OH$ | P63/m | Cr-O=1.66 (20) | $O_1$-Cr-$O_{(2)}$ = 115.0, 113.6, 103.6 and 106.4 | axis by the divalent Ca ions. The distortion of $[CrO_4]^{3-}$ tetrahedra is significantly greater than $[PO_4]^{3-}$ of spodiosite. | 73 |

TABLE 3

Electron Spin Resonance Data For Cr(V) Compounds

| Ion | Host | Temp. | $g_z$ $g\parallel$ | $\langle g \rangle$ | $g\perp$ | Line width (G) | $A_z$ $A\parallel$ | $A_x$ | $A\perp$ $A_y$ | Ref. |
|---|---|---|---|---|---|---|---|---|---|---|
| | | | | | | | $\times 10^{-4}$ cm$^{-1}$ | | | |
| Cr$^{5+}$ | 0.26% Cr as K$_3$CrO$_8$ in K$_3$NbO$_8$ | R.T. | 1.9434 ±0.0004 | | 1.9848 ±0.0004 | | | | | 66 |
| | K$_3$CrO$_8$, polycrystalline, | R.T. | 1.936 ±0.002 | | 1.983 ±0.002 | | | | | 65 |
| | K$_3$CrO$_8$ in H$_2$O$_2$ | 77°K | 1.951 | | 1.985 | | | | | 89 |
| | [CrO$_4$]$^{3-}$ : K$_2$(CrO$_4$) in KOH | 20°K | 1.98 | | 1.97 | | | | | 77 |
| | CrO$_3$ in SiO$_2$ | 20°K | 1.898 | | 1.975 | | | | | 78 |
| | Ca$_2$(PO$_4$,CrO$_4$)Cl | 77°K | 1.9936 ±0.0005 | | 1.9498 ±0.0002 | | 7.6 | 22.7 | 19.1 | 79 |
| | Ca(WO$_4$) with 1% Cr$^{5+}$ | 4.2°K | 1.988 | | 1.943 | | 4.0±1 | | 23±1 | 80 |
| | (Me$_4$N)CrOCl$_4$ pwdr. | R.T. | 1.985 | 1.990 | 1.996 | | | | | 90 |

TABLE 3 (continued)

| Ion | Host | Temp. | $g_z$ / $g\|$ | $\langle g \rangle$ | $g\perp$ | Line width (G) | $A_z$ / $A\|$ | $A_x$ (×10⁻⁴ cm⁻¹) | $A\perp$ / $A1$ | $A_y$ | Ref. |
|---|---|---|---|---|---|---|---|---|---|---|---|
| Cr⁵⁺ | KCrOF₄ in 48% HF | 77°K | 1.959 | 1.964 | 1.968 | | | +36.75 | | −22.97 | 93 |
| | K₂CrOF₅ in 48% HF | 77°K | 1.959 | 1.963 | 1.968 | | 1.74 | +36.75 | | −22.97 | |
| | (Et₄N)₂[CrOF₅] pwdr. | R.T. | 1.961 | 1.968 | 1.975 | | | | | | 90 |
| | Cr₂O₃ in Al₂O₃ | 77°K | | 1.96(γ) | | ~60 | | | | | 99,78 |
| | | R.T. | | 1.98 | | 40–50 | | | | | 100 |
| | CrO₃ in oleum | | 1.949 | | 1.975 | | 39 | | 9 | | 78 |
| | CrO₃ in 65% oleum | R.T. | 1.951 | 1.964 | 1.970 | | | | | | 97 |
| | | 77°K | 1.936 | 1.964 | 1.986 | | | | | | |
| | CrO₃ in 70% H₂SO₄ | 295°K | | 1.966 | 1.976 | 28 | | | | | 89 |
| | | 77°K | 1.969 | | | | | | | | |
| | K₂Cr₂O₇ in CF₃COOH | | 1.961 | | 1.977 | | | | | | 303 |
| | CrO₃ in 20% HF | 240°K | 1.960 | 1.960 | | | | | | | 89 |
| | CrO₃ in acetic acid | 77°K | 1.959 | | 1.970 | | | | | | 303 |
| [CrO]³⁺ | K₂CrOCl₅, acetic acid | 77°K | $2.008_5$ | | $1.977_3$ | | 36.1 | | 9.7 | | 91 |

| | | | | | |
|---|---|---|---|---|---|
| $Rb_2CrOCl_5$ powder | R.T. | 1.980 | 1.988 | 1.994 | 90 |
| $Rb_2CrOCl_5$ powder | R.T. | 1.978 | 1.986 | 1.990 | 92 |
| $Cs_2CrOCl_5$ powder | R.T. | 1.982 | 1.988 | 1.994 | 90 |
| $pyrHCrOCl_4$ soln. | 77°K | 2.008 | | 1.974 | 303 |
| $pyrHCrOCl_4$ powder | R.T. | $1.998_3$ | | $1.979_2$ | 91 |
| $(NH_4)_2CrOCl_5$ 20% HCl | 280°K | | 1.986 | | 89 |
| | 77°K | 1.995 | | 1.937 | |

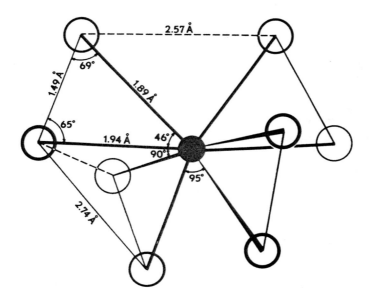

FIG. 5. The structure of the $[CrO_8]^{2-}$ anion [64].

The lower e becomes $a_1(d_{x^2})$ and $b_1(d_{x^2-y^2})$ while the upper $t_2$ level→ $b_2(d_{xy})$ and $e(d_{xz}, d_{yz})$. The electron spin resonance spectrum indicates that the single d-electron is located in a ground state of $b_1$ symmetry [66,65].

The optical spectrum of $K_3CrO_8$ in 30% $H_2O_2$, at room temperature displays one band at 2000 cm$^{-1}$ (f = 1.8 x 10$^{-2}$) which has been assigned to an electronic transition from a $b_1$ to an e molecular orbital [66]. The spectrum of the same material (<10$^{-3}$ molal) in KOH (temp.>77°K) displays two weak bands at 16900 and 18000 cm$^{-1}$, which have been assigned to the $b_1 \rightarrow a_1$ and $b_1 \rightarrow e$ transitions [65]. Approximation of the degree of Cr-O bond covalency, by calculation of crystal field parameters from the optical and esr data, is in agreement with the observed structure [65].

Cr(V) in tetrahedral coordination is found in a variety of alkali metal chromates. The earliest reports of pentavalent chromium were made by Klemm and Scholder with the high temperature preparation of micro-crystalline $Ba_3(CrO_4)_2$, $Sr_3(CrO_4)_2$, $Ba_5(CrO_4)_3OH$, and $Sr_5(CrO_4)_2OH$ [67-70]. X-ray powder photographs of these compounds, $Ca_3(CrO_4)_2$, $Ca_5(CrO_4)_3 \cdot OH$, and $[M_5(CrO_4)_3X]$, where M = Ca, Ba, or Sr and X = F or Cl [70,72,73,74,75] indicated that they were isomorphous with their phosphate analogs. The magnetic susceptibilities of compounds such as

$Ba_3(CrO_4)_2$ [70], $Na_3(CrO_4)_2$, and $Li_3(CrO_4)_2$ [71] yield magnetic moments $\approx 1.7$ B.M. Taken with the powder patterns these magnetic moments are strongly suggestive of the presence of the $Cr^{5+}$ ion. Recent complete structure determinations of $Ca_2CrO_4Cl$ [76] and $Ca_5(CrO_4)_3OH$ [72], and subsequent low-temperature measurements of the esr and optical spectra of these compounds, have further substantiated and characterized the Cr(V) ion in a tetrahedral site.

Unlike the esr of dodecahedral or octahedral Cr(V), the esr of the tetrahedral $[CrO_4]^{3-}$ species [77-81], is very broad or not evident at room temperature and has line widths of ~50 gauss only when $T < 80°K$. This is surprising as the electronic structure of a $d^1$ ion in $T_d$ symmetry should be comparable to that of a $d^9$ ion in $O_h$ symmetry. The room temperature esr spectrum of $Cu^{2+}$ in $O_h$ symmetry is easy to observe. It has been suggested this behavior may provide a clue to better understanding of the mechanism of spin-lattice relaxation [81].

The compounds $Ca_2CrO_4Cl$ and $Ca_2(CrO_4)_x(PO_4)_{1-x}Cl$ have been the subject of an intensive series of investigations. The crystal and molecular structure of both $Ca_2CrO_4Cl$ and $Ca_2PO_4Cl$ have been determined by Greenblatt et al. [76]. Both compounds exhibit distorted $[MO_4]^{3-}$ tetrahedra. The type of distortion is the same in both cases, namely a compression along one of the twofold axes. The distortion was found to be greater in the case of the $[CrO_4]^{3-}$ anion, and it was suggested that this was perhaps due to electronic effects. Comparison of the infrared spectra of $Ca_2VO_4Cl$, $Ca_2AsO_4Cl$, and $Ca_2CrO_4Cl$ [79] indicated again a more distorted tetrahedron for $[CrO_4]^{3-}$. These data were used to support the argument for the observation of the Jahn-Teller effect. These data are not conclusive, however, since the other $d^0$ $[MO_4]^{3-}$ species were also found to have distorted tetrahedral symmetry, indicating that crystal packing forces have a definite effect, while $d^1$ $[MnO_4]^{2-}$ species are known which show no distortion at all. The evidence for a ground state Jahn-Teller distortion in these systems is quite contradictory. The $[MnO_4]^{2-}$ tetrahedron in $K_2MnO_4$ appears, from X-ray data, to be regular within experimental error [82]. On the other hand, the absorption spectrum of $Cr^{5+}$ in $Li_3PO_4$ (the $[PO_4]^{3-}$ tetrahedra are quite regular) shows strong evidence of a distortion. In this system the distortion is clearly not a packing phenomenon, as the optical spectrum of $Mn^{5+}$ in the same host shows little or no evidence of a deviation from regular tetrahedral symmetry [83].

The electron spin resonance spectrum of $[CrO_4]^{3-}$ in $Ca_2PO_4Cl$ has been analyzed by Banks et al. [79]. The results of this work indicate that, in its ground state, the single d-electron occupies a $d_{z^2}$ orbital. The overall symmetry of the anion approximates $D_{2d}$, and there are two magnetically nonequivalent chromiums.

The optical spectrum of $Ca_2(CrO_4)_x(PO_4)_{1-x}Cl$ has been measured to

5°K [84,49]. This spectrum and its assignment are displayed in Figs.
6 and 7 [49]. This assignment suggests a d-orbital separation, $e_g$-$t_{2g}$,
of ~10,000 cm$^{-1}$, in agreement with the results of McGarvey [85].

The low temperature spectra of $Ca_2(CrO_4)_x(VO_4)_{1-x}Cl$ and $Sr_2(CrO_4)_x$
$(VO_4)_{1-x}Cl$ have also been measured [83] and are essentially the same as
that spectrum exhibited by $Ca_2(CrO_4)_x(PO_4)_{1-x}Cl$ with the exception that
many of the fine details stand out with much greater clarity when a
vanadate host is used.

Spodiosite analogs, e.g. $Ca_2CrO_4Cl$, as well as apatites, are best prepared by high temperature flux techniques. Typically one mixes $CaCl_2$
and a chromium compound of almost any oxidation state in approximately
a 20/1 ratio, heats the mixture to ~900°C for 24 hours, then cools the
resulting molten solution at a rate of 4°C per hour. This produces large
single crystals of $Ca_2CrO_4Cl$ which can then he removed from the mass
of flux by leaching with water. If the initial reaction temperature is
raised to ~1300°C and the molten mixture is quenched when the cooling
cycle reaches ~1200°C, the apatite phase is obtained. Using techniques
identical to these, it is possible to produce $Ca_2PO_4Cl$, $Ca_5(PO_4)_3Cl$ and
mixed chromium-phosphorus compounds. The latter are especially important for use as hosts for optical and esr measurements.

Banks et al. [86] have measured the esr and optical spectra of the apatite
analogs, $Ca_5(CrO_4)_x(PO_4)_{3-x}Cl$ and $Ca_5(CrO_4)_x(PO_4)_{3-x}F$. The results
of the electron spin resonance measurements indicate the presence of
$Cr^{5+}$ in three distinct sites in $Ca_5(CrO_4)_x(PO_4)_{3-x}Cl$ and in one site in
$Ca_5(CrO_4)_x(PO_4)_{3-x}F$.

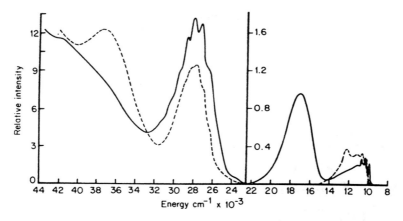

FIG. 6.  The polarized spectrum of $[CrO_4]^{3-}$ at 80°K [49].

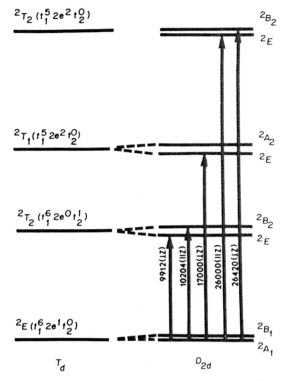

FIG. 7. The energy level diagram for $[CrO_4]^{3-}$ [49].

The 80°K spectra of the $F^-$ and $Cl^-$ species support the conclusion that three sites are present in the chloroapatite while only one is present in the fluoroapatite. These spectra are shown in Fig. 8. In the chloride spectrum the maxima at $\approx 9100$ cm$^{-1}$ are assigned to $t_1^6 2e^1 \rightarrow t_1^6 4t_2^1$ ($^2E \rightarrow ^2T_2$) transitions originating at two different molecular sites. Maxima at 13,300–13,850; 16,200–17,300; and 21,150 cm$^{-1}$ appear to be associated with charge transfer transitions occurring at three separate sites. The spectrum of the fluoroapatite shows no evidence for more than one type of chromophoric site.

There have been several attempts to rationalize the optical spectrum of $[CrO_4]^{3-}$ in terms of current theoretical approaches. Among the earliest efforts was that of Viste and Gray [39]. Employing a semiempirical approach they calculated the ordering of electronic levels for $[MnO_4]^-$, and used the result to interpret the reported spectrum of $[CrO_4]^{3-}$. Based upon the data of Bailey and Symons [87], the lowest-lying absorption maximum, $\sim 16,000$ cm$^{-1}$, was assigned as the one possible d-d transition.

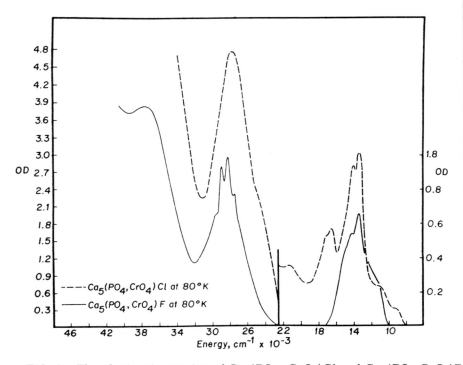

FIG. 8.  The electronic spectra of $Ca_5(PO_4, CrO_4)Cl$ and $Ca_5(PO_4,CrO_4)F$ [86].

There is some question as to the accuracy of their assignment for this anion inasmuch as the data mentioned earlier for $Ca_2CrO_4Cl$ indicates that the lowest-lying transition occurs at $\sim 10,000$ $cm^{-1}$ and not $16,000$ $cm^{-1}$. More will be said about this later.

Di Sipio et al. [88] have also applied a semiempirical molecular orbital approach to this problem, calculating the one-electron energy levels for $[CrO_4]^{3-}$ in $T_d$ and $D_{2d}$ symmetry. Using these calculated levels and employing an empirical correction for configuration interaction, they suggest a $\Delta$ ($Et_{2g}-E_{eg}$) of $10,650$ $cm^{-1}$, a value, remarkably, perhaps fortuitously, close to the experimental value as determined for $Ca_2CrO_4Cl$ [49,84,85].

Just as tetrahedral $[CrO_4]^{3-}$ species are found in highly alkaline solutions, a pseudo-octahedral configuration appears to be favored in neutral and acidic solutions. Characteristically for these species, the esr spectral line width is narrow and independent of temperature with $g||<g\perp$ in all but $M_2^I[CrOCl_5]$ and $M^I[CrOCl_4]$ (M = K, Rb, Cs, or $NH_4$) [89,90

91,92,93,94]. The esr results for a number of systems, both as solids and in solution, are presented in Table 3. Several sources of the reversal of g-anisotropy in the latter two structural types have been considered [80,90,92,95]. The esr and optical data for $[CrOCl_5]^{2-}$ and $[CrOF_5]^{2-}$ place the unpaired electron in a $b_2$ ($d_{xy}$) anti-bonding orbital. Recently, additional hyperfine structure, from the fifth fluorine in an axial position, has been observed [94]. The authors suggest that their data indicate extensive mixing of ground and excited states due to configuration interaction.

The electronic spectrum has been measured and assigned in several of the species. The reflectance spectrum of $(NH_4)_2 [CrOCl_5]$ at room temperature [95], $[CrOCl_5]^{2-}$ at room temperature [90] and 20°K [96], $K_2Cr_2O_4$ in 65% $H_2SO_4$, at room temperature [97] and $(Et_4N)_2[CrOF_5]$ [90] and several $[CrOCl_5]^{2-}$ and $[CrOCl_4]^-$ species in Nujol at 77°K [90] have all been reported (Reference 90 reports an extensive amount of data on these species.) Generally speaking, there are four band systems, I~12,000-13,000 cm$^{-1}$, II~18,000 cm$^{-1}$, III~22,000-24,000 cm$^{-1}$, and IV~25,000-40,000 cm$^{-1}$. The point symmetry in the $[CrOX_5]^{2-}$ compounds is $C_{4v'}$, having a tetragonally deformed octahedron with a chromyl bond, $[Cr=O]^{3-}$ along the symmetry axis. This removes the degeneracy of the $T_{2g}$ and $E_g$ levels, the $T_{2g}$ level splitting into $B_2(d_{xy})$, $E(d_{xz},d_{yz})$ components while the $E_g$ level produces components of $B_1(d_{x^2-y^2})$ and $A_1(d_{z^2})$ symmetry. (The type of metal orbital predominant in the m.o. giving rise to the spectroscopic state is indicated in parentheses.)

As can be seen in Table 4, there has been little agreement in assignment of observed transitions, with the exception of the assignment for band I.

The recent measurements of Ziebarth and Selbin [90] reveal vibronic coupling in band I, of the order of 600-800 cm$^{-1}$. This progression is assigned to a (Cr=O) stretching frequency which is 1000 cm$^{-1}$ in the ground state. Their overall assignment gives $10D_q = 18000$ cm$^{-1}$ and 22000 cm$^{-1}$ $[E(d_{x^2-y^2})-E(d_{xy})]$ in the oxo-chloro and oxo-fluoro ions respectively. A strong double bond between metal $d_{xz}$ and $d_{yz}$ and oxygen $2p_x$ and $2p_y$ and calculated spectral d-d transitions of $[Me^VOX_5]^{2-}$ where $Me^V$ = Cr, Mo, W or X = F, Cl, Br (using the simple $C_{\infty v}$ symmetry of the Me-O bond) are in agreement with observation made recently [98].

As noted in Table 3 the esr spectrum of $Cr^{5+}$ in $Al_2O_3$ has been measured [99,100]. Other less well-characterized or -studied $Cr^V$ oxygen-halogen compounds include $CrOCl_3$ [101], $CrOF_3$ [102], which does not appear to have been obtained in a pure form, $CrOF_4$ and $CrF_5$. The magnetic moment of $CrOF_4$ at R.T.=1.76 B.M. [102].

$CrF_5$, a red solid, mp 30°C, recently prepared with improved purity [103] has been shown to have an orthorhombic cell [61].

TABLE 4

Band Positions (in cm$^{-1}$) and Spectral Assignments for CrO$^{3+}$ Species

| Assignment | $(CrOCl_5)^{2-}$ [95] | $Cr^{5+}/65\%$ Oleum [97] | Species and energy | | | |
|---|---|---|---|---|---|---|
| | | | $(CrOCl_5)^{2-}$ [91] | $(CrOCl_5)^{2-}$ [93] | $(CrOCl_5)^{2-}$ [90] | $(CrOF_5)^{2-}$ [90] |
| $b_2(d_{xy}) \rightarrow e_\pi(d_{xz}, d_{yz})$ | 12,900 | 13,700 | 13,210 | | 12,500 | 8,350 |
| $b_2(d_{xy}) \rightarrow b_1(d_{x^2-y^2})$ | 23,500 | 18,100 | 23,200 | 23,500 | 18,000 | 23,200 |
| $e_\pi(\pi \text{ ox}) \rightarrow b_2(d_{xy})$ | 40,000 | | | | | |
| $b_1^b(\sigma_x) \rightarrow b_2(d_{xy})$ | | | 18,300 | 18,300 | | |
| $b_1^b(\sigma_x) \rightarrow b_1(d_{x^2-y^2})$ | | | | 40,000 | 22,500 | |
| $e_\pi^b(\pi \text{ Cl}) \rightarrow b_2(d_{xy})$ | | | | | 25,000 | 15,000 |
| $b_2(d_{xy}) \rightarrow a_1(d_{z^2})$ | | | | | 30,000 | |

## C. Chromium(IV)

Chromium (IV) oxide, $CrO_2$, is the only known transition metal oxide which is strongly ferromagnetic at room temperature. It has, therefore, been intensely studied from both the theoretical and practical points of view.

It was first isolated and associated with a magnetic component of chromium oxide mixtures in 1935 [104]. It was only obtained pure in the late 50's, however, This was achieved by thermal decomposition (350°C) of dry $CrO_3$ at an oxygen pressure of 60 atm [105]. A survey of methods used to grow single crystals appeared in 1967 [106]. Crystals between one and ten microns result from the thermal decomposition (400 - 500°C) of aqueous $CrO_3$ under high pressures (500 - 3000 atm) [107]. Crystals as large as 1.5 mm x 0.33 mm have been grown from $CrO_2$ at 900-1300°C and 60-65 K bar. Some large crystals have also been grown utilizing $Na_2CrO_4$ as a flux. Considerable intergrowth was reported, however. Details of the crystal structure appear in Table 5. The lattice parameters obtained from the single crystal [108] differ slightly from those obtained from the powder [109].

It has been observed that there is an anomalous change in the $\underline{c}$ axis with temperature [107-109], the dimensions of $\underline{c}$ increasing in the temperature range from -196 to ~100°C (~+.002Å), reaching a maximum value at ~0°C [109], ~-100°C [107] and then contracting as temperature is increased to 200°C (~0.006 - 0.01A). The ratio $c/a$ decreases smoothly, 0.005 units between -153 and +177°C (107). Darnell and Cloud [107] found the maximum slope of the decrease in the region of the Curie temperature. Reports of the value of the Curie temperature differ, however, by as much as 9°C: 124°C [109], 119°C [107], 121°C [110], 125°C [106], the most recent being 116°C [111].

The ferromagnetism of $CrO_2$ is especially interesting when one remembers that its two neighbors $VO_2$ and $MnO_2$ are paramagnetic and antiferromagnetic respectively, while the isoelectronic $MoO_2$ is diamagnetic. Obviously such differences in behavior must be rationalized in terms of a combination of local and collective electron effects and changes in local (i.e. point) and total (i.e. translational) crystal symmetry.

A discussion of the approach necessary for an understanding of this problem can be found in Refs. [110, 112, 113, 114].

The anomalous temperature dependence of the $CrO_2$ lattice parameters was at first tentatively attributed to a Jahn-Teller distortion of the 3d orbitals of $Cr^{4+}$ [107,109]. A distortion of the rutile anion octahedron, arising from a minimization of the elastic energy, rather than local electronic energy, perhaps having a local magnetic origin, was also considered to contribute in part [107].

TABLE 5

Crystal Structures of Cr(IV) Compounds

| Compound | Space group | Bond dist Å | Bond angle° | Comments | Ref. |
|---|---|---|---|---|---|
| $CrO_2$ | $P4/mnm$ | $O_{(1)}-O_{(2)}=2.57$ | $O_{(1)}-M-O_{(2)}=83$ | | 109 |
| | | $M-O_{(1)(2)}=1.94$ | | | |
| | | $M-O_{(3)(4)}=1.84$ | | | |
| | | $O_{(1)}-O_{(5)}=2.92$ | | | |
| | | $Cr-O$ (four) $=1.918$ | | The octahedral share | 108 |
| | | $Cr-O$ (two) $=1.882$ | | edges along the c | 110 |
| | | $Cr^{4+}-Cr^{4+}=2.918$ | | axis | |
| | | ($\parallel$ c-axis) | | | |
| $Sr_2CrO_4$ | $Pna_2$ | $T_1$: $Cr-O=1.83$, | $T_1$: O-Cr-O=2x112 | There are eight for- | 123 |
| | | 1.86, 1.83, 1.66 | 2x106, 109 and 110 | mula units in a unit | |
| | | $T_2$: $Cr-O=1.87$, | $T_2$: O-Cr-O=126, | cell and two types of | |
| | | 1.90, 1.95, 1.67 | 115, 97, 114 and | tetrahedra, $T_1$ and $T_2$, | |
| | | av. $Cr-O_{T_1}=1.80$ | 2x97 | one slightly more dis- | |
| | | av. $Cr-O_{T_2}=1.85$ | | torted than the other. | |
| $SrCrO_3$ | $Pm3m$ | | | Cr(IV) is in a cubic | 124 |
| | | | | persovskite lattice, | |
| | | | | a=3.818; Cr(IV) radius | |
| | | | | =0.57Å | |

Within a classification of sources of spontaneous crystallographic distortions [110], Goodenough has accounted for the magnetic ordering and moment and the metallic behavior (i.e. electrical conductivity) of $CrO_2$ [115]. Considering nearest neighbor interactions between sublattices in the rutile structure, each oxygen has three coplanar chromium near-neighbors. A cation-anion-cation 180° ($\pi$-type interaction between $p_\pi$ and $t_{2g}$ orbitals) ferromagnetic ordering, at an angle away from the $\underline{c}$ axis is, therefore, possible in addition to a cation-cation anti-ferromagnetic interaction directly along the $\underline{c}$ axis ($Cr^{4+}$-$Cr^{4+}$ 2.918Å). The elongation of the $\underline{c}$ axis as the temperature passes below the Curie point is attributed to electron repulsion between the half-filled d-orbitals, situated along the $\underline{c}$ axis, increasing as the spins become parallel. In $MoO_2$ the metal-metal distance is short enough (~2.50 A) [116] along the $\underline{c}$ axis to allow the cation-cation anti-ferromagnetic exchange to dominate. The observation, using neutron diffraction, of [100] planes of easy magnetization at ~40° angle to the $\underline{c}$ axis is consistent with Goodenough's scheme [107,108,117]. In his scheme the $\pi$-type overlap between the two metal $t_{2g}(\perp)$ orbitals and oxygen $p_\pi$ orbitals forms $\pi$ and $\pi*$ states continuously throughout the crystal, i.e. $\pi$ and $\pi*$ "bands." (It is helpful to remember that the 180° cation-anion-cation anti-ferromagnetic interaction of $KNiF_3$, for example, involves $\sigma$-type overlap, locally, with the $d_{x^2-y^2}$ electron traversing via the $p\sigma$ orbital to the $d_{x^2-y^2}$ orbital on the other metal.) $CrO_2$ is a relatively good electrical conductor, resistivity at room temperature, in powders ~ $4 \times 10^{-2}$-$2 \times 10^{-3}$ ohm-cm and ~$3 \times 10^{-4}$ ohm-cm in single crystals [107]. By associating the remaining electron (see $CrO_2$ energy band scheme, Refs. [113 and 115]), from the nonbonding localized $t_{2g}(||)$ orbital (pointing toward the next chromium on the $\underline{c}$ axis), with the half-filled $(\alpha)\pi*$ band, both the magnetic moment, $\mu = 2$ B.M. [108, 109], and conductivity can be accounted for [115,118].

A material like $CrO_2$, with both conductivity and ordered magnetic structure at room temperature is, of course, potentially useful in an electronic recording or storage device. The variation of electrical and magnetic properties with wavelength [111] and temperature [111,118] has begun to be studied.

There are two reports of electron spin resonance spectrum of Cr(IV), in $\alpha Al_2O_3$ and in Si (Table 6). Cr(IV) in $\alpha Al_2O_3$ is produced as bright orange crystals by heating Cr(III)/$\alpha Al_2O_3$ under an atmosphere of $O_2$ to 1400°C for sixteen hours [119]. Nitride ion, $N^{3-}$, substitution for $O^{2-}$, is used to provide charge compensation. The Cr(IV) here is ocatahedrally surrounded by 6 oxygens and has a $^3A_2$ ground state. Estimates of trigonal and lower symmetry field splitting were $D \simeq +7cm^{-1}$ and $E<0.05$ $cm^{-1}$. The spin lattice relaxation time, $<10^{-9}$ seconds exhibits a strong temperature dependence, the line width increasing until the signal is lost just below 77°K. The optical absorption spectrum, at 77° K, between 3,000 and 6,000 Å exhibits a strong peak at 4,600 Å and shoulder at 37,000 Å.

TABLE 6

Electron Spin Resonance Data for Cr(IV) Compounds

| Compound | Temp. | $g\|\|$ | $<g>$ | A | Ref. |
|---|---|---|---|---|---|
| $Cr^{4+}$ in $\alpha Al_2O_3$ | 4.2°K | 1.90 | | | 119 |
| $Cr^{4+}$ in Si | 4.2°K | | 1.9962 | 12.54 | 120 |

It was suggested that this spectrum is characteristic of the $\alpha Al_2O_3$ crystal and centers of anionic charge compensation than of the Cr(IV) ion [119].

Cr(IV) in silicon is tetrahedrally surrounded by four nearest neighbors. The four $^{53}Cr$ [natural abundance 9.54%] hyperfine lines were shown to be separated by $\simeq 13.5$ gauss. Each line is further split by interaction with neighboring $^{29}Si$ nuclei [120].

Lunkin et al. [121] have reported the preparation of $Cr^{4+}$ in an aluminum-lime glass. This was accomplished by inserting low concentrations of $Cr^{3+}$ in a melt consisting of 10(mole)% $SiO_2$, 30% $Al_2O_3$, and 60% CaO and treating the mixture with nitrates in an $O_2$ atmosphere. This procedure produced a blue material with absorption maxima at ~12,500, ~14,250, and ~16,000 cm$^{-1}$. No electron spin resonance spectrum was observed at room temperature. Irradiation of the glass with $\gamma$-rays produced a signal, g = 1.97, which the authors considered to be characteristic of $Cr^{5+}$. This was cited as supporting evidence for the presence of $Cr^{4+}$ in the original formulation.

The preparation of a number of compounds containing the $[CrO_4]^{4-}$ ion has been reported by Scholder [122,68,70]. Blue-black prisms of $Sr_2CrO_4$, stable in air and insoluble in hot acetic acid, were obtained by the reaction [122]:

$$Cr_2O_3 + SrCrO_4 + 5Sr(OH)_2 \xrightarrow[Ar]{900°C} 3Sr_2CrO_4 + 5H_2O \qquad (1)$$

The barium analog can be prepared in several different ways [68,70]:

$$Ba_3[Cr(OH)_6]_2 + Ba(OH)_2 \xrightarrow[N_2]{1000°} 2Ba_2CrO_4 + 6H_2O + H_2 \qquad (2)$$

$$BaCrO_4 + Cr_2O_3 + 5Ba(OH)_2 \xrightarrow[N_2]{1000°} 3Ba_2CrO_4 + 5H_2O \qquad (3)$$

$$BaCrO_4 + Ba(OH)_2 \xrightarrow[O_2]{400-500°C} Ba_2CrO_4 + H_2O + 1/2\,O_2 \qquad (4)$$

Measurement of the magnetic susceptibility of the bright green $Ba_2CrO_4$ yields a moment of 2.82 B.M. [67,70], consistent with two unpaired electrons.

Wilhelmi [123] has investigated the crystal structure of $Sr_2CrO_4$, noting that the structure of this compound is very similar to that of $Ba_2TiO_4$. His results indicate that the metal-oxygen bond lengths are rather unusual, being generally shorter than those found in $CrO_2$ and in one bond of each tetrahedron, identical with values found for Cr(V) and Cr(VI) oxygen bonds, cf. Table 5.

Black crystals of $SrCrO_3$, air stable, insoluble in dilute acids and alkali, have been prepared by reacting strontium oxide or chromate with $CrO_2$ at 800°C under 40,000 atm pressure [67,69]. The Cr(IV) is octahedrally coordinated to six oxygens, and observed metallic conductivity is ascribed to a band structure similar to one described for $CrO_2$ [124].

The barium analog, $BaCrO_3$, exhibits several polytypes [124,125]. All of these are prepared at 60-65 kbars pressure, but at a variety of temperatures ranging from 750°-1200°C. The overall reactions required to produce the various $BaCrO_3$ modifications are

$$BaO + CrO_2 \rightarrow BaCrO_3 \qquad (5)$$

$$Ba_2CrO_4 + CrO_2 \rightarrow 2BaCrO_3 \qquad (6)$$

The structural types produced include three hexagonal phases, three rhombohedral phases, one orthorhombic, one cubic, and one monoclinic phase. Details are given in Table 5.

In addition the lead analog $PbCrO_3$ [126,127], $Ba_3CrO_5$ [68], $Na_3CrO_4$ [68], and $Na_2CrO_3$ [128] have also been prepared. The preparation of the last compound involved the somewhat unusual procedure of reacting liquid sodium with $Cr_2O_3$. This produces a green hydroscopic powder which has a magnetic moment of 2.49-2.52 B.M. and appears to be isostructural with $Li_2TiO_3$.

Several fluorides of Cr(IV) are known. The binary fluoride, $CrF_4$, is best prepared by fluorination of the metal at 350°C [129,130]. This compound is a green solid, turning brown upon exposure to moisture, the vacuum sublimation product (+100°C) often having a blue hue. It is insoluble in organic solvents, fairly inert, with $\mu_{eff}$ = 3.02 B.M. at 294°K, $\theta$ = -70°. Though it is apparently amorphous, it has been suggested that the structure consists of chains of $CrF_6$ octahedra joined by shared edges [130].

Salts of the type $M_2^I CrF_6$ where M = Na, K, Rb, and Cs, can be prepared either by fluorinating a $CrCl_3$, MCl, or MF, $CrCl_3$ mixture with $F_2$ at 350-375°C [131,132] or $BrF_3$ at 130°C [130]. To obtain those salts of the type $MCrF_5$, where M = K, Rb, or Cs, a mixture of $CrF_4$ and MCl is reacted with $BrF_3$ [130]. The crystal habit is either cubic or hexagonal or both (polymorphic). The lattice parameters appear in Table 7.

The Cr(IV) radius in $Na_2CrF_6$ was determined to be 0.56 Å, very close to that found in $SrCrO_3$.

Many of the above salts and $CrF_4$ exhibit significant deviation from spin only (for two unpaired electrons) moments ($\mu_{eff}$ range 3.02-3.22 B.M.) and rather large negative Weiss constants ($\theta$ range -32° to -105°), indicative of spin orbit and/or anti-ferromagnetic interactions. Indeed the crystal and magnetic structures of Cr(IV) oxide and fluoride salts will very likely provide opportunities to test and practice our schemes for sorting out influences of localized and collective electrons in the solid state. In rutile-like structures, for example, the $\pi$-type overlap of $p\pi$ and metal $d\pi$ orbitals would be expected to be less in the less polarizable fluorides than oxides and crystalline d-band formation, and, therefore, electrical conductivity less likely [115].

### III.  MANGANESE

#### A.  Manganese (VII)

Attempts to account for the optical absorption spectrum of $[MnO_4]^-$ between 14,000 and 47,000 $cm^{-1}$ in terms of transitions within a known and consistent electronic structure have been pivotal in defining chemical bond covalency in the past fifteen years. In order to assign the absorption bands of $[MnO_4]^-$, a d° system; 18,000-23,000 $cm^{-1}$ (I), 24,500-30,000 $cm^{-1}$ (II), 30,000-38,000 $cm^{-1}$ (III), 38,000-46,00 $cm^{-1}$ (IV) [47], and ~14,400 $cm^{-1}$ [134], Fig. 9, an educated guess must be made as to the symmetries of the highest filled, predominantly ligand, orbitals.

As described in the section of Cr(VI), agreement that these orbitals are of $t_2$ and $t_1$ symmetries resulted from a great variety of semiempirical molecular orbital calculations, (for references, see discussion of $[CrO_4]^{2-}$). It was quickly realized, however, that band assignment was not possible solely on the basis of the ordered (by calculation) one-electron energy levels, and contributions from two or more transitions to a band were suggested here, as in the $[CrO_4]^{2-}$ spectra [41, 42,43,45].

TABLE 7

Lattice Parameters for Several Cr(IV) Fluorides

| Compound | Cubic | Hexagonal | Ref. |
|---|---|---|---|
| $Na_2CrF_6$ | | a = 9.14, c = 5.15 | 132 |
| $K_2CrF_6$ | a = 8.104 | | 130 |
| | a = 8.15 | a = 5.69, c = 9.34 | 133 |
| $Rb_2CrF_6$ | a = 8.506 | a = 5.94, c = 9.67 | 133 |
| $Cs_2CrF_6$ | a = 8.916 | | |
| | a = 9.004 | | |
| $KCrF_5$ | | a = 8.739, c = 5.226 | 130 |
| $RbCrF_5$ | | a = 6.985, c = 12.12 | 130 |
| $CrCrF_5$ | a = 8.107 | | 130 |

In two recent approximate calculations which take, in part, some excited state repulsions into account [41,43] the assignments in Table 8 were made.

Agreement between the observed and calculated sequence of spectroscopic states (in eV, $^1T_1$: 2.26, $^1T_2$: 3.50, $^1T_2$: 3.95 and $^1T_1$: 5.40) [47] supports the assignment of Brown et al. [43], while the observed similarity of

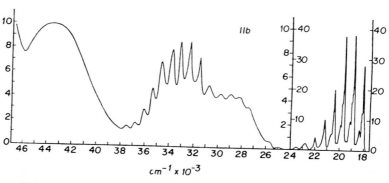

FIG. 9. The electronic spectrum of $KMnO_4$ at 4.2°K [47].

TABLE 8

The Electronic Structure of $[MnO_4]^-$

| Band No. | eV | Brown et al.[a] [43] | Dahl and Johansen [41] |
|---|---|---|---|
| I | $\simeq 2.27$ | $t_1 \to 2e\,(^1A_1 \to {}^1T_1)$ | $\begin{array}{l} t_1 \to 2e \\ 3t_2 \to 2e \end{array} (^1A_1 \to {}^1T_2)$ |
| II | $\simeq 3.5$ | $\begin{array}{l} t_1 \to 2e \\ 3t_2 \to 2e \end{array} (^1A_1 \to {}^1T_2)$ | $t_1 \to 4t_2\,(^1A_1 \to {}^1T_2)$ |
| III | $\simeq 4.0$ | $\begin{array}{l} t_1 \to 4t_2 \\ 3t_2 \to 4t_2 \end{array} (^1A_1 \to {}^1T_2)$ | $\begin{array}{l} 3t_2 \to 2e \\ t_1 \to 2e \end{array} (^1A_1 \to {}^1T_2)$ |
| IV | $\simeq 5.5$ | $\begin{array}{l} 3t_2 \to 2e \\ t_1 \to 4t_2 \end{array} (^1A_1 \to {}^1T_1)$ | $\begin{array}{l} 1e \to 4t_2 \\ 3t_2 \to 4t_2 \end{array}$ |

[a]Nomenclature $2t_2$ and $3t_2$ changed to $3t_2$ and $4t_2$ respectively for comparison with Dahl and Johansen who included 32 electrons, while Brown et al. used only 24.

vibrational progressions (the symmetric stretch $\nu_1(a_1) \simeq 770$ cm$^{-1}$ in the excited state) on bands I and III [47] tends to support Dahl and Johansen's assignment [41]. The approximate value of the crystal field splitting, $\Delta$, is $\simeq 14,000$ cm$^{-1}$ in the former and $\simeq 10,000$ cm$^{-1}$ in the latter calculation.

The Dahl and Johansen assignment, based on a calculation of one-electron energy levels, might be expected to be more tenuous than the Brown et al. CNDO calculation of energies of excited spectroscopic states. The great influence of the method of including ligand field corrections, in one-electron energy level calculations, on both the estimate of central metal charge and the order of filled molecular orbitals was recently demonstrated for this very ion, $[MnO_4]^-$ [40]. Nevertheless, it appears that the Dahl and Johansen assignment is more consistent with the most recent observations [134]. It will be remembered that $^3T_2$, $^3T_1$, $^1T_2$ and $^1T_1$ states result from the excited configuration $t_1^5e^1$, with the $^1A_1$ $^1T_2$ transition being the only totally (i.e., spin and electric dipole) allowed transition in tetrahedral symmetry. The $^1T_1$ state would be expected to be lower in energy than the $^1T_2$ [47,135]. So far this is consistent with the Brown et al. scheme, with the band intensity afforded perhaps through vibronic coupling. Vibronic structure would be expected to be evident on band IV in this case, however, and is not observed in [47,134,135]. More conclusive perhaps is the recent evidence of the $^1A_1 \rightarrow {}^1T_1$ band in the region of 14,400 cm$^{-1}$, anticipated in part by Ballhausen [135] from Teltow's data [33]. The manifold of the $t_1 \rightarrow 2e$ (predominant) transition at 2.27 eV is, therefore, $^1A_1 \rightarrow {}^1T_2$. Dahl and Johansen found that the order of molecular orbital levels remained essentially unchanged as the Mn-O bond length was varied from 1.552 to 1.689 Å. The experimentally determined bond distance, 1.629 Å [136] was used by Brown et al. [43]. Although a Mn-O distance of 1.59 A was used also in a recent ab initio calculation on [MnO$_4$]$^-$ [50], it probably did not affect the conclusion of the study, that consideration of interaction among excited configurations is necessary in any attempt to calculate transition energies. The approximate value of $\Delta$, ~12,000 cm$^{-1}$, which we believe reasonable [49], is apparently fortuitous, the energies between the $6a_1$ (highest filled) and all other (empty) levels being very close to this value and all transition energies being quite high (Table 9).

No doubt future experimental observations, such as study of magnetic circular dichroism of charge transfer bands will provide refinement of our concepts of electronic structure and covalency. (An excellent, up-to-date review of the evidence for and against the participation of (empty) 3d-orbitals in covalent bonding, in particular compounds, has been made by C. K. Jørgensen [137]).

Since the estimation of a $\nu_1(a_1)$ value in [MnO$_4$]$^-$ by Ballhausen [135], the fundamental frequencies of this ion have been measured in single crystals at 4.2°K [138], and have been compared with the fundamentals in the $^{18}$O labelled salt, (the unlabeled ion having $\nu_1 = 844$, $\nu_2 = 385$, $\nu_3 = 910$, and $\nu_4 = 407$ cm$^{-1}$ values) [139]. Variation in the positions and intensities of these fundamentals was observed in a recent report of the spectra (300-1000 cm$^{-1}$) of KMnO$_4$/MX (M = k, Rb and X = Cl, Br) taken over a concentration range of 1/100 to 1/6000 and at 40°C and 77°K [140]. The changes observed were interpreted as indicative of solid solution (i.e., MnO$_4^-$ occupying the X$^-$ site) formation. This occurred most readily in the RbBr. The ease of formation of such solid solutions was related to the polarizability of the alkali metal.

The Raman spectrum of [MnO$_4$]$^-$ is difficult to obtain reproducibly due to MnO$_2$ contamination from photochemical decomposition of the light-sensitive [MnO$_4$]$^-$ [141].

Although Mn$_2$O$_7$ has been known to be an unstable red-green oil, forming in sulfuric acid solutions of KMnO$_4$, it was first used as a powerful oxidant in the preparation of MnOCl$_3$, MnO$_2$Cl$_2$, and MnO$_3$Cl in 1968, and its spectrum, in CCl$_4$ at 25°C was reported at that time [142]. It reacts slowly (after a day at room temperature) with CCl$_4$, forming COCl$_2$,

TABLE 9

| Transition | Energy, $cm^{-1}$ |
|---|---|
| $6a_1 \rightarrow 8t_2$ | 12,570 |
| $6a_1 \rightarrow 7a_1$ | 12,100 |
| $6a_1 \rightarrow 7t_2$ | 11,600 |
| $6a_1 \rightarrow 2e$ | 10,400 |

$MnO_3Cl$ and $MnO_2$. Its spectrum is very similar to that of $MnO_3Cl$, but is less structured.

$MnO_3Cl$ is a very unstable violet-green gas at room temperature [143], forming a deep green-violet (almost black) [142] liquid at -30°C. It explodes above 0°C and hydrolyzes immediately to $[MnO_4]^-$ in water. Its spectrum (in $CCl_4$) shows a band analogous to the 18,000 $cm^{-1}$ band of $[MnO_4]^-$, but shifted up to ~20,000 $cm^{-1}$ [143] (~21,000 [142]), with some evidence of a band at ~15,000-16,000 $cm^{-1}$, no doubt the similarly shifted analog of that (at ~14,400 $cm^{-1}$) recently reported in $KMnO_4$ [134]. (The spectrum presented by Briggs [142] posses an odd shape below 15,000 $cm^{-1}$ and deserve reinvestigation preferably at reduced temperatures).

$MnO_3Cl$ and $MnO_3F$ can both be prepared by reacting $KMnO_4$ with $SO_3HX$, or with HX (X = F or Cl) and $H_2SO_4$ [144]. The $MnO_3F$ exists as dark green crystals up to -38°C and dark green liquid (and gas) with a boiling point (extrapolated) near 60°C. $MnO_3F$ is unstable above 0°C, but can be stored for months at dry ice temperature.

From the microwave spectra of $MnO_3F$ and $Mn(^{16}O)_2 \, ^{18}OF$, the structural parameters; Mn-O = 1.586 Å, Mn-F = 1.724 Å and O-Mn-F angle = 108°27', were determined [145]. The Mn-O distance is significantly shorter than in $[MnO \quad ]^-$.

## B. Manganese (VI)

Until the past year or two, the only known pure Mn(VI) compounds were the barium and potassium manganates, $K_2MnO_4$, and several hydrated forms of the sodium salt. Manganese dioxide dichloride, $MnO_2Cl_2$, was first prepared in 1968, as one of the products of the addition of $KMnO_4$ to an acid-organic solvent system at -60°C [142]. It is amber, in dilute $CCl_4$, dark brown as the pure liquid, unstable and sensitive to moisture, but apparently not explosive.

The potassium manganate can be prepared by oxidation of $MnO_2$ in fused KOH [146] or by decomposition of $[MnO_4]^-$ in excess base [147]. In 1969 the preparation of the pure sodium salt, by reaction of $MnO_2$ and NaOH at $400°-700°C$ under $O_2$ was reported in a Japanese patent [148]. The crystalline potassium, rubidium, and cesium salts were first prepared by reaction between the alkaline superoxides, $KO_2$, $RbO_2$ or $CsO_2$, and $MnO_2$ at $420°C$ [147].

The lattice parameters of pure $K_2MnO_4$, the isomorphous rubidium and cesium salts, and $BaMnO_4$ are given in Table 10.

The details of the refined structure of $K_2MnO_4$, reported in 1967 [82], appear in Table 11. The manganate is just as regular a tetrahedron as the permanganate. The slightly longer (0.03 Å) Mn-O distance in $[MnO_4]^{2-}$ may be the significant reflection of an electron now in a molecular orbital with both metal and ligand contribution [the $d^1$ electron in the $e_g$ ($\pi$) orbital]. The molecular packing is not determined by one or the other ion as in a 1:1 ionic solid here, however, but rather by cation-anion contacts. There are two types of oxygen polyhedra around potassium ions, one with 6 oxygen <2.9 Å and 3>2.9 Å distant, the other with 5>3.0 Å, 4~2.91 Å, and 1<2.9 Å distant from the potassium.

The most detailed polarized electronic spectrum of $[MnO_4]^{2-}$, at 20°K, published to date is the original by Teltow [33] on a single crystal of $K_2SO_4$ containing a small amount of $[MnO_4]^{2-}$. He observed four band systems, each with a great deal of fine structure, as follows (with the most intense peak in parenthesis): I, 16,125-20,250 cm$^{-1}$ [17,103]; II, 21,970-25,600 cm$^{-1}$ [22,800]; III, 26,430-32,440 cm$^{-1}$ [27,280], and IV, 33,120-36,030 cm$^{-1}$ [33,400]. The visible and uv spectra from 10,000-

TABLE 10

Lattice Parameters of Several $[MnO_4]^{2-}$
Containing Compounds

| | a, Å | b, Å | c, Å | Ref. |
|---|---|---|---|---|
| $K_2MnO_4$ | 7.66 | 10.33 | 5.89 | 149 |
| $Rb_2MnO_4$ | 7.997 | 10.670 | 6.044 | 147 |
| $Cs_2MnO_4$ | 8.360 | 11.052 | 6.247 | 147 |
| $BaMnO_4$ | 9.065 | 5.472 | 7.304 | 164 |

TABLE 11

Crystal Structure of $K_2MnO_4$

| Compound | Space group | Bond distance (Å) | Bond angles (°) | Comment | Ref. |
|---|---|---|---|---|---|
| $K_2MnO_4$ | Pnma | Mn-O = 1.647, 1.669, 1.661 Mn-O (average) = 1.659 K-O intermolecular 2.690- 3.230. O-O intermolecular 3.499. | O-Mn-O 109.5° ±0.6° | There are four molecules per unit cell in a regular tetrahedron | 82 |

35,000 cm$^{-1}$ of $[MnO_4]^{2-}$ in aqueous alkali at room temperature was reported in 1956 to have the following band peak positions: I, 16,530 cm$^{-1}$; II, 22,940 cm$^{-1}$; III, 28,490 cm$^{-1}$, and IV, 33,440 cm$^{-1}$ [150]. These have been the values of transition energies given a variety of assignments in the past ten years [39, 151-153 and references therein]. In addition a weak band around 12,000 cm$^{-1}$ was referred to and assigned by Carrington and co-workers [151,152].

The presence of this band in solid $Ba(SO_4, MnO_4)$ [154] and $K_2(SO_4, MnO_4)$ [134,154] has been confirmed and a definitive assignment presented, cf. Fig. 10.

All known manganates(VI) display a temperature dependent paramagnetism with effective moment consistent with one unpaired electron: $K_2MnO_4$ (155), $Rb_2MnO_4$, and $Cs_2MnO_4$ [147]. The earliest assignments of the optical transitions placed this electron in a $t_2$ orbital according to the 1952 Wolfsberg-Helmholz orbital scheme for $[MnO_4]^-$ and $[CrO_4]^{2-}$ [156]. Subsequent careful observation [157] and calculation [37], cf., Table 12, of the esr spectrum at 20°K of single crystals of $K_2CrO_4$ containing ~1% $MnO_4^{2-}$ indicated the unpaired electron to be in a doubly degenerate orbital, establishing the correctness of the alternate ordering $e(\pi)$ below the $t_2(\sigma,\pi)$ proposed by Ballhausen and Liehr [36]. Disagreement over

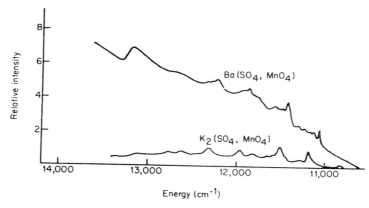

FIG. 10. The electronic spectrum of $[MnO_4]^{2-}$ at 4.2 °K [154].

the assignment of the "first", i.e., 16,530 cm$^{-1}$, transition was then expressed; some workers preferred a $t_1^6 e^1 \to t_1^6 t_2^1 (^2E \to {}^2T_2)$, "crystal field transition" [39] assignment, while others preferred a $t_1^6 e \to t_1^5 e_2^2 (^2E \to [^2T_{1+}{}^2T_2])$, ligand to metal charge transfer transition [152,153]. Only most recent experimentation has unequivocally established an absorption band on the red side of 16,530 cm$^{-1}$, 10,840-13,440 cm$^{-1}$ at 4.2 °K [134, 154]. With the $\simeq$12,000 cm$^{-1}$ band taken as the ligand field transition, the assignment of Carrington and Symons [151] is preferred. This assignment is also consistent with an estimation of band intensities [152] (using a dipole moment matrix value interpolated from $[MnO_4]^-$ and $[MnO_4]^{3-}$ values) and a consideration of the influence of spin-pairing energy on the first electron transfer bands of $[metal]^{6+}$ and $[metal]^{5+}$ tetroxo species [45]. The assignment is as indicated in Table 13 [151,158].

The fine structure of the unpolarized ligand field band of $[MnO_4]^{2-}$ in $K_2SO_4$ crystals [134] was rationalized on the basis of three origins ($^2T_2$ split in Cs site symmetry), at 10,840 cm$^{-1}$, 10,950 cm$^{-1}$ and 12,060 cm$^{-1}$, and combinations of vibrational overtones with intervals of the order of the asymmetric $\nu_4$ as seen in $[CrO_4]^{3-}$, and not the symmetric $\nu_1$ mode. The authors suggest this may be a part reflection of a Jahn-Teller distortion. It seems to us, however, that local static distortion in these mixed crystals removes any orbital degeneracy and would overwhelm the lowering of symmetry caused by the purely electronic effects of a Jahn-Teller distortion. Even stronger grounds for rejection of the hypothesis that this vibrational mode is somehow associated with a Jahn-Teller distortion is the recent observation of the presence of the $\nu_1$ mode on several band maxima in the spectrum of the $d^2$ system $(MnO_4)^{3-}$ [168].

The fundamental modes of $K_2MnO_4$, $\nu_1$: 810 cm$^{-1}$; $\nu_3$: 862 cm$^{-1}$; $\nu_4$: 328 cm$^{-1}$ have been measured at room temperature in Nujol mull and

TABLE 12

Electron Spin Resonance Data for $[MnO_4]^{2-}$

| Ion | Host | Temp. | $g_x$ | $g_y$ | $g_z$ | $A_x$ | $A_y$ | $A_z$ | Ref. |
|-----|------|-------|-------|-------|-------|-------|-------|-------|------|
| $Mn^{6+}$ | $K_2(Cr,Mn)O_4$ | $20°K$ | 1.970 | 1.966 | 1.938 | $0.25\epsilon$ $\times 10^{-2}$ cm$^{-1}$ | $0.33\epsilon$ $\times 10^{-2}$ cm$^{-1}$ | $1.35\epsilon$ $\times 10^{-2}$ cm$^{-1}$ | 37 |
| | | | | | | where $\epsilon = +1$ or $-1$ | | | 157 |

TABLE 13

Assignment of $[MnO_4]^{2-}$ Transitions

| Approximate band position $(cm^{-1})$ | Assignment |
| --- | --- |
| 12,000 | $t_1^6 e^1 \to t_1^6 \, t_2^1 (^2E \to {}^2T_2)$ |
| 16,530 | $t_1^6 e^1 \to t_1^5 e^2 (^2E \to {}^2T_1$ or ${}^2T_2)$ |
| 22,940 | $t_1^6 e^1 \to t_1^5 e^2 (^2E \to {}^2T_2$ or ${}^2T_1)$ |
| 28,490 | $t_1^6 e^1 \to t_1^5 et_2 (^2E \to {}^2T_1$ or ${}^2T_2)$ |
| 33,440 | $t_1^6 e \to t_1^5 et_2 (^2E \to {}^2T_2$ or ${}^2T_1)$ |

aqueous solutions [160]. These were used to estimate force constants for $[MnO_4]^-$, $[MnO_4]^{2-}$, $[MnO_4]^{3-}$ and $RuO_4$, $[RuO_4]$ $[RuO_4]^{2-}$ series. In the manganates especially, the metal-oxygen stretching force constants increased linearly from 4.5 mdyne/Å to ~5.7 mdyne/Å as the formal Mn oxidation state increased from +5 to +7 [161].

## C. Manganese(V)

With the exception of the recently reported compound, $MnOCl_3$, pentavalent manganese is known only as the tetroxo anion, $[MnO]^{3-}$. The hydrated sodium salt, $Na_3MnO_4 \cdot 10H_2O$, was the first of the $[MnO_4]^{3-}$-containing compounds to be prepared [162]. Intense blue crystals were obtained by extraction, using concentrated NaOH, of the product of a molten salt reaction involving $MnO_2$, $NaNO_2$ and $Na_2O$ or $Na_2O_2$. Other dry-melt preparations of the crystalline sodium hydrate include oxidation of MnO by $KNO_3$ or $MnO_2$ by atmospheric oxygen or $KMnO_4$ in alkaline melts at ~320°C [162]. The permanganate alone, when fused with NaOH evidently evolves oxygen and forms the blue anhydrous $Na_3MnO_4$. This compound has been isolated and analyzed as the hydrate [162]. Scholder has shown [70] that the deep blue needles crystallizing out of a NaOH melt were the adduct compound $Na_3MnO_4 \cdot NaOH$.

The $[MnO_4]^{3-}$ ion can be prepared in strongly alkaline solutions by reduction of $[MnO_4]^-$ with $Na_2SO_3$ [161] or KI [163]. Upon dilution or warming, the $[Mn^VO_4]^{3-}$ disproportionates into $[Mn^{VI}O_4]^{2-}$ and $Mn^{IV}O_2$. Between $0°C$ and room temperature, $KMnO_4$, in alkaline solution, is converted to $Mn^{VI}$ and $MnO_2$, passing through a $Mn^V$ intermediate state and forming mixed crystals of the hypomanganate in the presence of the arsenate, $Na_3(MnO_4, AsO_4)$ [162]. Alkali metal manganate (V) compounds are a characteristic deep blue or blue green, in contrast to the violet Mn(VII), green Mn(VI), and brown Mn(IV) oxides. Presumably all alkali salts $M_3[MnO_4]$ can be prepared by using the appropriate starting materials in the above reactions. Reference to preparations of the anhydrous $M_3[MnO_4]$, M = K, Rb or Cs, in a dissertation of 1957, indicated contamination with Mn(VI) was a problem (cited in Ref. [163]). The anhydrous $K_3MnO_4$ and $Na_3MnO_4$ have been prepared from the disproportionation of manganate(VI) at $520°C$, under a stream of nitrogen. The reaction proceeds principally according to the equation [68, 146].

$$3A_2MnO_4 = 2A_3MnO_4 + MnO_2 + O_2 \qquad (7)$$

The blue sodium hydrate becomes intense green when dried over $P_2O_5$ [70]. The preparation of blue-green $Li_3MnO_4$, by reaction of $LiMnO_4$ with LiOH at $124°C$ was also reported at this time [70]. This compound is stable in a LiOH solution at $0°C$ or in absolute methanol, but disproportionates in pure water instantly forming a violet solution.

In the earliest observations, the pentavalence of the Mn was indicated by a combination of the empirical formula and visual isomorphism of coprecipitated $Na_3[MnO_4]$ and $Na_3[MO_4]$, where M = As, P, or V [162]. The variation of molar magnetic susceptibility of $Na_3MnO_4 \cdot 10H_2O$ with temperature (at $90°$, $195°$, and $273°K$) was then seen to parallel values calculated for the $Mn^{5+}$ case rather than those for a $Mn^{4+}$ - $Mn^{6+}$ mixture. A value of $\mu = 2.8$ B.M., $\theta \sim 0$, close to that expected for the two unpaired electrons of Mn(V) was obtained [67].

The emerald green $Ba_3[MnO_4]_2$ can easily be prepared by oxidation of any Mn(II)-Mn(IV) oxide with BaO, $Ba(OH)_2$, or $BaO_2$ in molar ratio Mn:Ba = 1:1.5 in an $O_2$ stream at $800°$-$900°C$ [70]. Two specific equations are

$$2BaMnO_4 + Ba(OH)_2 \xrightarrow{900°C} Ba_3[MnO_4]_2 + H_2O + 1/2\ O_2 \qquad (8)$$

$$BaMnO_3 + Ba_2MnO_4 + 1/2\ O_2 \xrightarrow{900°C} Ba_3[MnO_4]_2 \qquad (9)$$

It has also been prepared by heating intimate mixtures of $Ba(NO_3)_2$, $MnO_2$ and a trace of $Na_2CO_3$ at $600°C$ for several hours [164]. The preparative methods of the early 1950's were recently criticized and a purer $Ba_3[MnO_4]_2$ reported from reaction between barium metamanganite, $BaMnO_3$ and $Ba(OH)_2$ at $700°C$ under a stream of dry oxygen [165].

The blue-green $Sr_3[MnO_4]$ is best obtained by oxidation of the hydroxyl salt $Sr_2[Mn(OH)_6]$ with $O_2$ at 250°C - 350°C [70,144]. Both Ba and Sr compounds can also easily be obtained from their alkaline solutions, by reduction of $BaMnO_4$ or $KMnO_4$ with ethanol [68].

A blue barium manganate(V) of the hydroxyapatite type, $Ba_5(MnO_4)_3OH$ can be prepared by heating $Mn_2O_3$ and $BaO_2$, finely powdered and intimately mixed, in a stream of moist oxygen for 20 hours at 800°-1,000°C [167]. If the $O_2$ stream is dry, the emerald green tertiary compound, $Ba_3[MnO_4]_2$, is obtained. Each form easily converts to the other simply by heating in the dry or moist oxygen flow [166]. The tertiary $Ba_3[MO_4]$ 's where M = P, As, V, Cr, or Mn are hexagonal isomorphs, with the ratios $c/a$ = 3.75, 3.67, 3.68, 3.72, and 3.76 respectively. The hydroxy-apatites, $Ba_5(MO_4)_3OH$, where M = P, V, Cr, or Mn, and $Sr_5(MnO_4)_3OH$ also have isomorphous hexagonal structures with the respective $c/a$ ratios = 0.745, 0.761, 0.751, and 0.747. On the basis of these X-ray data the ionic radii of the $V^{5+}$, $Cr^{5+}$, and $Mn^{5+}$ were estimated to be 0.48, 0.46, and 0.45 Å respectively [166].

The reported effective moments of the tertiary barium and barium and strontium hydroxyapatite-Mn(V) compounds are, as for $Na_3MnO_4 \cdot 10H_2O$, almost identical with that expected for two unpaired electrons [163]. The $\mu_{eff}$ values of $M_3[MnO_4]$ where M = K, Rb or Cs are low (2.57-2.68 B.M.) possibly due to Mn(VI) contamination. To our knowledge only three attempts to study the electron spin resonance spectrum have been reported [157]. The esr of single crystals of $Na_3VO_4 \cdot 12H_2O$ containing 1-2% $[MnO_4]^{3-}$, grown from a 12 M solution of KOH, was measured at 20°K and 90°K. Evidently the variety of orientations of the $[MnO_4]^{3-}$ anions to the magnetic field in these crystals (twelve molecules per unit cell) gave rise to a complex spectrum, which although consistent with S = 1 (paired sets of hyperfine structure, with isotropic splitting = 0.00625 cm$^{-1}$) was not further analyzed [157]. Similar problems have been encountered in the measurement of the esr spectrum of $Ca_2(PO_4,MnO_4)Cl$ [167,168]. Further efforts to interpret the esr spectrum of the $[MnO_4]^{3-}$ ion would be well worthwhile. A host such as $Li_3PO_4$, where the $[MO_4]^{3-}$ tetrahedra are regular, might help alleviate many of the problems caused by multiple orientations of the Mn(V) magnetic axes.

Two new species of manganate(V) were recently reported, $Sr_2(MnO_4)OH$ and $MnOCl_3$. Blue-green $Sr_2(MnO_4)OH$ and its dark green dihydrate are prepared by addition of $SrCl_2$ to an aqueous NaOH solution of $KMnO_4$ [169]. This compound has a hexagonal structure, a = 9.97 Å, c = 7.48 Å, $c/a$ = 0.750, isomorphous with the $A_5[BO_4]_3OH$ hydroxyapatites referred to above and has $\mu_{eff}$ = 2.84 B.M. at room temperature. The room temperature reflectance spectrum of $Sr_2[MnO_4]OH$ has a single strong absorption at 490 m$\mu$ (20, 408 cm$^{-1}$) close to values reported for other Mn(V) salts, Table 14. A reinvestigation of the spectra of the manganese salts listed in Table 14 is in order as the first transition in $Ca_2(PO_4,MnO_4)Cl$ occurs at much lower energies than those reported here [170].

TABLE 14

Reflectance Spectra of Some Mn(V)
Containing Compounds

| | $m\mu$ | $cm^{-1}$ | Ref. |
|---|---|---|---|
| $Ba_3[MnO_4]_2$ | 526 | 19,000 | 169 |
| $Ba_5[MnO_4]_3OH$ | 519 | 19,231 | 169 |
| $Sr_5[MnO_4]_3OH$ | 503 | 20,000 | 169 |
| $Na_3MnO_4 \cdot 10H_2O$ | 485-495 | 20,408 | 169 |

Manganese oxide trichloride, $MnOCl_3$, is a volatile mint green liquid at 0°C and decomposes above this temperature. It is an "ozone-like" smelling solid at -68°C, completely miscible in $CCl_4$, $CCl_3F$, slightly soluble in cold 100% $H_2SO_4$ and hydrolyzes vigorously, but apparently without explosion. Dry $CCl_4$ solutions keep for about a day at ambient temperature. It is prepared by reducing $KMnO_4$ with finely ground sucrose in a chloroform solution of $HSO_3Cl$ while maintaining the temperature between -30°C and 0°C [142]. The visible spectrum in $CCl_4$ at 25°C displays featureless peaks at 25,000 $cm^{-1}$, 16,000 $cm^{-1}$, and 13,000 $cm^{-1}$ with intensity ratio ~1.5:0.25:0.5. This spectrum differs from that of $[MnO_4]^{3-}$ in alkaline solution (6-12 M). Here the absorption maxima occur at ~30,800 $cm^{-1}$ and 14,800 $cm^{-1}$ with a small shoulder at 13,000 $cm^{-1}$, intensity ratio ~1.7:0.4:0.3 [168,150,152,45].

The ground state of the $[MnO_4]^{3-}$ is $^3A_2$, transitions to levels of $^3T_1$ symmetry being the only spin and electric dipole allowed excitations. The two bands, 14,000 $cm^{-1}$ and 30,800 $cm^{-1}$, were originally assigned according to the Ballhausen-Liehr one-electron level ordering as $^3A_2 \rightarrow {}^3T_1$ $(t_1^5 e^3)$ and $^3A_2 \rightarrow {}^3T_1 (t_1{}^5 e^2 t_2{}^1)$, respectively [152]. They were reassigned a year later as $e \rightarrow t_2$ (14,800 $cm^{-1}$) and $t_1 \rightarrow e$ (30,800 $cm^{-1}$) solely on the basis of expected crystal field splitting value [45]. Two reports of low temperature electronic spectra of the calcium apatite, $Ca_5(PO_4,MnO_4)_3Cl$ [167], and spodiosite $Ca_2(PO_4,MnO_4)Cl$ [167,170] mixed crystals have appeared since, providing data for a more reasoned assignment of the $[MnO_4]^{3-}$ transitions.

The 14,800 $cm^{-1}$ band reported the solution spectrum is split, at room temperature, into two maxima, at 17,500 $cm^{-1}$ and 13,800 $cm^{-1}$ in the spodiosite. In the compound with the apatite structure, two bands (with shoulders) are again observed. These occur at 16,000 (18,500 sh) and 14,500 (13,000 sh) $cm^{-1}$. At low temperatures, 2°K [167] and 80°K [170], further fine structure appears; sharp low intensity maxima (oscillator strengths $\simeq 10^{-6}$ - $10^{-7}$) are observed around 8,400 $cm^{-1}$ - 8,700 $cm^{-1}$ in both compounds while the spodiosite exhibits a distinct vibrational progression in the region 21,500 $cm^{-1}$ - 25,000 $cm^{-1}$. As can be seen in Fig. 11, three distinct maxima on the 13,800 $cm^{-1}$ and on the 17,500 $cm^{-1}$ bands were also resolved in the spodiosite [167,170]. On the basis of oscillator strengths the $\simeq 14,000$ $cm^{-1}$, $\simeq 17,000$ $cm^{-1}$, and $\simeq 33,000$ $cm^{-1}$ bands were assigned by Kingsley et al. [167] as $^3A_2 \rightarrow ^3T_1$ transitions. The broad weak 11,000 $cm^{-1}$ band was assigned as a transition to a $^3T_2$ state. This is in accord with the ordering of states predicted by crystal field theory for a $d^2$ ion in Td symmetry $[^3A_2 < ^3T_2 < ^3T_1(F) < ^3T_1(P)]$. Because the sum of the energies of the first two bands roughly equalled that of the third (33,000 $cm^{-1}$) it was deduced that the spectra could be rationalized in terms of the one-electron excitations from $^3A_2(t_1^6 e^2)$, Table 15.

The $Ca_2PO_4Cl$ structure having been determined [76], the polarized spectrum of $Ca_2(PO_4, MnO_4)Cl$ was measured at 80°K and an interpretation attempted in terms of $D_2d$ symmetry [170]. In particular the band at 17,500 $cm^{-1}$ showed no component $\vec{E} || Z$ and, therefore, cannot have $^3T_1$ parentage. Surprisingly the observed polarizations are best fit by an ordering of levels $^3T_1(F) < ^3T_2(F) < ^3T_1(P)$. Discounting this possibility, the band at 17,500 $cm^{-1}$ is assigned as having $^3T_2$ parentage $[^3T_2(Td) =$

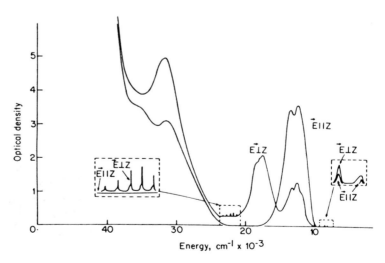

FIG. 11. The polarized spectrum of $[MnO_4]^{3-}$ at 80°K [170].

TABLE 15

Electronic Structure of $[MnO_4]^{3-}$

| Excited electronic state | In $Ca_5(PO_4)_3Cl$ | In $Ca_2(PO_4)Cl$ |
|---|---|---|
| $^3T_1(t_1^5e^2t_2)$ | 33,500 | 32,400 |
| $^1T_2(t_1^6et_2)$ | | 25,000–21,500 |
| $^3T_1(t_1^6et_2)$ | 16,700 | 17,500 |
| $^3T_1(t_1^5e^3)$ | 14,800 | 13,800 |
| $^1A_1(t_1^6e^2)$ | 13,600 | |
| $^3T_2(t_1^6et_2)$ | 11,000 | 11,000 |
| $^1E(t_1^6e^2)$ | 8,764 | 8,703 |
| | 8,598 | 8,410 |

$(^3B_2 + {}^3E)(D_2d)$ and $^3B_1 \xrightarrow{xy} {}^3E$, $^3B_1 \nrightarrow {}^3B_2]$ and the strongly xy and z allowed
components centered around 13,800 cm$^{-1}$ as a mixture of transitions to
$^3E$ and $^3A_2$ states from the $^3T_2(t_1^6et_2)$, $^3T_1(t_1^6et_2)$, and $^3T_1(t_1^5e^3)$ mani-
folds. The polarization behavior in the two areas of spin forbidden peaks
is also not in accord with that expected from $^1E(\sim 8,500$ cm$^{-1})$ and $^1T_2$
$(\sim 22,000$ cm$^{-1})$ parentages. Both peaks are observed in both polarizations
in the former while only one, $\Gamma_4 \rightarrow \Gamma_3$ should be present when $\vec{E}||Z$. For a
transition to a $^1T_2$ parent the allowed $\Gamma_5 \rightarrow \Gamma_5$ should be, and is not, ob-
served when $\vec{E}||Z$. It should be very useful to look at the low temperature
spectra of single crystals of other $[MnO_4]^{3-}$ species to aid in sorting out
the assignment of the electronic spectra of $[MnO_4]^{3-}$.

It is pertinent at this point to interject some general comments pertaining
to the electronic structure of tetrahedral oxyanions in which the central
ion is a transition metal. It has been common practice to calculate a
crystal field splitting energy for these highly covalent anionic complexes.
This value of $\Delta$ has generally been arrived at as one of the results of a
Hückel type molecular orbital calculation and has been used more often
than not as a criterion of the validity of certain types of semiempirical
approaches. Unfortunately, in almost every case, the experimental data

collected before 1965, which was used as a basis of comparison for the theoretical calculations, was incomplete or inaccurate. This applies to the oxyanions $[MnO_4]^-$, $[MnO_4]^{2-}$, $[MnO_4]^{3-}$, and $[CrO_4]^{3-}$ and may well be found to apply to other oxyanions as experimentation continues. While the consequences of this inadequacy have undoubtedly had a direct effect upon the development of a calculational treatment which is useable for inorganic systems it has also led to the tacit assumption of the violation of the Principle of Electroneutrality as proposed by Pauling [171]. A specific case in point is the suggestion by Viste and Gray [39] that $\Delta$ (which can be taken as the separation of the metal $e_g$ and $t_{2g}$ orbitals, but should not be construed to mean $\Delta$ in the normal crystal field sense) for tetrahedral oxyanions is a rapidly increasing function of the charge of the central metal ion. Keeping in mind the Principle of Electroneutrality and that the separation of the $e_g$ and $t_{2g}$ orbitals is very much a function of metal-ligand charge separation the above suggestion is somewhat surprising. One might expect on the contrary, that a point might be reached beyond which $\Delta$ would not increase (charge equalization occurring through $\pi$-interactions). A recent interpretation based upon the more modern and accurate data bears out the validity of this latter concept, cf. Table 16.

The infrared spectra of sodium and barium manganates(V) have been observed at room temperature in Nujol mulls and KBr discs. The vibrational modes are assigned in Table 17. As may be seen the symmetric $\nu_1$ vibration is anomalously absent in the spectrum of $Ba_3(MnO_4)_2$ [166]. This vibration is also anomalously absent in the $Ba_3[CrO_4]_2$ and $Ba_3(VO_4)_2$. It is, however, observed in $Sr_2(MnO_4)OH$ where $\nu_1 = 864$ and $\nu_3 = 772$ cm$^{-1}$ [167]. Vibrational modes of $[MnO_4]^{3-}$ were observed in the spodiosite mixed crystals [168] to be $\nu_1 = 750$ cm$^{-1}$ ($A_1$), $\nu_4 = 460$ cm$^{-1}$ ($T_2$), and $\nu_2 = 227$ cm$^{-1}$ and 264 cm$^{-1}$(E). The observed shift of $\nu_3$ (weighted mean) to higher frequencies in the $Ba[Mn^{VI}O_4]$ and $Ba(Mn^{VII}O_4)_2$ compounds (~840 cm$^{-1}$ and ~890 cm$^{-1}$, respectively) is consistent with the metal 3d orbitals participating in $\sigma$ as well as $\pi$ bonding [165].

## D.   Mn(IV)

The known oxide and fluoride compounds containing tetravalent manganese which will be described include: $MnO_2$, $Mg_6MnO_8$, $M_2^{II}MnO_4$ ($M^{II} = Sr$, Ca, Zn, Mg), $M^{II}MnO_3$ ($M^{II} = Ca$, Cd, Zn), $CaMn_3O_7$, $Cd_2Mn_3O_8$, $(NH_4)_6MnMo_9O_{32}\cdot 8H_2O$, $Na_{12}MnNb_{12}O_{38}\cdot 5H_2O$, $MnF_4$ and $M_2^I[MnF_6]$ ($M^I = $ Li, Na, K, Rb, Cs and $M^{II}[MnF_6]$($M^{II} = Ba$, Sr, Ca, Mg). In addition Mn(IV) appears in many mixed valence oxides such as $Mn_5O_8$, (Al, Li)$MnO_2(OH)_2$, $ZnMn_3O_7$, $BaFeMn_7O_{16}$, $Mn_7O_{13}$, $CuMnO_4$, $(La_{1-x}Ca_x)MnO_3$, and $M^{III}Mn_2O_5$; M = Dy and Bi.

The dioxide, the common mineral form of manganese, occurs in two crystalline phases, the tetragonal "$\beta$" or pyrolusite form and the rarer orthorhombic ramsdellite phase. There also exists a series of combinations

TABLE 16

Formal Charge vs. Crystal Field Splitting Energy

| Complex | Metal formal charge | $\Delta,^a\text{cm}^{-1}$ | $\Delta,^b\text{cm}^{-1}$ |
|---------|---------------------|---------------------------|---------------------------|
| $CrO_4^{3-}$ | 5 | 16,000 | 10,000 |
| $CrO_4^{2-}$ | 6 |  | ~9,000 |
| $MnO_4^{6-}$ | 2 |  | 4,500 |
| $MnO_4^{5-}$ | 3 |  | 7,000 |
| $MnO_4^{4-}$ | 4 |  | 10,000 |
| $MnO_4^{3-}$ | 5 | 11,000 | 11,000 |
| $MnO_4^{2-}$ | 6 | 19,000 | ~10,000 |
| $MnO_4^{-}$ | 7 | 26,000 | ~10,000 |

a. Ref. 39.  b. Ref. 49.

TABLE 17

Assignment of the $[MnO_4]^{3-}$ Vibrational Modes

| | $\nu_1\ \text{cm}^{-1}$ | $\nu_3\ \text{cm}^{-1}$ | $\nu_4\ \text{cm}^{-1}$ | Ref. |
|---|---|---|---|---|
| $Na_3[MnO_4]$ | 863 | 770 | 348 | 160 |
| $Ba_3[MnO_4]_2$ |  | ~760 |  | 166 |

of these phases, which have come to be collectively referred to in the literature as the "$\gamma$" form [172]. The gray-black solid $MnO_2$ is insoluble in water, easily reduced in acids and forms manganate (IV) salts on fusion or reaction with oxides and alkalis. The unit cell parameters of the tetragonal, $\beta$, (a,c = 4.396, 2.871Å [172]; 4.398, 2.8738 [173]) and the ramsdellite phases $(a, b, c, = 4.533, 9.27, 2.866$ [174]; 4.46, 9.32, 2.85 [175]) were first reported about 20 years ago, but details of structure are approximate and scattered, Table 18, perhaps due to variations in mineral samples.

Single crystals of the $\beta$ form, black needles, about 1 mm long, were first grown in 1969 [173]. A sample of high purity $MnO_2$ was dissolved in 6N HCl which contained a trace of $KClO_3$. This was sealed in a gold tube for 24 hours at 700°C and 3,000 atm.

The Mn-O bond lengths quoted by Abrahams and Bernstein for $D_yMn_2O_5$ [176] show much more variety than those found for $\beta$-$MnO_2$ [172-176].

The $Mn^{4+}$-O bond in the periodate, $Na_7H_4[Mn^{4+}(IO_6)_3].17H_2O$ [178], appears to offer an example of an ion-dipole, $Mn^{4+}$-$OIO_5$, interaction. Similarly, a preliminary report indicates that Mn(IV) is probably octa-hedrally coordinated to six oxygens in the Na and $KMn^{IV}(IO_6)$ periodates [179].

The $\gamma$ form of $MnO_2$ can be formed by oxidation of $\alpha MnOOH$ by $O_2$ or $HNO_3$ [180]. Heating $\gamma$-$MnO_2$ to 480°C converts it to $\beta$-$MnO_2$ [181]. The variety of powder patterns of $\gamma$-$MnO_2$ samples has been accounted for by a model of mixed, layered, single (pyrolusite) and double (ramsdellite) chains, each phase characterized by a "p" (pyrolusite concentration) value [173,181].

In the early 1950's, $Mg_6MnO_8$ was synthesized from a mixture of MgO and $MnCO_3$, in an 8:1 molar ratio, held at 1100°C for 5 hours in an $O_2$ atmosphere [182]. The Mn-O distances are slightly longer than expected from the sum of ionic radii, perhaps due to the large vacancy left by the Mg. The octahedra around the Mn ions are regular, while those around the Mg ions are distorted. Were the material sufficiently magnetically dilute, it would be interesting to compare the esr spectrum of $Mg_6MnO_8$ with the recently reported spectrum of $Mn^{4+}$ in MgO [183] in which a $[Mn^{4+}$-hole] paired structure is proposed.

The brown-black strontium and barium manganates $M_2^{II}MnO_4$, are pre-pared by heating the Mn(VI) manganate at 500°C - 1,000°C under nitrogen and in the presence of the corresponding hydroxide [184]. Evidently only the powder pattern of the strontium salt could be obtained at the time of the original preparation (tetragonal and isomorphous with $K_2NiF_4$, $I4/mmm$, $a, c = 3.79, 12.43$ Å) [185]. The Ca, Zn and Mg analogues, $M_2^{II}MnO_4$, the $M^{II}MnO_3$ (M = Ca, Cd, Zn), $CaMn_3O_7$ and $Cd_2Mn_3O_8$ compounds were prepared during a comprehensive study of solid state reactions between $MnO_2$ and Ca, Zn, Mg or Cd oxides [186]. The few cubic lattice parameters reported sensitively reflect changes in cation size (Ca- and $ZnMnO_3$, $a_o = 7.46$ and $8.35$ Å; Ca- and $MgMnO_4$, $a_o = 8.53, 8.38$). Studies of collective electron phenomena (conductive and magnetic) in these compounds should be interesting. The compound $Mg_6MnO_4$ is empirically a variant of the $Mg_2MnO_4$ with 4MgO units. Both have $a_o = 8.38$ Å and correspond to different ratios of $MnO_2$ in MgO, $MnO_2.2MgO$ and $MnO_2.6MgO$.

## TABLE 18

### The Crystal Structures of Mn(IV) Compounds

| Compound | Space group | Bond distance Å |
|---|---|---|
| Pyrolusite ($\beta$MnO$_2$) | P4/mmm | 2Mn–O, 1.878 |
| | | 4Mn–O, 1.891 |
| | | Mn$^{IV}$–Mn$^{IV}$, 2.871 |
| | | across a shared edge |
| | | 4Mn–O, 1.91 |
| | | 2Mn–O, 1.85 |
| Ramsdellite | Pbnm | Mn–O, 1.87–1.94Å |
| | | 1.89 average |
| Na$_7$H$_4$[Mn$^{IV}$(IO$_6$)$_3$]. 17H$_2$O | Pbcn | Mn–O = 1.90 |
| Mg$_6$MnO$_8$ | Fm3m | 6Mn–O, 1.928 |
| | | 2Mg–O, 2.095 |
| | | 4Mg–O, 2.102 |
| (NH$_4$)$_6$MnMo$_9$O$_{32}$. 8H$_2$O | R32 | rh. a = 10.08 |
| | | $\alpha$ = 104°24' |
| | | hex. a = 15.94, |
| | | c = 12.38 |
| Na$_{12}$MnNb$_{12}$O$_{38}$. 50H$_2$O | P2/n | Mn–O, 1.87 |
| | | (Nb)$_2$–O, 1.88–2.05 |
| | | (bridge) |

| Bond angles(°) | Comments | Ref. |
|---|---|---|
| O-Mn-O 82.5° | The slightly distorted $MnO_6$ octahedra form infinite single chains in c-direction, sharing 2 opposite edges. Neighboring chains share corners. | 172,173, 175,176 174 |
| | Neighboring chains share 2 edges forming infinite <u>double</u> chains in c-direction. | 109 |
| | The $Mn^{IV}$ is at the center of a regular octahedron of three pairs of oxygen atoms, each pair shared by a different $IO_6$ octahedron. | 178 |
| | Four chemical formula units per unit cell, a MgO type structure with one fourth $Mg^{2+}$ replaced by the pair $[Mn^{4+}-$ hole] causing an ordering of holes. | |
| | Mg-O from ionic radii = 2.102. | 182 |
| | Nine $MoO_6$ octahedra surround a central $MnO_6$ unit, per cell. | 205 |
| O-Mn-O between $[Nb_6O_{19}]$ anions = 96.7° | The $Mn^{IV}O_6$ octahedron joins 2 $[Nb_6O_{19}]^{8-}$ groups and is elongated along a pseudo 3-fold axis approximately in a,b plane. b-axis $\parallel$ to rod. | |

C. ROSENBLUM AND S. L. HOLT

TABLE 18 continued

| Compound | Space group | Bond distance Å |
|---|---|---|
| | | Nb-O, 1.75-1.79 (terminal) |
| | | $(Nb)_6$-O, 2.4 (central) |
| | | $Nb_2$-Mn-O, 2.10 |
| | | O-O in $[Nb_6O_{19}]^{8-}$, 2.7-2.9 |
| $CuMn_2O_4$ | Fd3m | Cu-O, 2.04 |
| | | Mn-O, 1.95 |
| | | O-O, 2.54, 2.96, 3.35 |
| $BiMn_2O_5$ | Pbam | $Mn^{3+}$-$Mn^{3+}$, 2.9 |
| | | $Mn^{4+}$-$Mn^{4+}$, 2.8 |
| | | $Mn^{3+}$-$Mn^{4+}$, 3.71 |
| | | 3.48 |
| $DyMn_2O_5$ | Pbam | $Mn^{4+}$-$Mn^{4+}$, 2.898 |
| | | $Mn^{3+}$-$Mn^{3+}$, 2.870 |
| | | $2Mn^{3+}$-O, 1.926 basal |
| | tetrag. pyram. | $2Mn^{3+}$-O, 1.911 |
| | | $1Mn^{3+}$-O, 2.02 |
| | octahed. | $4Mn^{4+}$-O, 1.940 |
| | | $2Mn^{4+}$-O, 1.873 |
| $Mn_2^{II}Mn_3^{IV}O_8$ | C2/m | a = 10.34, b = 5.72, c = 1.85, $\beta$ = 109°25' |
| $Li_{1-x}Al_x(Mn^{II}Mn^{IV})$ $O_2(OH)_2$ | C2/m | 4Mn-O, 1.93 |
| | | 2Mn-O, 1.97 |
| | | 4Al-OH, 1.95 |
| | | 2Al-OH, 1.93 |

| Bond angles(°) | Comments | Ref. |
|---|---|---|
| | length. Anions are separated by water molecules. | 206 |
| Mn–O–Mn $\simeq 90°$<br>$Mn^{3+ \text{ or } 4+}$-O-Cu 125° | | 218 |
| $Mn^{3+}$-O-$Mn^{3+}$ 93.3<br>$Mn^{4+}$-O-$Mn^{4+}$<br>$Mn^{3+}$-O-$Mn^{4+}$ 129° | | 221 |
| O-$Mn^{4+}$-O 83.3-96.6°<br>O-$Mn^{4+}$-O 176.7°<br>O-$Mn^{3+}$-O 83.7-100.7°<br>O-$Mn^{3+}$-O 163.3° | The Dy is surrounded by a polyhedron of 8 oxygen atoms. The $Mn^{IV}$ lie along infinite chains ‖ c. These are linked by O-$Mn^{III}$-O-$Mn^{III}$-O paths. Four molecules per unit cell. 5.6 wt% lead impurity found. | 176 |
| | | 212 |
| | Infinite 2 dimensional sheets of $MnO_6$ edge joined octahedra alternate with $(Al,Li)(OH)_6$ octahedra in layers. Hydroxyl bonds join the O and OH. | 213 |

TABLE 18 continued

| Compound | Space group | Bond distance Å |
|---|---|---|
| $ZnMn_3O_7 \cdot 3H_2O$ | Pl | Mn-O, 1.94 |
| | In monoclinic | Mn-O, 1.96 |
| | subcell space | Zn-O, 1.93 |
| | group = C2/m. | Zn-O, 1.98 |
| | | Zn-OH$_2$, 2.14, |
| | | 2.15 |

| Bond angles(°) | Comments | Ref. |
|---|---|---|
| | layers on top of one another. Empty octahedral sites are 7.54 Å apart in a trigonal grouping | 214 |

Interest in $Mn^{4+}$ ion has been focused on two aspects: one, the magnetic and electrical properties of $MnO_2$ in comparison with neighboring dioxides, $VO_2$, $CrO_2$, and $MoO_2$ and the other, its spin resonance and optical spectra, as a member of the family of $d^3$ ions, $V^{2+}$, $Cr^{3+}$, and $Mn^{4+}$. Indeed, efforts to sort out and estimate the influences of metal charge and local symmetry on the parameters associated with zero field ground and excited state splittings and the crystal field strength form a literature of the $d^3$ family, from which we shall provide only a few guiding references.

$MnO_2$ is antiferromagnetic, with a Néel temperature, recently redetermined on powder and single crystals, of 94°K [173]. The Néel temperature (84°K) reported in 1951 [187] was probably for a single crystal of mixed phases ($\gamma$) grown as it was from MnOOH heated at 300°C. Neutron diffraction data on $MnO_2$ powder and crystals can be elegantly accounted for by an unusual spin structure, having separate corner and body center antiferromagnetic spin sublattices, each with spin direction spiralling twice (perpendicular to the c-axis) within seven chemical unit cells [245].

Electrical conductivity in rutile-type transition metal dioxides is believed to occur in energy bands arising either from direct metal-metal d-orbital overlap [M-M<Rc (Rc = critical interaction distance)] or metal(d)-oxygen (sp)-orbital overlap. Observed overall trends in the first, second and third row dioxides were recently summarized and rationalized by considering both variation of the crystallographic c/a ratios and the number of d-electrons per cation [173]. In both $CrO_2$ and $MnO_2$ the critical interaction distance is exceeded by the actual M-M distance (2.92 and 2.87 A, respectively) and the filled $t_{2g}$ orbitals $||$ c do not overlap. Conduction presumably occurs, therefore, in a $M_{t_{2g}}$ -$O_{p\pi}*$, cation-anion band which is empty in $CrO_2$ and has one electron in $MnO_2$. The resistivity of the $MnO_2$, which is $10^3$-$10^4$ times greater than that of $CrO_2$, is not, however, accounted for by this model, which attributes the highly metallic behavior of the dioxides of Ru, Ir, Os and Rh as consistent with partial filling of the M-$O_\pi*$ band.

The six-line esr spectra (I - 5/2, $^{55}$Mn) of both tetragonally ($SrTiO_3$, MgO, $TiO_2$, $SnO_2$) and trigonally ($Al_2O_3$, $[Mo_9O_{32}]^{10-}$ and $[Nb_{12}O_{38}]^{16-}$) distorted $Mn^{IV}O_6$ octahedra have been summarized in Table 19.

The $SrTiO_3$ host undergoes a cubic to tetragonal transition below 100°K, which is evidently not reflected in altered g or $|A|$ values, but is accompanied by an increase in the zero field splitting, 2D, from 0.013 cm$^{-1}$ to 0.02 cm$^{-1}$ (1.4 - 2 gauss) [188]. The g value of Mn(IV) is greater than that of Cr(III) (g = 1.9788) in this host. This is consistent with more covalency and a stronger crystal field, $\Delta$ as it appears in g = 2.0023 - 8 $\zeta_o/\Delta$, at the Mn(IV) than at the Cr(III) site [183]. The g and $|A|$ values of the lithium and non-lithium doped manganese - MgO systems are very similar to those of the $SrTiO_3/Mn^{4+}$ system. Mn(II) is promoted to Mn(IV) in the MgO powder by addition of Li$^+$ ion (as the carbonate [189], fluoride [190] or chloride [183]) before firing of the mixture. The tetragonal center is the paired $[Mn^{4+}O^{2-}\text{-}Li^+O^{2-}]$ or lattice charged $[Mn^{2+}\text{-}Li^-]$ species [183, 190]. Irradiation of Mn(II)-MgO crystals (~70kV, 1 hour (183)) results in $Mn^{4+}$ formation $([Mn^{4+}]/[Mn^{2+}] \approx 1/500)$. The source of tetragonal distortion here is believed to be a $[Mn^{4+}\text{- hole}]$ or lattice charged $[Mn^{2+}\text{-hole}^{2-}]$ structure. The g and $|A|$ values of the octahedral $[Mn^{2+}]$ in MgO (2.0014 $cm^{-1}$ and 81 x $10^{-4}$ $cm^{-1}$ [191]) are not very different from those of the tetragonal spectrum. This, and reasonable agreement between the observed values of the zero field splitting (2D = 0.106 $cm^{-1}$ for $Mn^{4+}$) and a value calculated without assuming effects of a tetragonal field at the ligands, led the authors [183] to propose the point defect (hole) along the z-axis as the source of tetragonal distortion. This is interesting because considering only the radii of cations $[Mn^{4+}$ (~0.53 Å), $Mg^{2+}$ (0.73 Å), or $Ti^{4+}$ (0.63 Å)] involved, one would imagine either the oxygen octahedron around $Mn^{4+}$ to be distorted or the $Mn^{4+}$ not in its center. This is seen not to be the case in the $Mg_6MnO_8$ (i.e. $MnO_2 \cdot 6MgO$) structure, which has regular Mn and distorted Mg octahedra and ordered $Mn^{4+}$-hole pairs.

The combined influence of this cation-hole tetragonal distortion and spin-orbit coupling is greater on the ground, $^4A_2$, and less on the excited, $^2E$, state splittings of $Cr^{3+}$ than for $Mn^{4+}$ in MgO (for $Cr^{3+}$ in MgO, 2D = ~0.164 $cm^{-1}$ [192], $\Delta E$ = 94 $cm^{-1}$ [193]). The 2D values of $Mn^{4+}$ in rutile powder [194] and crystal [195] and in $SnO_2$ [196] are greater than those found in the MgO system. The g and $|A|$ values exhibit distinct angular dependence in these two hosts. In addition super hyperfine structure is seen in both, with $a_x = a_y = 0.456$ x $10^{-4}$ $cm^{-1}$, $a_z = 0.93$ x $10^{-4}$ $cm^{-1}$ in $TiO_2$ [197] and $a_x$, $a_y$, $a_z$ = 28.36, 27.43 and 31.71 x $10^{-4}$ $cm^{-1}$ in $SnO_2$ [196]. In both hosts the substitutional Mn(IV) ion has two cations closer along the c-axis (Mn-Ti = 2.959 Å, Mn-Sn = 3.185 Å) than the eight at the unit cell corners (Mn-Ti, 3.569 Å [197], Mn-Sn, 3.708 Å [196]). Overlap of the $Mn^{4+}$ orbitals with these gives rise to the SHFS and is very much stronger in the $SnO_2$. It is a curious and, as far as we know, still unexplained observation that while the zero field ground state splitting for $Mn^{4+}$ in $SnO_2$ is about twice that in $TiO_2$, the values for $Cr^{3+}$ in both these hosts are about equal [196] (1.16 $cm^{-1}$ and 1.1 $cm^{-1}$, respectively).

Single crystals of $\alpha Al_2O_3$ containing $Mn^{4+}$ at some $Al^{3+}$ sites can be grown from fluxes to which MgO has been added, the $Mg^{2+}$ providing charge compensation [198]. The g-value of $Mn^{4+}$ is greater than that for $Cr^{3+}$ (g = 1.984) in this host, and as in MgO reflects the effects of increased charge. The zero field ground state splitting is almost equal to that of $Cr^{3+}$ in $Al_2O_3$ (0.38 $cm^{-1}$ [199]) while the excited state splitting is more than twice that of $Cr^{3+}$ in $Al_2O_3$ (29 $cm^{-1}$ [199]). The influence of the greater trigonal distortion of the $Mn^{4+}$ (-$K_{Mn4+} \approx 700$ $cm^{-1}$, -$K_{Cr3+} \approx 330$ $cm^{-1}$ [200] on the D value may be offset by the the higher Dq value [198] and greater covalency in the $Mn^{4+}$. This was the essence of a recent suggestion [201] to account for the variation of $^4A_2$ and $^2E$

## TABLE 19

### The Electron Spin Resonance Data for Mn(IV) Compounds

| Ion | Host | Temp. | $g_z$ / $g\parallel$ | $\langle g \rangle$ |
|---|---|---|---|---|
| $Mn^{4+}$, 0.01% | $SrTiO_3$, xstal | 300°K | | 1.994 |
| | | 77°K | | |
| $Mn^{4+}$, Li/Mg 5% | MgO, powder | 300°K | | 1.990 |
| | | 4.2°K | | |
| 1%LiF | MgO, powder | R.T. | | 1.995 |
| LiCl | MgO, powder | R.T. | | 1.9941 |
| | MgO, xstal | R.T. | 1.9931 | |
| $Mn^{4+}$, 0.01% | $TiO_2$, powder | R.T. | | 1.990 |
| | , xstal | 77°K | 1.990 | |
| $Mn^{4+}$, 0.04% | $SnO_2$ | 77°K | 1.9907 | |
| $Mn^4$, 0.1-0.001% | $Al_2O_3$, xstal, | 1.6-295°K | $g\parallel \simeq 1.9937 \simeq g\perp$ | |
| $Mn^{4+}$ | $(NH_4)_6Mn$ | | | |
| | $Mo_9O_{32}\cdot$ | R.T. | 1.9920 | |
| | $8H_2O$ | | | |
| | aqueous soln. | 77°K | | 3.8 |
| | $Mn^{IV}[Nb_6O_{19}]_2$ | 77°K | | 3.8 |

| $g_x$ ($g_\perp$) | $g_y$ | $A_z$ ($A\|$) | $A_x$ $\times 10^{-4} cm^{-1}$ | $A_y$ ($A_\perp$) | $\Delta(^2E)$ $cm^{-1}$ | $2D$ $cm^{-1}$ | Ref. |
|---|---|---|---|---|---|---|---|
| | | | 70.0 | | | 0.013 | 188 |
| | | | | | | 0.020 | |
| | | | 72.0 | | 69 | | 189 |
| | | | 70.85 | | | 0.056 | 183 |
| | | | 70.82 | | | | |
| 1.9940 | | 71.1 | 70.6 | | | 0.106 | 183 |
| | | | 72.0 | | | 0.806 | 194 |
| 1.995, | 1.991 | 72.7 | 72.3 | 70.3 | | 0.816 | 195 |
| 1.9879, | 1.9870 | 74.99 | 72.26 | 70.12 | | 1.76 | 196 |
| | | | 70.00 | | 80 | 0.39 | 198 |
| | | | | | 85 | | |
| 1.9880 | | 76.0 | 68.4 | | | 1.772 | 208 |
| | | | | | | | 210 |
| | | | | | | | 210 |

splittings (2D &$\Lambda$) in the $d^3/Al_2O_3$ series, with mechanical compression ($O$ kg/cm$^2$ - 600 kg/cm$^2$) along the c-axis [200]. Feher and Sturge observed the rate of change of D with stress to be almost uniform for $V^{2+}$, $Cr^{3+}$ and $Mn^{4+}$, while that of $\Lambda$ differed widely. They suggested a model with the ground state splitting more influenced by long-range interactions and the excited state, more by local interactions. A model assuming only increased metal-oxygen overlap with increasing charge, however, roughly accounts for the observed trends of the trigonal parameters v and v' with stress [201]. More data from $d^3$ ions in a number of hosts and from other isoelectronic series are needed.

The ratios of trigonal deformation parameter to the magnitude of the $^2E$ state splitting ($-K/\Delta E$) in $Mg_2TiO_4(Mn^{4+})$ and $MgTiO_3(Mn^{4+})$ were recently observed to be the same order of magnitude as those for $Cr^{3+}$ and $Mn^{4+}$ in $Al_2O_3$ [202]. A very highly resolved emission spectrum from $MgTiO_3(Mn^{4+})$ indicated a trigonal distortion, $\Delta(^2E) = 31$ cm$^{-1}$ in this host [203].

The spin lattice relaxation times within the metastable $^2E$ states of $V^{2+}$, $Cr^{3+}$ and $Mn^{4+}$ decrease in inverse proportion to $[\Delta(^2E)]^3$ [204].

Mn(IV) is also in a trigonal environment in both molybdenum, $(NH_4)_6$ $Mn^{(4+)}Mo_9O_{32}.8H_2O$ [205] and niobium, $(Mn^{4+}[Nb_6O_{19}]_2)^{12-}$ [206] heteropoly complexes. The three layers of three $MoO_6$ octahedra per unit cell of the Mo complex have no center or plane of symmetry but have a $C_3$ axis through the central Mn. The optical spectrum [207] very closely resembles that of $Mn^{4+}$ in $Al_2O_3$, Table 20, while the zero field splitting [208] is about four times as large. The low temperature polarized spectrum of these easily prepared [205] strongly dichroic orange-red rhombohedral rods certainly ought to be measured. There is uncertainty in the assignment of both the $^2T_2$ and $^4T_2$ transitions and the estimate of trigonal distortion in both $Al_2O_3(Mn^{4+})$ [198, 204] and $[MnMo_9O_{32}]$ [207, 209, 210]. If the 20,600 cm$^{-1}$ and 21,300 cm$^{-1}$ absorptions arise from trigonal splitting of the $^4A_{2g} \rightarrow ^4T_{2g}$ transition, then the analogous distortion in the niobium complex (20,750 cm$^{-1}$ and 21,900 cm$^{-1}$) is even greater [209]. On the other hand the absorptions around 20,700 cm$^{-1}$ may be the $^4A_2 \rightarrow ^2T_2$ transition, made intense from having coupled with the allowed transition to $^4T_2$ state.

It is interesting, of itself, that manganese (IV) appears in so many mixed valence solids, often naturally occurring minerals. $Mn^{II}$ substitutes for Cd in $Cd_2Mn_3O_8$ forming $Mn_2^{II}Mn_3^{IV}O_8$. This compound might be viewed as a phase of $MnO_2$ (i.e. $Mn_{10}O_{16}$) [211]. Both octahedra in the mineral lithiophorite which has a unit cell $2[Al_{0.68}Li_{0.32}Mn^{2+}_{0.17}Mn^{4+}_{0.82}O_3]$ 1.00 $H_2O$ are distorted, each sharing six shortened (O-O, 2.56 Å) edges with a neighbor [212]. Chalcophanite, $ZnMn_3O_7.3H_2O$, has a similar layered structure, except with a Mn vacancy every seventh site in the infinite 2-dimensional sheet of distorted $MnO_6$ octahedra. The Zn atoms,

TABLE 20

Electronic Spectrum of Some $Mn^{4+}$ Compounds

| Assignment | in $Al_2O_3$ (204) $Cr^{3+}$ ($cm^{-1}$) | $Mn^{4+}$ | $[MnMo_9O_{32}]^{6-}$ (207) ($cm^{-1}$) | $K_2MnF_6$ (225) ($cm^{-1}$) |
|---|---|---|---|---|
| $^4A_2 \rightarrow {}^4T_1$ | 25,000 | 25,000 | 25,500–26,000 | 28,600 |
| $^2T_2$ | 21,000 | 21,200 | 20,600 (?) | 19,300 |
| $^4T_2$ | 18,000 | 20,800 | 21,300 | 22,200 |
| | | 21,300[a] | | |
| $^2T_1$ | 15,050 | 15,500 | 15,100 | |
| | | | | 14,000 |
| $^2E$ | 14,418 | 14,782 | 14,300 | |

[a] [198].

above and below these vacancies, have three short bonds to the Mn oxygens (as in tetrahedrally coordinatee Zn) and three long bonds to water layer oxygens (as in octahedrally coordinated Zn) [213]. There are two molecules per unit cell and the Mn is predominantly tetravalent ($Mn^{4+}{}_{3-\delta}Mn^{2+}{}_{\delta}$) $O_7$, $0.0 < \delta < 0.25$ [213]. The same is true for hollandite $2[Ba, FeMn^{4+}{}_{6.26}Mn^{2+}{}_{0.63}O_{16}]$ [214].

A double layered manganate, $Mn_7O_{13}.5H_2O$, probably analogous to the $ZnMn_3O_7$, with ordered holes, but with mixed $Mn^{4+}$-$Mn^{3+}$ states was recently described [215], as was a spinel-type $Mn_3X^{2+}Li_2O_8$ (X = Mg, Ni and Co) [216]. The spinel-type $Cu^{1+}{}_{tet}(Mn^{3+},Mn^{4+})_{oct} \rightarrow (Mn^{3+},Mn^{4+})_{oct}O_4{}^{2-}$ structure of copper manganite is believed to develop from an irreversible electron transfer between $Mn^{3+}{}_{tet}$ and $Cu^{2+}{}_{oct}$ in $Mn^{3+}{}_{tet}$ $[Cu^{2+},Mn^{3+}]_{oct}O_4{}^{2-}$ [217]. The distortion from cubic symmetry decreases in these compounds with increasing $[Mn^{4+}]$. This is also true of a series of mixed valence perovskite-type crystals $(La_{1-x}A_x)MnO_3$, A = Ca, Si or Ba [218]. In these manganites the effect of both variation of lattice parameters (upon changing cation "A") and variation of $Mn^{4+}$ concentration on magnetic behavior and $T_{Curie}$ has been accounted for by Goodenough [219] in terms of transitions between covalent and ionic Mn-O-Mn bonds.

The $BiMn_2O_5(Mn^{3+}, Mn^{4+})$ mixed crystal also shows unusual magnetic behavior. It becomes antiferromagnetic, with a magnetic cell double the chemical unit cell and no longer having an inversion center, below $42°K$ [220]. A ferroelectric moment is, therefore, being sought [221]. An antiferromagnetic ordering of the Mn ions in $DyMn_2O_5$ may be masked by the paramagnetism of the dysprosium ($DyMn_2O_5$, $T_{Néel} = 8°K$) [176].

Blue, hygroscopic $MnF_4$ can be prepared in a few minutes by the action of fluorine gas on the metal powder at $200°-800°C$ [222]. The yellow $M_2^I MnF_6$ ($M^I$ = Li, Na, K, Rb and Cs) salts occur in cubic, hexagonal and trigonal modifications, with the K and Rb ones exhibiting polymorphic changes with temperature [223]. The $M^{II} MnF_6$ compounds ($M^{II}$ = Mg, Ca, Sr, Ba) occur only in hexagonal form [224]. The pentafluoromanganates(IV), $M^I MnF_5$, ($M^I$ = K, Rb, Cs) result upon fluorination of the $M^I Mn^{II} F_3$ salts $450°-500°C$. All have magnetic moments, $\mu \simeq 3.9$ B.M., close to the spin only moment expected for $Mn^{4+}$ [223].

A well resolved reflectance spectrum of $K_2MnF_6$ at $300°K$ and $77°K$ was reported, Fig. 12, and assigned recently [225], Table 20, with Dq = 2,200 $cm^{-1}$, B = 587 $cm^{-1}$, and $\beta = 0.55$. Vibrational structure, spacing $475\pm50$ $cm^{-1}$, was observed on the 22.2 kK band. That seen on the analogous emission ($^4T_{2g} \rightarrow {}^4A_{2g}$) band in $Cs_2MnF_6$ had a spacing of $\simeq 520$ $cm^{-1}$ [226]. The vibrational frequencies in $K_2MnF_6$ [227] and $Cs_2MnF_6$ [226] were recently assigned. Disagreement exists about whether the absorption in the 16 kK region (16,080, $K_2MnF_6$ and 16,025 $cm^{-1}$ $Cs_2MnF_6$) is due to decomposition [225] or a vibronically coupled spin forbidden $^4A_{2g} \rightarrow {}^2T_{1g}$, $^2E_g$ [226,227].

## IV. IRON

### A. Fe(VI)

Iron(VI) has been observed only in oxide salts, $[FeO_4]^{2-}$ of Na, K, Cs, or Sr. Anodic oxidation of the metal [228] or oxidation of $Fe_2O_3$ by $Cl_2$ in concentrated alkaline solutions yields the red-purple $M_2^I[FeO_4]$ salts [228,239]. The $M^{II}[FeO_4]$ species are presumably prepared in an analogous manner. Reported studies of the barium [230,231] and strontium [231] salts include no details of preparative method. The red-brown barium salt has been obtained by heating $K_2FeO_4$ and $Ba(CH_3CO_2)_2$ [70]. An alternate preparation of the $K_2FeO_4$ salt involves heating $K_2O$ and $Fe_2O_3$ in a ratio of 2:1 at $150°C$ under oxygen [163]. Above $210°C$, a mixture of $K_3FeO_4$ and $KFeO_2$ results. The violet-black $Cs_2FeO_{3.93}$ is formed by heating $CsO_{1.93}$ with $Fe_2O_3$ at $260°C$ under $O_2$ for 200-300 hours [231]. Only a mixture of $Rb_3FeO_4$ and $RbFeO_2$ is reported to result from the heating of $Rb_2O$ and $Fe_2O_3$, $250°-350°C$ [163].

The $[FeO_4]^{2-}$ ion, an even more powerful oxidizing agent than $[MnO_4]^-$, is easily reduced in acid and, though fairly stable in base, does decompose

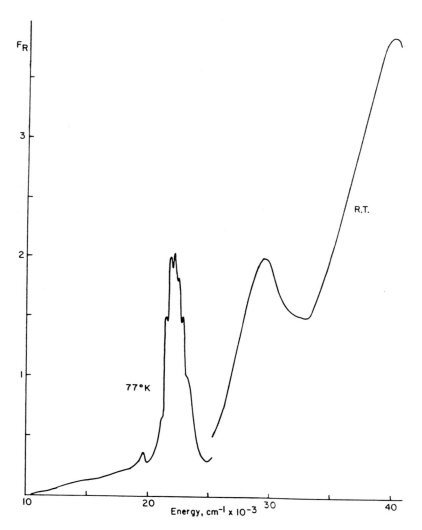

FIG. 12. The electronic spectrum of $K_2MnF_6$ [225].

slowly. Contamination with $Fe^{3+}$ is probably the cause of the temperature dependence and rather high extrapolated moment of $K_2FeO_4$, $\mu = 3.06$ B.M., reported in 1950 [229]. When the final precipitate of $K_2FeO_4$ with associated contaminants KCl and KOH, was washed only once with cold methanol [163], a negligible temperature dependence and values of $\mu = 2.8$-$2.9$ B.M. ($3d^2$ Spin only $\mu = 2.83$), $\theta \approx 0°$ were obtained. The value for $\mu$, for $Cs_2[FeO_4]$ is 2.78 B.M., $\theta = -25°$ [163].

In $[FeO_4]^{2-}$ the iron atom is evidently tetrahedrally surrounded by oxygen, as $K_2FeO_4$ is isomorphous with $K_2SO_4$ and $K_2CrO_4$ [157]. No crystal structure of an iron(VI) species has been reported. The observed esr spectrum of single crystals of $K_2CrO_4$ containing $[FeO_4]^{2-}$, grown from solution, is consistent with an S = 1, tetrahedral ground state, the two electrons being in a doubly degenerate orbital, Table 21 [157]. This observation caused a revision of an early assignment of the electronic spectrum of the $[FeO_4]^{2-}$ ion. The original interpretation of the spectrum of $[FeO_4]^{2-}$, Fig. 13, suggested the assignment of electronic energy levels shown on the left of Table 22. This assignment was based upon the

TABLE 21

Electron Spin Resonance Data for $[FeO_4]^{2-}$

| host | $<g>$ | $cm^{-1}$ |
|------|-------|-----------|
| $K_2CrO_4$ | 2.000 | $D_x = 0.0504$ |
|  |  | $D_y = 0.0180$ |
|  |  | $D_z = 0.0684$ |

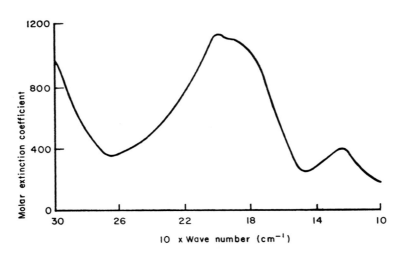

FIG. 13. The electronic spectrum of $[FeO_4]^{2-}$.

TABLE 22

The Electronic Spectrum of $[FeO_4]^{2-}$

| Energy (cm$^{-1}$) | Ref. 156 | Assignment Ref. 232 | Ref. 39 |
|---|---|---|---|
| 12,720 | $3t_2 \rightarrow 2a_1$ | $e \rightarrow t_2$ | $2e \rightarrow 4t_2$ |
| 17,800 | $t_1 \rightarrow 3t_2$ | $t_1 \rightarrow e$ | |
| 19,600 | $3t_2 \rightarrow 2e$ | $t_1 \rightarrow e$ | $2e \rightarrow 4t_2$ |

assumption that the ordering of one electron energy levels proposed by Wolfsberg and Helmholz for $[MnO_4]^-$ [35] was the correct one. A later interpretation by Symons [232] based upon the Liehr-Ballhausen one electron ordering [36] concluded that the lowest energy transition is d-d in nature while the upper two maxima are charge transfer in origin. Viste and Gray [39] have disagreed with both these interpretations suggesting that both the 12720 and 19600 cm$^{-1}$ bands arise from the d-d transition ($2e \rightarrow 4t_2$). Based upon the results of the recent reinvestigations of the $[MnO_4]^-$ [47, 134], $[CrO_4]^{3-}$ [49], $[MnO_4]^{2-}$ [134,154], and $[MnO_4]^{3-}$ [170] it would be surprising if any of the interpretations presented thus far is completely correct. Of paramount importance to the entire question of the electronic structure of $[FeO_4]^{2-}$ is the acquisition of a high quality solid state absorption spectrum.

Three reports of the Mössbauer effect isomer shift, using $Co^{57}$ in stainless steel [234,233] or $Co^{57}$ in Pt [230] source have appeared. The results are summarized in Table 23. (The difference between values given in Ref. 230 and those of 233 and 234 arises from the difference in source host.) The isomer shift in all cases is much smaller (i.e., less toward the negative) than would be expected for the $3d^2$ case based upon a rough extrapolation from shifts observed in $3d^8$ - $3d^5$ cases [235]. Since the magnitude of shift is proportional to total s-electron density at the nucleus, increased 3d-electron density, via covalency enhanced by a high iron charge, may cause expansion of 3s and 4s orbitals and reduction of s-electron density at the nucleus. This could qualitatively account for the smallness of the isomer shift [230,233,234]. This idea is consistent with recently observed rather small internal fields in $BaFeO_4$ (115kOe at 4.2°K) [234], $K_2FeO_4$ (130KOe at 2°K) [234] and $K_2FeO_4$ (142KOe at 1.6°K) [233].

$BaFeO_4$ and $K_2FeO_4$ are both antiferromagnetic, with a broad Néel

TABLE 23

Mössbauer Data for $[FeO_4]^{2-}$
mm/sec - Isomer Shift

| Temp, °K | $K_2FeO_4$ | $BaFeO_4$ | $BaFeO_4 \cdot H_2O$ | Ref. |
|---|---|---|---|---|
| 298 | -0.79 | | | |
| 78 | -0.69 | | | 234 |
| 298 | -1.269 | -1.213 | -1.258 | |
| 178 | -1.215 | | | 230 |
| 153 | | | | |
| 141 | | -1.162 | -1.200 | |
| 293 | -0.88 | | | |
| 1.6 | -0.99 | | | 233 |

temperature inflection around 10°K for the former [231] and two values $T_N = 3.6°K$ [233] and $T_N \simeq 5°K$ [231] reported for the latter. Antiferromagnetic exchange, therefore, certainly contributes to the reduced internal field at the temperatures reported. $SrFeO_4$ appears to be paramagnetic down to 2°K [231].

## B. Fe(V)

The $3d^3$ ion, iron(V), has been studied very little. The black $K_3FeO_4$ was first prepared in 1954 by heating a 3:1 mixture of potassium and iron oxides at 450°C for two days [166]. Its magnetic susceptibility is temperature independent (range 90°K, 195°K, and 294°K) $\mu_{eff} = 3.5$-$3.6$ B.M. [236]. $Rb_3[FeO_4]$ is presumably prepared in an analogous manner, $\mu_{eff} = 3.6$-$3.7$ [163]. Both values are somewhat smaller than an expected spin only value of 3.87 B.M.

## C. Fe(IV)

The strontium and barium orthoferrates, $M_2^{II}FeO_4$, were first prepared in 1909 by reaction of $Fe(NO_3)_3 + Sr(NO_3)_2$ and $Fe(OH)_3 + Ba(OH)_2$ at 400°C - 600°C under oxygen [237]. The iron/oxygen ratio was observed to indicate non-stoichiometry, however. Indeed, it has only been in the past six years that conditions necessary to obtain 100% $Fe^{4+}$ in $SrFeO_3$

(high $O_2$ pressure and firing at 1400 °C for sixteen hours) have been determined [239,238]. Reaction between ferric hydroxide salts, $M_3^{II}$ [Fe(OH)$_6$]$_2$ and $M^{II}$(OH)$_2$, at 800 °C - 900 °C in air yields a compound of the composition $M^{II}FeO_3 + M^{II}(OH)_2$ [70]. These "components" are inseparable (without decomposition) and presumably the "compound" is identical, with $M_2^{II}FeO_4 \cdot H_2O$. Hydrated and anhydrous, black and gray-black $M^{II}FeO_3$ (M = Ba, Sr) species were reportedly obtained by thermal decomposition of aqueous $M^{II}FeO_4$ [70].

The mass of Fe(IV) literature to date (with the exception of the [Fe$^{IV}$ (diars)$_2X_2$]$^{2+}$ species) centers around Sr(Fe$^{4+}_{1-2x}$Fe$^{3+}_{2x}$)$O_{3-x'}$, $0 < x < 0.5$ or $SrFeO_{2.5-3.0}$. A $Ba_8Fe_8O_{21}$ mixed oxide (100% Fe$^{4+}$ would be $Ba_8Fe_8O_{24}$) was reported in 1946 [240], but it has been the strontium salt which has been studied in detail.) The reported changes in lattice parameters with changing amount of Fe$^{4+}$ are summarized in Table 24. These structures are perovskite analogues. They are antiferromagnetic with a range of Néel temperatures shown in Table 25. Reported values of magnetic moments are anomalous and unaccounted for quantitatively, Table 26. Based upon a $d^4$-$d^5$ ratio of 58/42 the expected moment is 5.32 B.M., instead of the observed 4.38 B.M. Even with 100% of the $d^4$ species present the normal moment should be 4.90 B.M. A strong negative Fe$^{3+}$-Fe$^{3+}$ interaction masking a positive Fe$^{3+}$-Fe$^{4+}$ one has been suggested [242]. Some electron exchange mechanism between Fe$^{3+}$-Fe$^{4+}$, causing preferential alignment of the Fe$^{3+}$ spins evidently increases to a

TABLE 24

Lattice Parameters versus Fe$^{4+}$ Concentration
for $SrFeO_x$

| | %Fe$^{4+}$ | Lattice Parameter Å | Ref. |
|---|---|---|---|
| $SrFeO_{2.5}$ | 0 | a=5.671, b=15.59, c=5.528 | 238 |
| $SrFeO_{2.75}$ | 50 | a=3.853 | 241 |
| $SrFeO_{2.80}$ | 58 | a=3.85, c=3.87 | 242 |
| $SrFeO_{2.84}$ | 72 | a=3.851, c=3.867 | 239 |
| 2.86 | 75 | a=3.859 | 243 |
| | | a=3.869 | 244 |
| $SrFeO_{2.93}$ | 86 | a=3.869 | 218 |
| $SrFeO_{3.0}$ | 100 | a=3.850 | 239 |

TABLE 25

Néel Temperatures for $SrFeO_x$

|  | $T_N$, °K | Ref. |
|---|---|---|
| $SrFeO_{2.86}$ | 40-50 | 243 |
|  | 50 | 244 |
| $SrFeO_{2.84}$ | 80 | 239 |
| 2.97 | 130 | 239 |
| 3.0 | 130 | 239 |
| 3.0 | 134 | 238 |

TABLE 26

Magnetic Moments for $Fe^{4+}$-Containing Species

|  | $Fe^{4+}\%$ | $\mu$ B.M. | Ref. |
|---|---|---|---|
| $SrFeO_{2.79}$ | 58 | 4.38 | 242 |
| 2.86 | 72 |  |  |
|  |  | 8.06 | 244 |
| 2.93 | 75 |  |  |
| 3.0 | 100 | 6.3 | 239 |

maximum, then decreases, between 58% and 100% $Fe^{4+}$. The lattice spacings, symmetry and number of anion vacancies also change with changing $Fe^{3+}/Fe^{4+}$ ratio. The mechanism of magnetization, or para-magnetic moment enchancement, may involve a metal-oxygen-metal super-exchange since the $T_{Néel}$ increases with increasing anion vacancies.

Mössbauer study of $SrFe_{1-x}Cr_xO_{3-y}$, x<0.75, 0<y<0.3 indicates the presence of some $Fe^{4+}$ [246]. This system differs from the $SrFeO_3$ and $Sr(FeTi)O_3$ systems, however, in that the perovskite cubic cell parameter decreases on either side of x = 0.3 and the $T_{Néel} \approx 230$°K.

The electrical resistivity is lower in the $SrFeO_{3.0}$ than in the mixed valence species; 58% $Fe^{4+}$, $\rho = 3.3 \times 10^{-1}$ ohm·cm (243); 72% $Fe^{4+}$, $\rho \simeq 10^1$-$10^{-2}$ ohm·cm [239,243]; 100% $Fe^{4+}$, $\rho \simeq 10^{-3}$ ohm·cm [239]. As the cation-cation distance is large, 5.49 Å, in these materials, conduction is believed to occur in covalently mixed metal ($t_{2g}$)- oxygen ($p\pi$) collective electron states (bands) [239]. This would be consistent with earlier observations in the series $Sr[Fe_{1-x}Ti_x]O_3$ [243]. At x<0.8 a steep increase in conductivity coincides with a decrease in the cubic cell parameter, $a_o$, from 3.905 Å, the normal lattice parameter for $Sr[Fe_{1-x}Ti_x]O_3$ where x>0.8. Evidently the structure begins to collapse around the $Fe^{4+}$ ion at this concentration ($r_{Fe^{4+}} = 0.59$Å - 0.60Å [238,242], $r_{Fe^{3+}} = 0.67$Å, $r_{Ti^{4+}} = 0.69$Å).

Two Mössbauer studies of isomer shift in Fe(IV) at 298°K are in approximate agreement, Table 27. The observed shifts, about -0.1 mm/sec, relative to stainless steel are in the right direction but, as in $Fe^{6+}$, much smaller than expected for a $3d^4$ configuration, $\simeq$ -1.2mm/sec relative to steel [235]. They are closer to values expected for $3d^5$, again perhaps due to participation of the 3d orbitals in covalent bonding with oxygen. This would be consistent with observed increase in negative shift with increasing anion vacancies. Contribution from increasing $Fe^{3+}$ ($I_s$ = +0.55 mm/sec [244] relative to steel) would tend to decrease the observed value. The real shift may, therefore, be more negative in the $SrFeO_{2.6}$ sample.

Fe(IV) plays a role in photochromic changes recently observed in $CaTiO_3$: Fe-Mo [247] and $SrTiO_3$:Fe-Mo [248]. Color (absorption band 4000-5000Å) can be reversibly induced in both of these materials by excitation with light ~4,500 Å or exposure to a 25 keV electron beam. The proposed models involve holes and electrons, trapped at $Fe^{3+}$-$Mo^{6+}$ centers, respectively, forming the colored $Fe^{4+}$-$Mo^{5+}$ species.

TABLE 27

Mössbauer Data for $Fe^{4+}$
Is/mm/sec Relative to Copper

|  |  | Ref. |
|---|---|---|
| $SrFeO_{3.2}$ | -0.171 | 238 |
| $SrFeO_{2.93}$ | -0.22 | 244 |
| $SrFeO_{2.86}$ | -0.204 | 238 |
| $SrFeO_{2.6}$ | -0.507 | 238 |

Conversion of $Fe_2O_3$ to a perovskite, $Fe^{II}Fe^{IV}O_3$, by 130 Kbar pressure at 1000°C may have been effected recently [249]. The estimated $Fe^{IV}$-O distance, 1.77 Å at the octahedral, B, sites is significantly different from that estimated in $SrFeO_{2.93}$ ($Fe^{4+}$ 86%, 1.94Å [218]).

## V. COBALT

### A. Co(V)

The compounds $M_3CoO_4$ (M = K [250], Na, Rb, and Cs [251]) believed to contain Co(V), have been prepared fairly recently. The blue-black $K_3CoO_{\simeq 4}$ forms upon heating a mixture of $KO_x$ and $Co_3O_4$, in a ratio of 3:1 for a period up to 23 hours, at 460-550°C under an oxygen atmosphere [237,250]. Estimation of 1.4-1.65 "active" oxygen atoms per cobalt (i.e., by titration of $I_2$ formed in the presence of KI and HCl) and the ease of isomorphic doping into the colorless $K_3PO_4$ host, to form intense blue crystals supported the proposed pentavalence. The measured magnetic moments were also consistent with the presence of four (diluted sample) or two (undiluted sample) unpaired electrons. Both temperature and cobalt concentration dependence were exhibited, however. In the undiluted crystal, the moment was found to decrease with temperature, $\mu_\beta$ = 2.1 B.M. (293°K), 1.7 B.M. (193°K), and 1.2 B.M. (90°K). In the $K_3PO_4$ host the moment decreased with increasing cobalt concentration, $\mu$ = 5.1 B.M. (3.2% $Co^{5+}$), 4.3 B.M. (8.1% $Co^{5+}$), and 2.6 B.M. (5.6% $Co^{5+}$) [250]. Two species of $K_3CoO_{\simeq 4}$ having different powder patterns were observed. The one having a pattern more analogous to the $K_3MO_4$, M = P, V, Mn, and Fe series is unstable and transforms to the other in about two days [250]. The Rb and Cs salts are presumably prepared like the K salt [251]. An earlier attempt to prepare the Cs salt, however, by reacting $Co_3O_4$ and $CsO_2$ at 350°C for 72 - 92 hours, resulted in a mixed valence, $CsCo^{III}Co^{IV}O_4$, compound ($\mu$ = 0.68 - 1.14 B.M. in a temperature range between 90 and 295°K [236]). The red-violet $Na_3CoO_4$ results from reaction of $Co_2O_3$ and $Na_2O$ at 450°C under 100 atm $O_2$ [251]. Co(V) is also present in $KCoO_3$, prepared by reacting $KO_{0.7}$ and CoO for 24 hours at 300-400°C under ~50 atm $O_2$ [167].

### B. Co(IV)

A brown-black $Co^{IV}O_2$ and an olive-green solid, suggested to be $H_2Co^{IV}O_4$, were reported in an early study of higher valent cobalt oxides [252]. The better characterized red-brown microcrystalline $BaCo^{IV}O_4$ results from reaction of the barium and cobalt(II) hydroxides under $O_2$ at 1050°C [70].

Magnetic and electrical properties of Co(IV) containing crystals $SrCoO_x$ (59.3% $Co^{4+}$, a = 3.85Å) and $La_{.5}Sr_{.5}CoO_x$ (61% $Co^{4+}$, a = 3.83Å) are unusual [242]. The resistivities (3.9 and 6.9 x $10^{-4}$ ohm.cm respectively) are very much lower than either the Co(III) only ($8.9 \times 10^6$ and $4.9 \times 10^2$ ohm.cm) or $Fe^{III}$-$Fe^{IV}$ ($3.3 \times 10^{-1}$ and $4.9 \times 10^{-2}$ ohm·cm) containing analogues. These ~60% $Co^{IV}$ (40% Co(III)) crystals also exhibit spontaneous magnetization below room temperature and magnetic moments (3.32 and 3.65 B.M.) far below spin only values (5.50 and 5.52 B.M.), which would be estimated on the basis of Co(IV)-Co(III) content [242]. Goodenough has suggested a model with $Co^{IV}$ in a $^2T_2$ ground state and Co(III) having $^1A_1$ and $^5T_2$ ground states of comparable energies. The two magnetic sublattices form in parallel (111) planes [253].

The electron spin resonance spectrum of Co(IV) in an $Al_2O_3/MgO$ single crystal [254] displays an eight line hyperfine pattern with $g_\perp = 2.58 \pm 0.01$, $A = 0.01 \pm 0.003$ cm$^{-1}$, and $B = 0.0198 \pm 0.0002$ cm$^{-1}$. This result is consistent with a low spin $^2T_2$ ground state.

The cesium salt, $Cs_2CoF_6$, the only Co(IV) fluoride compound known, can be prepared by direct fluorination of either $Cs_2CoCl_4$ or $Cs_2Co(SO_4)_2$ at 300°C. The brown-gold crystals produced by this reaction are isomorphous with $K_2PtCl_4$ (Fm3m), with cell dimensions, a = 8.905Å (255) or a = 8.92Å [256]. The electronic spectrum, measured at 295 and 77°K [257] is shown in Fig. 14. The assignment of this spectrum, Table 28, is based upon the assumption of a $^2T_2$ ground state (the magnetic moment varies from 2.46 B.M. at 90°K to 2.97 B.M. at 294°K [255,256]). Allen and Warren point out that the temperature dependence of the magnetic moment could indicate the necessity of analyzing the absorption spectrum in terms of a ground state with variable spin multiplicity. The spectrum,

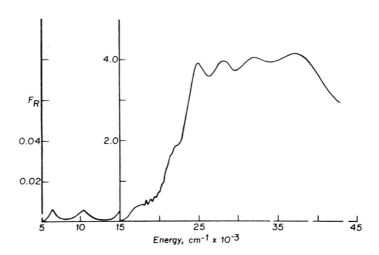

FIG. 14.   The electronic spectrum of $Cs_2CoF_6$ [257].

## TABLE 28

### Electronic Spectrum of $Cs_2CoF_6$ [a]

| Band position, cm$^{-1}$ | $\epsilon$ | $f$ | Assignment |
|---|---|---|---|
| 6,400 | 0.5 | $2.3 \times 10^{-6}$ | $^2T_{2g} \to {}^4T_{1g}$ |
| 10,300 | 0.5 | $4.4 \times 10^{-6}$ | $^2T_{2g} \to {}^4T_{2g}$ |
| 17,300 | 32 | $3.0 \times 10^{-4}$ | $^2T_{2g} \to {}^2A_{2g}, {}^2T_{1g}$ |
| 18,900 | 34 | $2.2 \times 10^{-4}$ | $^2T_{2g} \to {}^2T_{2g}$ |
| 21,400 | 106 | $1.3 \times 10^{-3}$ | $^2T_{2g} \to {}^2E_g$ |
| 24,400 | 198 | $3.2 \times 10^{-3}$ | $^2T_{2g} \to {}^2T_{1g}$ |
| 28,300 | Strong | | $\pi \to t_{2g}$ |
| 32,200 | | | $^2T_{2g} \to {}^2E_g$ |
| 37,100 | Strong | | $\pi \to t_{2g}$ |

[a]There has been a rather imaginative use of Gaussian analysis in some areas of this spectrum; therefore, care should be exercised in placing too much faith in the absolute reliability of values of $\epsilon$, $f$, and band position, especially for the higher energy absorptions.

however, shows no indication of an anomaly within the temperature range studied. The crystal field parameters derived from the $Cs_2CoF_6$ spectrum are $Dq = 2030$ cm$^{-1}$, B = 635 cm$^{-1}$, $\beta = 0.54$. These values are consistent with a strong-field complex, intermediate in covalency between the Mn$^{4+}$ and Ni$^{4+}$ analogs. The demonstration of a decreasing stability of the metal$^{4+}$ formal valence state reflected in the increasing participation of the extremely electronegative fluorine in covalent bonding with the metal is an important feature of this series.

It would appear that further work is in order with regard to the temperature dependence of both the spectrum and magnetic moment of $Cs_2CoF_6$.

Tetravalent cobalt may form heteropoly complexes with tungstic and molybdic oxides, similar to those of Mn(IV). The only report, however, of preparative work and suggestive empirical formulae, e.g., $3K_2O.CoO_2$. $9MoO_3 + 6 \ 1/2 \ H_2O$ (from chemical analysis) dates from 1907 [258] with apparently no subsequent studies of this system to date.

## VI. NICKEL

### A. Ni(IV)

The structure and spectra of a number of Ni(IV) alkaline mixed metal oxides, periodate, heteropoly salts and alkaline fluoride ionic solids have been reported in recent years. The anhydrous binary oxides of Ni(III) or Ni(IV) have not been observed. The crystal size and composition of Ni(II) oxide and hydroxide change upon oxidation with hypochlorite [259] or bromine [260] to hydrated species of unknown valence. Although a stoichiometric excess of oxygen can be determined, the powder patterns do not differ appreciably from that of NiO, except in being somewhat more amorphous [261].

The Ni(IV) salt, $BaNiO_3$, has been reported to result from reaction of NiO with barium hydroxide [262] or oxide [261] at about 700°C, in the presence of oxygen. Approximate details of crystal structure of the latter compound, which was obtained only as a black powder appear in Table 29. It is like a perovskite, with 12 oxygens coordinated to the barium, but the nickel octahedra apparently share faces [261]. The estimated Ni-Ni distance appears to be short enough for metal-metal interaction. The moment was reported to be equivalent to 0.8 unpaired electron [261]. A black product analyzed as $BaNiO_{2.5}$, prepared by a reaction between barium nitrate and nickel nitrate or carbonate, in air, at 680°C, had the same hexagonal cell parameters, a = 5.58, c = 4.832, and density [263] as the above "$BaNiO_3$". This may or may not imply an error in characterization of $BaNiO_3$. A distinct product which analyzed as $BaNiO_{2.5}$ had been prepared in the earlier study [261]. Further studies have not been reported and would clearly be useful.

Ni(IV) is likely to be present in the $M_2^INiO_3$ ($M^I$ = Na, K, Rb) compounds, in $RbNiO_4$ and in $Rb_3NiO_4$. These higher oxides result from firing NiO with an alkaline peroxide, ($M^IO_x$, 0<x<2) in an oxygen atmosphere [264]. Speculation on the valence of nickel in these solids was based on fitting observed with calculated moments and varying the amount of ($O^{2-}$) and ($O_2^{2-}$) necessary to obtain fit. For example the compounds $M_2^INiO_3$ had moments of 2.0 B.M. and could be formulated as $Ni_2^{III}Ni^{IV}(O^{2-})_7(O_2^{2-})$. The moments of $CsNiO_{4.5}$ and $Cs_3NiO_5$, 2.9 B.M., implied all nickel present as the tetravalent species.

Tetravalent nickel also appears in alkali nickel periodates. The purple-black, metallic crystals $NaNi[IO_6].H_2O$ or $KNi[IO_6]0.5H_2O$ result either from persulfate oxidation of a mixture of the alkali per-iodic salt, $Na_3H_2IO_6$, or the reaction of $KIO_4$ and nickel sulfate heptahydrate [266].

### TABLE 29

Structural Data for Ni(IV) Containing Compounds

| Compound | Space group | Bond distance Å | Bond angles | Comments | Ref. |
|---|---|---|---|---|---|
| $BaNi^{IV}O_3$ | C6mc | 6Ni-O 2.01<br>6Ba-O 2.90<br>6Ba-O 2.79<br>Ni-Ni 2.42 | | Hexagonal a = 5.580, c = 4.832 two molecules per unit cell. | 261 |
| $KNi^{IV}[IO_6]$ | P312 | Ni-O 2.10<br>I-O 1.85<br>K-O 2.56 | $O_1-Ni-O_2$ 172°<br>$O_1-I-O_4$ 179°<br>4 $O_1-Ni-O_x$ 80-94°<br>4 $O_1-I-O_x$ 87-94° | Isomorphous with $PbSb_2O_6$, both iodine and nickel atoms are octahedrally surrounded by oxygen, two octahedra sharing an edge. | 265 |
| $NaNi^{III}O_2$ | C2/m | 4Ni-O 1.95<br>2Ni-O 2.17<br>4Na-O 2.29<br>2Na-O 2.34 | | In the monoclinic form, below 220°C, two molecules per unit cell. Each oxygen is coordinated to 2 nickel atoms at 1.95 Å, one at 2.17 Å, two sodium atoms at 2.29 Å and one at 2.34 Å. | 284 |

The moments, 1.11 and 1.08 B.M., at 31°C, are low, but not zero as expected for low spin Ni(IV). This is perhaps due to some spin ordering or due perhaps to contamination by Ni(II). The magnetic moment of every known Ni(IV) and Ni(III) compound, as we shall see, indicates a low spin structure. Whether or not this spin quenching is caused by the strong crystal field alone or is invariably accompanied by increased covalency, as detectable by electronic and spin resonance spectra, is the focus of many studies of the Ni(III) and Ni(IV) oxides and fluorides today.

The Ni(IV) analogues of the Mn(IV) - Mo, Nb and V - heteropoly complexes have been prepared and all are diamagnetic. Fine purple-black crystals of $[NH_4]_6[NiMo_9O_{32}]$ or $3[(NH_4)_2O].NiO_2.9MoO_3.yH_2O$ form upon persulfate [267] or peroxydisulfate [268] oxidation of a boiling solution of nickel sulfate and ammonium paramolybdate. Precipitation with $BaCl_2$ forms the brown crystalline $3[BaO]$ salt [267]. Based upon viscosity measurements and similarity between solution and reflectance spectra [268], it appears that the $[NiMo_9O_{32}]^{6-}$ ion, which oxidizes $H_2O_2$ and $H_3O^+$, is only slightly solvated in solution. A band observed at 17,600 $cm^{-1}$ was assigned to the $^1A_{1g} \rightarrow {}^1T_{2g}$ transition. The ammonium salt is isomorphous with the Mn(IV) analog [268], the $M^{4+}$ ion being surrounded by an octahedron of oxygen atoms [209].

Ni(IV) is also believed to be octahedrally coordinated in the niobate, $K_8Na_4NiNb_{12}O_{38}.21H_2O$ [268]. Strongly pleochroic, dark maroon needles, $Na_{12}NiNb_{12}O_{38}.$ (48-50 $H_2O$) result from the reaction of $NiSO_4.7H_2O$ and $Na_7HNb_6O_{19}.15H_2O$ in the presence of excess NaOBr. Cherry red aqueous solutions of this compound absorb broadly around 27,700 $cm^{-1}$ (log $\epsilon \approx 4.2$) and decompose slowly at room temperature. Small cherry red crystals of $K_7[NiV_{13}O_{38}].18H_2O$ form upon reacting $NiSO_4.7H_2O$, $KVO_3$, $HNO_3$ and potassium peroxydisulfate [269]. Oxidation number studies support the proposed tetravalence of the nickel ion. No maxima have been detected in either the ultraviolet or visible spectra. The observed diamagnetism of the Ni(IV) ion has been used to support a proposed structure which is described as having each of the twelve edges of a central nickel octahedron shared with a $[VO_6]$ octahedron, the thirteenth $[VO_6]$ residing on one of the fourfold axes [269].

Although binary fluorides of Ni(III) and (IV) are unknown, $M_2^INiF_6$ (M = Na, K, Rb, Cs) have been reported [270]. The red crystalline potassium salt [271] and the rubidium and cesium salts [133] are generally prepared by fluorination of a mixture of the corresponding alkali and nickel(II) chlorides at about 275°C, though some modifications have been recently tried [270,272,273]. The sodium salt has been prepared by fluorination of $NiF_2$ and NaCl or NaF at 350°C and 350 atm for 10 hours [270]. Treatment of "$Na_2NiO_3$" with fluorine at 280°C, 120 atm for several days will also produce the alkali fluoride [274]. $Na_2NiF_6$ decomposes extremely easily (even under dry argon or sealed in silica tubes) and both preparations are reportedly impure, with up to 10% monoclinic $Na_3NiF_6$ and 10% orthorhombic $NaNiF_3$ [274] present.

Unlike a number of tetravalent transition metal hexafluorides, the nickel salts exhibit no polymorphism [133,274] the K and Rb compounds notably remaining cubic for as long as 60 hours at 100°-250°C [133]. It does seem, therefore, in agreement with recent suggestions, that the $[NiF_6]^{2-}$ anion is sufficiently small and/or covalently bound to absorb a change in cation size without an accompanying change in type of packing [274]. The cell parameters for the K, Rb and Cs salts respectively are $a_o$ = 8.11 [271,272], 8.44 [133,270] and 8.92Å [133].

Four independent reports of the electronic spectrum of these salts have appeared in the last two years. All four reports contain measurements made at room temperature; in HF solution [270,272], Nujol mull [270], KBr disc [272] and as a powder [274,275] while one report presents a spectrum obtained at 80°K (in a mineral oil mull) [273], Fig. 15. The observations and assignments are in reasonable agreement with one another, $K_2NiF_6$: $^1A_1 \rightarrow {}^1T_1$ 19080 [272], 19000 [275], 19230 [273] cm$^{-1}$; $Na_2NiF_6$: 19000 cm$^{-1}$ [274,275]; $Cs_2$, $Rb_2$ and $(NO)_2NiF_6$: 18116, 18018 and 18248 cm$^{-1}$ respectively [270]. Vibrational fine structure is present, $\nu$ = 492 [272], 486 - 490 [273,275] cm$^{-1}$ on this band, and has tentatively been assigned to arise from the Ni-F stretch ($A_{1g}(\nu_1)$ 560 cm$^{-1}$ [275] or $E_g(\nu_2)$ 520 cm$^{-1}$ [276] in the ground state.) The second allowed transition, $^1A_1 \rightarrow {}^1T_2$, has been assigned as a shoulder at 25300 cm$^{-1}$ [275] or 25830

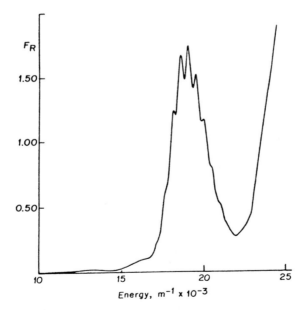

FIG. 15.  The electronic spectrum of $K_2NiF_6$ [275].

$cm^{-1}$ [273]. The spin forbidden transitions to the $^3T_{1g}$ and $^3T_{2g}$ states were observed at 12900 and 16000 $cm^{-1}$ [274,275], while an intense charge transfer band was reported at 37037, [273] 30800 [275] and 32785 $cm^{-1}$ [273]. The positions of spin allowed bands are remarkably close to predictions made by Jørgensen [277,278] and emphasize the usefulness of $\Delta$, $\beta$ (of the spectrochemical and nephalauxetic series) and $\chi_{opt}$ as empirical parameters. The calculated 10Dq, B and $\beta$ for the $K_2NiF_6$ were found to be 20900 $cm^{-1}$, 485 $cm^{-1}$ and 0.40 [273] and 20100 $cm^{-1}$, 515 $cm^{-1}$ and 0.43 [275]. Values of 19900 $cm^{-1}$ and 450 $cm^{-1}$ for 10 Dq and B were also calculated [273] from a reflectance spectrum reported in 1965 [279]. Although generally referred to as "diamagnetic" in the literature [273, 279] the susceptibility of $K_2NiF_6$ at 298, 195 and 90°K is 2.16, 1.68 and 1.04 B.M., perhaps due to interaction with interstitial $O^{2-}$ [280].

A study of the $^{19}F$ nuclear magnetic resonance spectra of $K_2NiF_6$ and $Rb_2NiF_6$ indicate highly shielded $^{19}F$ nuclei [279]. When compared to the chemical shifts observed for compounds such as $[TiF_6]^{2-}$, $[SiF_6]^{2-}$, $[SnF_6]^{2-}$ etc. it is found that the shift (upfield) for $[NiF_6]^{2-}$ is at the very least three times as great as that observed for the aforementioned anions. No suitable explanation has been presented for this behavior. The cell dimensions and structures of a number of $[MF_6]^{2-}$ have been reported by Bode and Voss [281].

All reported ground state vibrational modes of the $M_2NiF_6$ salts have collected in Table 30.

### B. Ni(III)

In addition to the hydrated higher "valent" nickel oxide and the several salts which contain some Ni(III), trivalent nickel appears in $M_2Ni_2O_5$ (M = $K^+$, $Rb^+$ and $Ba^{++}$), $M^INiO_2$ (M = Na and Li), and $LaNiO_3$. The black K- and $RbNiO_{2.5}$ (tetragonal cells a,c = 4.93, 6.10 Å and 5.02, 6.58 Å respectively) can be isolated around 310-400°C in the reaction which produces $M_2NiO_3$ oxides at 450-500°C [264]. The moments (1.6 and 1.7 B.M.) can be accounted for by the formulation $M_2^INi_2^{III}(O^{2-})_3(O_2^{2-})$. A black compound, roughly formulated as "$Ba_2Ni_2O_5$" can be prepared by fusion of the barium and nickel oxides in barium chloride, hydroxide or carbonate at ≃1000°C. This material has a complex powder pattern, low resistivity and a magnetic moment of 1.24 B.M. at room temperature [261,262].

The crystal type, magnetic behavior and conductivity of NiO is altered upon heating with $LiCO_3$ (800°C) [282] or $Li_2O_2$ (1000°C) [283] to form $Li^+Ni_{1\pm2x}^+Ni_x^{+++}O$. The cubic lattice parameter, $a_o$, decreases with increasing lithium content (0-5 wt%) [283]. Below x = 0.3, $Li_xNi_{1-x}O$ is cubic and antiferromagnetic. In the range 0.3<x<0.5 it is rhombohedral and ferromagnetic [282]. The compounds $NaNi^{III}O_2$ and $LiNi^{III}O_2$ are prepared by bubbling $O_2$ through molten alkali hydroxides in a nickel tube

TABLE 30

Ground State Vibrations for $[NiF_6]^{2-}$

| $M^I$ in $M_2^I NiF_6$ | $\nu_1$ | $\nu_2$ | $\nu_3$ | $\nu_4$ | $\nu_5$ | Lattice | Ref. |
|---|---|---|---|---|---|---|---|
| K | 562 | 520 | 658 | 345 | 310 | 178, 138 | 276 |
|  |  |  | 654 |  |  |  | 281 |
|  |  |  | 663 | 349 |  |  | 272 |
|  | 560 |  | 654 |  |  |  | 275 |
|  |  |  | 662 | 350 |  | 138 | 270 |
| Na |  |  | 658 | 350 |  | 200 | 270 |
|  |  |  | 669 |  |  |  |  |
| Rb |  |  | 654 | 343 |  | 110 | 270 |
| Cs |  |  | 654 | 337 |  | 106 | 270 |
| (NO) |  |  | 647 | 352 |  | 101, 166 | 270 |

at 800°C [284]. The observed Ni-O and Na-O distances, Table 29, are close to those expected from sums of ionic radii, 2.0 and 2.35 Å. Above 220°C $NaNiO_2$ has a hexagonal habit, a = 2.96, c = 15.77 Å, and may be a mixed valence compound [264]. Each $Li^+$ and $Ni^{3+}$ is at the center of a regular oxygen octahedron in rhombohedral $LiNiO_2$ (hexag. a = 2.88 Å, c = 14.2 Å [284].

$LaNiO_3$, prepared by heating a stoichiometric mixture of the oxides or carbonates at 900°C [285] or the oxalates at 800°C [286], is also rhombohedral, a = 7.676 Å $\alpha$ = 90°41', with low spin $Ni^{3+}$ at the center of a regular octahedron. The absence of evidence of magnetic order down to 10°K and the metallic conductivity in $LaNiO_3$, between 200°C and 300°C, are consistent with the presence of a partially filled $\sigma*$ band (of $e_g$ orbitals) [286]. In light of this it would be interesting to determine the presence, or lack, of magnetic order and conductivity in pure $LiNiO_2$. From the size of $Li^+$ and probable presence of undistorted low spin $Ni^{3+}$, one would expect this to be a metallic rather than magnetic solid, even though the resistivity of the semiconducting mixed valence oxide levels off above 2 wt% $Li^+$ [287].

A summary of electron spin resonance data for $Ni^{3+}$ in oxide hosts appears in Table 31. A trigonal distortion of the octahedral, $d^7$, low spin, $^2E_1$ ground state of $Ni^{3+}$ in the $Al^{3+}$ site in $Al_2O_3$ was originally estimated as ~1000 $cm^{-1}$ from polarized optical spectra taken at 35 and 78°K [288]. A study of the esr spectrum appears to indicate, however, [289] that while the ground state is $^2E$ there is also a Jahn-Teller distortion present. The isotropic g-value of $Ni^{3+}$ in $SrTiO_3$ is also believed to arise from a dynamic Jahn-Teller effect [290]. Since this crystal undergoes a transition from cubic to tetragonal below 110°K, at which temperature the axial spectrum develops, it is not obvious how much anisotropy is contributed either by a static Jahn-Teller effect or a crystal packing induced site distortion [290,291]. The Jahn-Teller splitting of the ground state of $Ni^{3+}$, separates the two lowest energy levels by an estimated 665 $cm^{-1}$ in $SrTiO_3$ and 1530 $cm^{-1}$ [291] in $Al_2O_3$. The 10Dq in $Al_2O_3$ is about 17000 [291] - 18000 [288] $cm^{-1}$.

There are two characteristic esr spectra of nickel in $TiO_2$. One arises from low spin $Ni^{3+}$ in an interstitial site, and has line widths of 2 and 1.2 gauss at 292 and 78 - 4.2°K [292]. (Although the $Ni^{3+}$ is small enough (~0.65 Å) to fit at a substitutional site (radii <0.73Å), the $Ni^{2+}$ (~0.78 Å) from which the 7 - 10% $Ni^{3+}$ arises is not.) A second spectrum, perhaps with $Ni^{3+}$ or $Ni^+$ in a substitutional site (g[110] = 2.272, g[c] = 2.237 and g[1$\bar{1}$0] = 2.050) develops at 4.2 and 78°K upon irradiation with light (~5000 Å, 2.48 eV). It is unlikely that this spectrum is due to an electron-hole pair as the incident visible radiation has less than known band gap energy for electron-hole pair production (~3.2 eV).

A similar observation of blue light (4000 Å, 3.1 eV) generated esr signals which are isotropic in ZnS and anisotropic in ZnO, has been attributed to $Ni^{3+}$ [293]. A very similar anisotropic three line spectrum, which broadens above 77°K and disappears above ~150°K has been attributed to high spin $Ni^{3+}$ in $K^+$ sites in $KTaO_3$ [294]. Another narrow ($\simeq$10 gauss) anisotropic spectrum, $g_{||} > g_\perp$, is attributed to low spin $Ni^{3+}$ in $Ta^{5+}$ sites. As $g_{||} > g_\perp$ it appears that the axial distortion is compressional. This suggests that the distortion is due to a Jahn-Teller effect as an adjacent oxygen vacancy would produce an elongation of the octahedron. No change was observed at or below the $KTaO_3$ ferroelectric Curie temperature, 60°K, though any change may have been masked due to the presence of an already very strong axial distortion [294]. The second low spin spectrum, $g_{||}$ = 2.169 and $g_\perp$ = 2.086, develops only on addition of sodium. The $K_{1-x}Na_xTaO_3$ also has altered phase transition temperatures. What relationship, if any, exists between these two phenomena is not known.

The crystal field and ground state splitting energies of low spin $Ni^{3+}$ in various hosts, estimated from g values and Tanabe-Sugano diagrams are summarized in Table 32. As would be expected $Ni^{3+}$ at a $K^+$ site exhibits a relatively small crystal field splitting, 10Dq = 7400 $cm^{-1}$ [294]. A most unusual spectrum, tentatively assigned to $Ni^{3+}$ in SrO, has both g values close to 4 [295].

## TABLE 31

Electron Spin Resonance Results for $Ni^{3+}$ or $Cu^{3+}$ Ions

| Ion | Host | Temperature | $g\parallel$ | $\langle g \rangle$ | $g\perp$ | | Ref. |
|---|---|---|---|---|---|---|---|
| $Ni^{3+}$ | $Al_2O_3$ | 50 - 290°K | | 2.146 | | | 289 |
| | $SrTiO_3$ | 203°K | | 2.180 | | | 290 |
| | | 80 | 2.172 | | 2.184 | | |
| | | 20 | 2.136 | | 2.202 | | |
| | | 4.2 | 2.110 | | 2.213 | | |
| | $TiO_2$ | 4.2°K, 290°K | 2.254 ($g_z$) | | 2.084 ($g_x$), 2.085 ($g_y$) | | 292 |
| | ZnS | 1.3°K | | 2.148 | | | 293 |
| | ZnO | 4.2 - 1.3°K | 2.1426 | | 4.3179 | | |
| | $KTaO_3$ | 77°K | 2.216 | | 4.423 | $K^+$ site, high spin | 294 |
| | | | 2.234 | | 2.111 | $Ta^{5+}$ site, low spin | |
| | | | 2.169 | | 2.086 | | |
| $Cu^{3+}$ | $Al_2O_3$ | 1.8°K | 2.0788 | | 2.0772 | | 289 |

TABLE 32

Crystal Field Parameters for $Ni^{3+}$

| | 10Dq (cm$^{-1}$) | $\Delta(^2E)$ cm$^{-1}$ | Ref. |
|---|---|---|---|
| $Al_2O_3$ | 18,000 | 1,000 | 288 |
| | 17,000 | 1,530 | 291 |
| $SrTiO_3$ | | 665 | 291 |
| $K_3NiF_6$ | 16,200 | | 296 |
| $KTaO_3$ | | | |
| $Ta^{5+}$ sites | 18,000 | 1,400 | 294 |
| $K^+$ sites | 7,400 | | 294 |

Unlike either yellow Ni(II) or red Ni(IV) complex fluorides, the Ni(III) fluorides $M_3NiF_6$, M = K, Na are violet-rose and $BaNiF_5$, violet-gray. The former are made by fluorinating the alkali chlorides and nickel chloride or sulfate at 300-400°C [167, 281, 256]. These fluorides are cubic. Attempts to make $Rb_3NiF_6$ and $Cs_3NiF_6$ appear to have failed [256]. The barium salt has been prepared by fluorination of $BaNi(CN)_4$ at 300°C. A brown modification, also $BaNiF_5$, evidently develops upon further reaction at 500°C [255].

The $M_3NiF_6$(M=K and Na) compounds are not high spin, as are the iso-electronic Co(II) fluoro complexes. They are not fully spin paired either, however, having anomalous, temperature dependent "intermediate" moments ($K_3NiF_6$, $\mu$ = 2.12, 2.30, and 2.54 B.M. at 90, 195 and 294°K [167, 256,280]). The magnetic behavior of mixed crystals $K_3NiF_6/K_{2.8}AlF_{5.8}$ indicates $O^{2-}$ may replace $F^-$ and cause an oxidation of $Ni^{3+}$ to $Ni^{4+}$ in this system [280]. A low spin - high spin - ($^2E_g$ - $^4T_{1g}$) ground state equilibrium could also account for the moment which lies between 1 and 3 unpaired electrons. This latter is consistent with a recent fit of the optical spectrum using a Dq/B value slightly above the $^2E_g$-$^4T_{1g}$ crossover in a $d^7$, $D_{4h}$ diagram (Dq = 1620, B = 703 cm$^{-1}$ [296]). The tetragonal ground state splitting, $^2E_g(A_1) \rightarrow {}^2E_g(B_1)$, was found by optical measurements to be 6,800 and 6,100 cm$^{-1}$ in the K and Na salts. The ligand field bands $^2A_1 \rightarrow {}^2E(^2T_{1g})$, $^2E(^2T_{2g})$ and $^2A_2(^2T_{1g})$ were assigned at 12300, (12800

in the sodium analog) 15700 and 19200 cm$^{-1}$. These bands and assignments from the room temperature and 77°K reflectance spectra are in
agreement with earlier observations [256]. Charge transfer bands were
reported at 32,000 and 37,000 cm$^{-1}$ [256,296]. The high degree of covalency, unusual for fluorides, indicated here by $\beta = 0.63$ [296], is nonethe-
less, smaller than that observed in the more unstable Ni(IV) or the Cu(III)
analogues. Vibrational fine structure has not been observed for the
Ni(III) alkali fluorides [272], however, low temperature single crystal
polarized spectra might allow observation of such. See Fig. 16.

The $\nu_3$, M–F stretching vibration narrows and increases in energy with
increasing formal nickel charge: Ni(II), (III), and (IV), 445, 580, and 663
cm$^{-1}$.

## VII. COPPER

### A. Cu(III)

The chemistry of Cu(III) is at present somewhat limited, but will probably receive increased attention in the near future because of improved
handling techniques. The crystal structures of the oxides, periodates and
tellurates and fluorides have not been reported, though the compounds
have been known for some 20-30 years. The electronic spectrum of a
fluoride was first reported in 1969.

The potassium or sodium cuprate(III), $MCuO_2$, is obtained by heating the

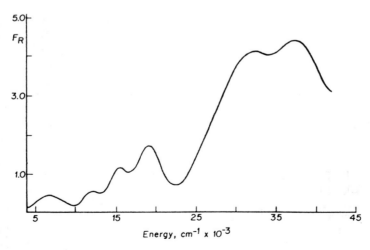

FIG. 16.  The electronic spectrum of $K_3NiF_6$ [297].

oxides (400-500°C) [297] or treating the hydroxides with alkaline hypobromite [298]. The potassium compound, $KCuO_2$, is a steel blue, diamagnetic solid [297]. A "$Ba(CuO_2)_2$ x $H_2O$" species forms upon treatment of the red-brown $NaCuO_2$ solution with $BaCl_2$ [298] or upon heating BaO and CuO [297].

The periodate or tellurate ions stabilize Cu(III) after oxidation of alkaline Cu(II) by persulfate or hypochlorite ion. It has been suggested [266] that in diamagnetic $Mn[Cu(IO_6)_2]$ x $H_2O$, the Cu(III) ion is coordinated, in a square planar configuration, to 2 pairs of oxygens, each from an $[IO_6]^{5-}$ octahedron. The tellurates $M_5^IH_4[Cu(TeO_6)_2]$ (M = K or Na) have been widely used as titrants [299] (for sugars, proteins, glycerol and organic acids), but not studied for the nature of the Cu(III) ion.

Copper(III) is present in $Al^{3+}$ sites in single crystals of $Al_2O_3$ grown from the oxide fluxes [289,300]. Measurements of the axial electron spin resonance spectrum indicate hyperfine splitting constants, A and B (for $Cu^{63}$), to be $-64.6 \times 10^{-4}$ and $-60.2 \times 10^{-4}$ cm$^{-1}$ respectively. The g values found for Cu(III)/$Al_2O_3$, Table 32, are closer to the free electron value than those of the isoelectronic $Ni^{II}$ ion (g|| = 2.196 and g = 2.187), reflecting the stronger crystal field surrounding the more highly charged copper ion [289]. From the optical spectrum the $^3A_2(t_2^6e^2) \rightarrow {}^3T_2(t_2^5e^3)$ transition, which is equal to 10 Dq, is found to occur at 21000 cm$^{-1}$ [289]. (This is compared to an energy of 10,000 cm$^{-1}$ for $Ni^{II}$). Bands at 15,000 and 30,000 cm$^{-1}$ were assigned to an excitation to the $^1E$ and $^3T_1$ states. The trigonal field splitting of the $^3T_2$ state ($^3E, {}^3A_1$) is roughly estimated as ~515 cm$^{-1}$ [289, 300].

The light green $K_3CuF_6$ or brown $Cs_2[Rb$ or $K](CuF_6)$ are obtained by fluorinating a mixture of the chlorides at $\simeq 250$°C [301]. The latter are cubic, of the $K_2NaCrF_6$ crystal type, with a = 8.92 and 9.08 Å respectively. A mixed valence red brown $Cs_{1.67}Cu_{0.67}^{II}[Cu^{III}F_6]$, a = 8.82 Å, is also known [301]. The diffuse reflectance spectrum of $Cs_2KCuF_6$, Fig. 17, exhibits maxima at 9,600, 14,100, 16,400, and 20400 cm$^{-1}$. These have been assigned to transitions from the $^3A_{2g}$ ground state to the $^1E_g$, $^3T_{2g}$, $^1A_{1g}$ and $^3T_{1g}$ excited states [296]. Strong $\pi \rightarrow e_g$ charge transfer bands fall at 29,500 and 37,200 cm$^{-1}$. $B_{35}$ and $\beta_{35}$ were estimated at 641 cm$^{-1}$ and 0.55, the latter being the lowest value known for first row transition metal(III) fluorides.

## ACKNOWLEDGMENTS

The authors wish to express their gratitude to the National Science Foundation for financial support, to Miss Elizabeth DeVore and Mrs. Cathy Isernhagen for technical assistance, and to Mrs. Margaret Fowler, Librarian, Chemistry Department, Imperial College.

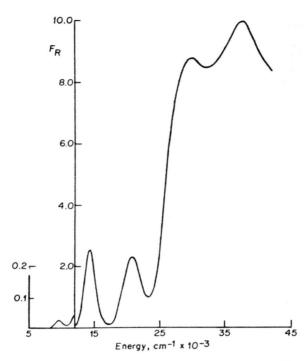

FIG. 17. The electronic spectrum of $Cs_2KCuF_6$ [297].

## REFERENCES

1. A. Bystrom and K. A. Wilhelmi, <u>Acta Chem. Scand</u>., <u>5</u>, 1003 (1951).

2. B. Gossner and F. Mussgnug, <u>Z. Krist</u>. <u>72</u>, 476 (1930).

3. J. A. Campbell, <u>Acta Cryst</u>. <u>9</u>, 192 (1956).

4. U. Klement, <u>Naturwiss</u>. <u>45</u>, 83 (1958).

5. K. A. Wilhelmi, <u>Arkiv Kemi</u> <u>26</u>, 149 (1966).

6. J. K. Brnadon and J. D. Brown, <u>Can. J. Chem</u>. <u>46</u>, 993 (1968).

7. J. S. Stephen and D. W. J. Cruickshank, <u>Acta Cryst</u>. <u>B26</u>, 437 (1970).

8. C. J. H. Schutte and A. M. Heyns, <u>J. Mol. Structure</u> <u>5</u>, 37 (1970).

9. J. Jaffray, <u>Compt. Rend</u>. <u>241</u>, 1114 (1955).

10. J. J. Rush, T. I. Taylor and W. W. Havens, Jr., <u>J. Chem. Phys</u>. <u>35</u>, 2265 (1961).

11. J. J. Rush, T. I. Taylor and W. W. Havens, Jr., J. Chem. Phys. 37, 234 (1962).

12. K. A. Wilhelmi, Acta Chem. Scand. 12, 1965 (1958).

13. K. A. Wilhelmi, Arkiv Kemi 26, 131 (1966).

14. K. A. Wilhelmi, Arkiv Kemi 26, 141 (1966).

15. K. A. Wilhelmi, Acta Chem. Scand. 19, 165 (1965).

16. K. A. Wilhelmi, Chem. Comm. 437 (1966).

17. J. S. Stephens and D. W. J. Cruickshank, Acta Cryst. B26, 222 (1970).

18. A. Bystrom and K. A. Wilhelmi, Acta Chem. Scand. 4, 1131 (1950).

19. F. Hanic and D. Stempelova, Chem. Zvesti, 14, 156 (1960).

20. L. Helmholz and W. R. Foster, J. Am. Chem. Soc. 72, 4971 (1950).

21. R. W. G. Wyckoff, Crystal Structures, Vol. 3, Interscience, New York (1965).

22. J. A. A. Ketelaar and E. Wegerif, Rec. Trav. Chim. 58, 948 (1939).

23. H. Fischmeister, Acta Cryst. 7, 776 (1954).

24. W. H. Zachariasen, and G. E. Ziegler, Z. Krist. 80, 164 (1931).

25. H.W. Smith, Jr. and M. Y. Colby, Z. Krist., 103A, 90 (1941).

26. J. J. Miller, Z. Krist. 99A, 32 (1938).

27. Reports of Standard X-Ray Diffraction Powder Patterns, National Bureau of Standards (1963).

28. D. I. Bujor, Z. Krist. 105A, 364 (1944).

29. J. H. Clouse, Z. Krist. 83, 161 (1932).

30. C. W. F. T. Pistorius and M. C. Pistorius, Z. Krist. 117, 259 (1962).

31. H. G. K. Westenbrink, Rec. Trav. Chim. 46, 105 (1927).

32. K. Brandt, Arkiv Kemi, Mineral. Geol. 17A, No. 6 (1943).

33. J. Teltow, Z. Physik, Chem., B43, 198 (1939).

34. J. Duinker and C. J. Ballhausen, Theoret. Chem. Acta 12, 325 (1968).

35. M. Wolfsberg and L. Helmholz, J. Chem. Phys., 20, 837 (1952).

36. C. J. Ballhausen and A. D. Liehr, J. Mol. Spectry. 2, 342 (1958).

37. D. Schonland, Proc. Roy. Soc. (London), A254, 111 (1960).

38. J. P. Dahl and C. J. Ballhausen, Adv. Quant. Chem. 4, 170 (1968).

39. A. Viste and H. B. Gray, Inorg. Chem. 3, 1113 (1964).

40. R. M. Canadine and I. H. Hillier, J. Chem. Phys. 50, 2984 (1969).

41. J. P. Dahl and H. Johansen, Theoret. Chim. Acta 11, 26 (1968).

42. L. Oleari, G. DeMichelis, and L. Disipio, Mol. Phys. 10, 111 (1966).

43. R. D. Brown, B. H. James, M. F. O'Dwyer, and K. R. Roby, Chem. Phys. Letters 1, 459 (1967)

44. P. E. Best, J. Chem. Phys. 44, 3248 (1966).

45. A. Carrington and C. K. Jørgensen, Mol. Phys. 4, 395 (1960).

46. H. Basch, A. Viste, and H. B. Gray, J. Chem. Phys. 44, 10 (1966).

47. S. L. Holt and C. J. Ballhausen, Theoret. Chim. Acta 7, 313 (1967).

48. J. P. Dahl and H. Johansen, Theoret. Chim. Acta 11, 8 (1968).

49. C. Simo, E. Banks, and S. L. Holt, Inorg. Chem. 9, 183 (1970).

50. I. H. Hillier and V. R. Saunders, Chem. Comm., 1275 (1969).

51. T. M. Dunn and A. H. Francis, J. Mol. Spectry. 25, 86 (1968).

52. J. A. Campbell, Spectrochim. Acta 21, 1333 (1965).

53. R. G. Darrie, W. P. Doyle, and I. Kirkpatrick, J. Inorg. Nuc. Chem. 29, 979 (1967).

54. A. K. Muller, E. J. Baran, and P. J. Hendra, Spectrochim. Acta 25A, 1654 (1969).

55. M. S. Mathur, C. A. Freuzel, and E. B. Bradley, J. Mol. Struct. 2, 429 (1968).

56. R. Mattes, Z. Naturforschg. 241, 772 (1969).

57. W. H. Hartford and M. Darrin, Chem. Rev. 58, 1 (1958).

58. W. T. Smith, Angew. Chem. Intern. Ed. Eng. 1, 467 (1962).

59. W. E. Hobbs, J. Chem. Phys. 28, 1220 (1958).

60. G. D. Flesch and H. J. Svec, J. Am. Chem. Soc. 81, 1787 (1959).

61. A. J. Edwards, Proc. Chem. Soc. 1963, 205.

62. O. Glemser, H. Rocsky, and K. H. Hellberg, Angew. Chem. Intern. Ed. Eng. 2, 266 (1963).

63. K. H. Hellberg, A. Muller, and O. Glemser, Z. Naturforsch. 21B, 118 (1966).

64. R. Stomberg and C. Brosset, Acta Chem. Scand. 14, 441 (1960).

65. J. D. Swalen and J. A. Ibers, J. Chem. Phys. 37, 17 (1962).

66. B. R. McGarvey, J. Chem. Phys. 37, 2001 (1962).

67. W. Klemm, Angew. Chem. 63, 396 (1951).

68. R. Scholder, Angew. Chem. 65, 240 (1953).

69. R. Scholder, Angew. Chem. 70, 583 (1958).

70. R. Scholder and W. Klemm, Angew. Chem. 66, 461 (1954).

71. R. Scholder and H. Schwartz, Z. Anorg. Allgem. Chem. 326, 1, 11 (1963).

72. K. A. Wilhelmi and O. Jonsson, Acta Chem. Scand. 19, 177 (1965).

73. W. Johnson, Mineral Mag. 32, 408 (1960).

74. R. Scholder and E. Suchy, Z. Anorg. Allgem. Chem. 308, 295 (1961).

75. E. Banks and K. L. Jaunarajs, Inorg. Chem. 4, 78 (1965).

76. M. Greenblatt, E. Banks, and B. Post, Acta Cryst. 23, 166 (1967).

77. A. Carrington, D. J. E. Ingram, D. Schonland, and M. C. R. Symons, J. Chem. Soc. 1956, 4710.

78. L. L. van Reijen and P. Cossee, Disc. Farad. Soc. 41, 277 (1966).

79. E. Banks, M. Greenblatt, and B. R. McGarvey, J. Chem. Phys. 47, 3772 (1967).

80. K. V. Lingan, P. G. Nair, and B. Venkataraman, Proc. Ind. Acad. Sci. 70A, 29 (1969).

81. L. L. van Reijen, P. Cossee, and H. J. van Haren, J. Chem. Phys. 38, 572 (1963).

82. G. J. Palenik, Inorg. Chem. 6, 507 (1967).

83. J. Milstein, S. L. Holt, and B. R. McGarvey, Inorg. Chem. 10, (1972).

84. E. Banks, M. Greenblatt, and S. L. Holt, J. Chem. Phys. 49, 1431 (1968).

85. B. R. McGarvey, Electron Spin Resonance of Metal Complexes 1, Plenum Press, New York (1969).

86. E. Banks, M. Greenblatt, S. L. Holt, and B. R. McGarvey, to be published.

87. N. Bailey and M. C. R. Symons, J. Chem. Soc. 203 (1957).

88. L. DiSipio, G. DeMichelis, E. Tondello, and S. Oleari, Gazz. Chim. Ital. 96, 1775 (1966).

89. N. S. Garif'Yanov, Proc. Acad. Sci. USSR, Phys. Chem. 155, 249 (1964).

90. O. V. Ziebarth and J. Selbin, J. Inorg. Nuc. Chem. 32, 849 (1970).

91. H. Kon and N. E. Sharpless, J. Chem. Phys. 42, 906 (1965).

92. C. R. Hare, I. Bernal, and H. B. Gray, Inorg. Chem. 1, 831 (1962).

93. P. T. Monoharan and M. T. Rogers, J. Chem. Phys. 49, 5510 (1968).

94. N. S. Garif Yanov, E. Kamenev, and I. V. Ovchianikov, Russ. J. Phys. Chem. 43, 611 (1969).

95. H. B. Gray and C. R. Hare, Inorg. Chem. 1, 363 (1962).

96. H. L. Krauss, M. Leder, and G. Munster, Chem. Ber. 96, 3008 (1963).

97. H. C. Mishra and M. C. R. Symons, J. Chem. Soc. 4490 (1963).

98. E. Wendling, Rev. Chim. Minerale 4, 425 (1967).

99. D. G. Howard and R. H. Lindquist, J. Chem. Phys. 38, 573 (1963).

100. J. Deven, J. Haber, and S. Kosek, Bull. Acad. Poln. Sci. Ser. Sci. Chem. 13, 21 (1965).

101. R. B. Johannesen and H. L. Krauss, Chem. Ber. 97, 2094 (1964).

102. H. C. Clark and Y. N. Sadana, Can. J. Chem. 42, 50, 702 (1964).

103. T. A. O'Donnell and D. F. Stewart, Inorg. Chem. 5, 1434 (1966).

104. A. Michel and J. Bernard, Compt. Rend. 200, 1316 (1935).

105. K. A. Wilhelmi and O. Jonsson, Acta Chem. Scand. 12, 1532 (1958).

106. B. L. Chamberland, Mat. Res. Bull. 2, 827 (1967).

107. F. J. Darnell and W. H. Cloud, Bull. Soc. Chem. France, 1164 (1965).

108. W. H. Cloud, D. S. Schreiber, and K. R. Babcock, J. Appl. Phys. 33, 1193 (1962).

109. K. Siratoni and S. Iida, J. Phys. Soc. Japan, 15, 2362 (1960).

110. J. B. Goodenough, Mat. Res. Bull. 2, 37 (1967).

111. A. M. Stoffel, J. Appl. Phys. 40, 1238 (1969).

112. J. B. Goodenough, Magnetism and the Chemical Bond, Interscience-Wiley, N. Y. 1963.

113. A. Nussbaum, Solid State Physics, 18, 165 (1966).

114. P. F. Bongers, Phillips Tech. Rev. 28, 13 (1967).

115. J. B. Goodenough, Bull. Soc. Chem. France, 1200 (1965).

116. A. Magneli and G. Andersson, Acta Chem. Scand. 9, 1378 (1955).

117. J. B. Goodenough, Mat. Res. Bull. 2, 165 (1967).

118. D. S. Chapin, J. A. Cafalas, and J. M. Honig, J. Phys. Chem. 69, 1402 (1965).

119. R. H. Hoskins and B. H. Soffer, Phys. Rev. 133, A490 (1964).

120. G. W. Ludwig and H. H. Woodbury, Solid State Phys. 13, 265, 271 (1962).

121. S. P. Lunkin, D. G. Galimov, and D. M. Yudin, Optics and Spectroscopy, 25, 323 (1968).

122. R. Scholder and G. Sperka, Z. Anorg. Allgem. Chem. 285, 49 (1956).

123. K. A. Wilhelmi, Arkiv Kemi 26, 157 (1967).

124. B. L. Chamberland, Solid State Comm. 5, 663 (1967).

125. B. L. Chamberland, Inorg. Chem. 8, 286 (1969).

126. W. L. Roth and R. C. DeVries, J. Appl. Phys. 38, 951 (1967).

127. W. L. Roth and R. C. DeVries, J. Am. Ceram. Soc. 51, 72 (1968).

128. C. C. Addison and M. G. Barker, J. Chem. Soc. 1965, 5534.

129. H. Von Wartenberg, Z. Anorg. Allgem. Chem. 247, 135 (1941)

130. H. C. Clark and Y. N. Sadana, Can. J. Chem. 42, 50 (1964).

131. E. Huss and W. Klemm, Z. Anorg. Allgem. Chem. 262, 25 (1950).

132. D. H. Brown, K. R. Dixon, R. D. W. Kemmitt, and D. W. A. Sharp, J. Chem. Soc. 1965, 1559.

133. H. Bode and E. Voss, Z. Anorg. Allgem. Chem. 286, 136 (1956).

134. P. Day, L. DiSipio, and L. Oleari, Chem. Phys. Letters 5, 533 (1970).

135. C. J. Ballhausen, Theoret. Chim. Acta 1, 285 (1963).

136. G. J. Palenik, Inorg. Chem. 6, 503 (1967).

137. C. K. Jorgensen, Structure and Bonding, 6, 94 (1969).

138. R. A. Schroeder, E. R. Lippincott, and C. E. Weir, J. Inorg. Nuc. Chem. 28, 1397 (1966).

139. S. Pinchas, D. Samuel, and E. Petreanu, J. Inorg. Nuc. Chem. 29, 335 (1967).

140. P. Manzelli and G. Taddei, J. Chem. Phys. 51, 1484 (1969).

141. P. J. Hendra, Spectrochim. Acta 24A, 125 (1968).

142. T. S. Briggs, J. Inorg. Nuc. Chem. 30, 2866 (1968).

143. D. Michel and A. Diowa, Naturwiss. 53, 129 (1966).

144. A. Engelbrecht and A. V. Grosse, J. Am. Chem. Soc. 76, 2042 (1954).

145. A. Javan and A. Engelbrecht, Phys. Rev. 96, 649 (1954).

146. R. Scholder and H. Waterstradt, Z. Anorg. Allgem. Chem. 277, 172 (1954).

147. G. Duquenoy, C. R., Acad. Sci. Paris, Ser. C, 268, 828 (1969).

148. A. Hiroichi, S. Suej, and M. Kunikichi, Japanese Patent 6821179
     (Cl 115P17) Chem. Abstracts 70, 116728t (1969).

149. F. H. Herbstein, Acta Cryst. 13, 357 (1960).

150. A. Carrington and M. C. R. Symons, J. Chem. Soc., 3373 (1956).

151. A. Carrington and M. C. R. Symons, J. Chem. Soc., 889 (1960).

152. A. Carrington and D. S. Schonland, Mol. Phys. 3, 331 (1960).

153. R. F. Fenske and C. C. Sweeney, Inorg. Chem. 3, 1105 (1964).

154. C. A. Kosky and S. L. Holt, Chem. Comm. 11, 668 (1970).

155. K. A. Jensen and W. Klemm, Z. Anorg. Allgem. Chem. 237, 47
     (1938).

156. A. Carrington, D. S. Schonland, and M. C. R. Symons, J. Chem. Soc.
     659 (1957).

157. A. Carrington, D. J. E. Ingram, K. A. K. Lott, D. S. Schonland, and
     M. C. R. Symons, Proc. Roy. Soc. (London) A254, 101 (1960).

158. A. Carrington and M. C. R. Symons, Chem. Rev. 63, 443 (1963).

159. W. P. Griffith, J. Chem. Soc. A, 1467 (1966).

160. B. Krebs, A. Muller, and H. W. Roesky, Mol. Phys. 12, 469 (1967).

161. H. Lux, Z. Naturforsch, 1, 281 (1946).

162. B. Jezowska-Trzebiatowska, J. Nawojska, and M. Wronska, Roczniki
     Chem. 25, 405 (1951).

163. W. Klemm, C. Brendl, and G. Wehrmeyer, Chem. Ber. 93, 1506
     (1960).

164. F. Jellinck, J. Inorg. Nuc. Chem. 13, 329 (1960).

165. E. J. Baran and P. J. Aymonino, Spectrochim. Acta 24A, 291 (1968).

166. W. Klemm, Angew. Chem. 66, 468 (1954).

167. J. D. Kingsley, J. S. Prener, and B. Segall, Phys. Rev. 137A, 189
     (1965).

168. J. Milstein, Thesis, Polytechnic Institute of Brooklyn (1970).

169. E. J. Baran and P. J. Aymonino, Monatsh. Chem. 100, 1674 (1969).

170. J. Milstein and S. L. Holt, Inorg. Chem. 8, 1021 (1969).

171. L. Pauling, J. Chem. Soc. 1461 (1948).

172. P. M. DeWolff, Acta Cryst. 12, 341 (1959).

173. D. B. Rogers, R. D. Shannon, A. W. Sleight, and J. L. Gillson, Inorg. Chem. 8, 841 (1969).

174. A. M. Bystrom, Acta Chem. Scand. 3, 163 (1949).

175. Y. Kondrashev and A. I. Zaslavsky, Izv. Akad. Nauk USSR Ser. Fiz. 15, 179 (1951).

176. S. C. Abrahams and J. L. Bernstein, J. Chem. Phys. 46, 3777 (1967).

177. K. Siratori and S. Iida, J. Phys. Soc. Japan 15, 2362 (1960).

178. A. Linek, Czech. J. Phys. B13, 398 (1963).

179. I. Reimer and M. W. Lister, Can. J. Chem. 39, 2431 (1961).

180. R. Giovanoli and U. Leuenberger, Helv. Chim. Acta 52, 2333 (1969).

181. J. H. A. Landy and P. M. DeWolff, Appl. Sci. Res. B10

182. J. S. Kasper and J. S. Prener, Acta Cryst. 1, 246 (1954).

183. J. J. Davies, S. R. P. Smith, and J. E. Wertz, Phys. Rev. 178, 608 (1969).

184. R. Scholder, Z. Elektrochem. 56, 879 (1952).

185. D. Balz and D. Plieth, Z. Elektrochem. 59, 545 (1955).

186. H. Toussaint, Rev. de Chim. Min. 1, 141 (1964).

187. H. Bizette, J. Phys. Radium, 12, 161 (1951).

188. K. A. Muller, Phys. Rev. Letters 2, 341 (1959).

189. M. Nakada, K. Awanzu, and S. Ibuki, J. Phys. Soc. Japan 19, 781 (1964).

190. B. Henderson and T. P. P. Hall, Proc. Phys. Soc. (London) 90, 511 (1967).

191. W. Low, Phys. Rev. 105, 793 (1957).

192. J. E. Wertz and P. Auzins, Phys. Rev. 106, 484 (1957).

193. G. F. Imbusch, A. L. Schawlow, A. D. May, and S. Sugano, Phys. Rev. 140A, 830 (1965).

194. H. G. Andersen, Phys. Rev. 120, 1606 (1960).

195. H. G. Andersen, J. Chem. Phys. 35, 1090 (1961).

196. U. H. From, R. B. Dorain, and C. Kikuchi, Phys. Rev. 135A, 710 (1964).

197. E. Yamaka and R. G. Barnes, Phys. Rev. 135A, 144 (1964).

198. S. Geshwind, P. Kislink, M. P. Klein, J. P. Remeika, and D. L. Wood, Phys. Rev. 126, 1684 (1962).

199. N. Laurance and J. Lambe, Phys. Rev. 132, 1029 (1963).

200. E. Feher and M. D. Sturge, Phys. Rev. 172, 244 (1968).

201. G. E. Stedman, J. Chem. Phys. 51, 4123 (1969).

202. A. Louat, F. Gaume-Mahn, C. R. Acad. Sci., Paris, 266B, 1128 (1968).

203. G. Villela, Y. Saikali, A. Louat, J. Paris, and F. Gaume-Mahn, J. Phys. Chem. Solids 30, 2599 (1969).

204. G. F. Imbusch, S. R. Chinn, and S. Geschwind, Phys. Rev. 161, 295 (1967).

205. J. L. T. Wangh, D. P. Shoemaker, and L. Pauling, Acta Cryst. 7, 438 (1954).

206. C. M. Flynn, Jr. and G. D. Stucky, Inorg. Chem. 8, 335 (1969).

207. L. C. W. Baker and T. J. R. Weakley, J. Inorg. Nuc. Chem. 28, 447 (1966).

208. G. R. Byfleet, W. C. Lin, and C. A. McDowell, Mol. Phys. 18, 363 (1970).

209. B. W. Dale and M. T. Pope, Chem. Comm. 792 (1967).

210. B. W. Dale, J. M. Buckley, and M. T. Pope, J. Chem. Soc. A, 301 (1969).

211. H. R. Oswald, W. Feitknecht, and M. J. Wampetich, Nature 207, 72 (1965).

212. A. D. Wadsley, Acta Cryst. 5, 676 (1952).

213. A. D. Wadsley, Acta Cryst. 8, 165 (1955).

214. B. Mukheyes, Acta Cryst. 13, 164 (1960).

215. R. Giovanoli, E. Stahli, and W. Feitknecht, Chimia 23, 264 (1969).

216. J. Preudhomme, C. R., Acad. Sci. Paris, 267C, 1632 (1968).

217. A. P. B. Sinha, N. R. Sanjana, and A. B. Biswas, J. Phys. Chem. 62, 191 (1958).

218. H. L. Yakel, Acta Cryst. 8, 394 (1955).

219. J. B. Goodenough, Phys. Rev. 100, 564 (1955).

220. E. F. Bertaut, G. Buisson, S. Quezel-Ambrunaz, and G. Quezel, Solid State Comm. 5, 25 (1967).

221. S. Goshen, D. Mukamel, H. Shaked, and S. Shtrikman, J. Appl. Phys. 140, 1590 (1969).

222. H. W. Roesky, O. Glemser, and K. H. Hellberg, Angew. Chem Intern. Ed. Eng. 4, 1098 (1965).

223. R. Hoppe, W. Liebe, and W. Dahne, Z. Anorg. Allgem. Chem. 307, 276 (1961).

224. R. Hoppe, J. Inorg. Nuc. Chem. 8, 437 (1958).

225. G. C. Allen, G. A. M. El-Sharkarwy, and K. D. Warren, Inorg. Nuc. Chem. Letters 5, 725 (1969).

226. N. A. Matwiyoff and L. B. Asprey, Chem. Comm., 75 (1970).

227. C. D. Flint, Chem. Comm. 482 (1970).

228. H. Lux, Handbook of Prep. Inorg. Chem. (G. Brauer, Ed.) Vol. 2, 1504, Academic Press, N. Y. 1965.

229. H. J. Hrostowski and A. B. Scott, J. Chem. Phys. 18, 105 (1950).

230. W. Kerler, W. Neuwirth, E. Fluck, P. Kuhn, and B. Zimmerman, Z. Phys. 173, 321 (1963).

231. T. Sinjo, T. Ichida, and T. Takada, J. Phys. Soc. Japan, 26, 1547 (1969).

232. M. C. R. Symons, Advances in the Chemistry of Coord. Cpds., 430 (1961), MacMillan Co., New York.

233. A. Ito and K. Ono, J. Phys. Soc. Japan, 26, 1548 (1969).

234. G. K. Wertheim and R. H. Herber, J. Chem. Phys. 36, 2497 (1962).

235. L. R. Walker, G. Wertheim, and V. Jaccarino, Phys. Rev. Letters 6, 98 (1961).

236. K. Wahl, W. Klemm, and G. Wehrmeyer, Z. Anorg. Allgem. Chem. 285, 322 (1956).

237. L. Moeser and H. Borck, Chem. Ber. 42, 4279 (1909).

238. P. K. Gallagher, J. B. MacChesney, and D. N. E. Buchanan, J. Chem. Phys. 41, 2429 (1964).

239. J. B. MacChesney, R. C. Sherwood, and J. F. Potter, J. Chem. Phys. 43, 1907 (1965).

240. M. Erchak, I. Fankuchen, and R. Ward, J. Am. Chem. Soc. 68, 2085 (1946).

241. C. Brisi, Ric. Sci. 24, 1858 (1954).

242. H. Watanabe, J. Phys. Soc. Japan 12, 515 (1957).

243. T. R. Clevenger, Jr., J. Am. Ceram. Soc. 46, 207 (1963).

244. G. Shirane, D. E. Cox, and S. L. Ruby, Phys. Rev. 125, 1158 (1962).

245. A. Yoshimori, J. Phys. Soc. Japan, 14, 807 (1959).

246. E. Banks and M. Mizushima, J. Appl. Phys. 40, 1408 (1969).

247. Z. J. Kiss and W. Phillips, Phys. Rev. 180, 924 (1969).

248. B. W. Faughnan and Z. J. Kiss, J. Quant. Electron. 5, 17 (1969).

249. A. F. Reid and A. E. Ringwood, J. Geophys. Res. 74, 3238 (1969).

250. C. Brendel and W. Klemm, Z. Anorg. Allgem. Chem. 320, 59 (1963).

251. R. Scholder, Bull. Soc. Chim. France 1112 (1965).

252. A. Metzl, Z. Anorg. Allgem. Chem. 86, 358 (1914).

253. J. B. Goodenough, J. Phys. Chem. Solids 6, 287 (1958).

254. M. G. Townsend and O. F. Hill, Trans. Farad. Soc. 61, 2597 (1965).

255. R. Hoppe, Rec. Trav. Chim. 75, 569 (1956).

256. W. Klemm, W. Brandt, and R. Hoppe, Z. Anorg. Allgem. Chem. 308, 179 (1961).

257. G. C. Allen and K. D. Warren, Inorg. Chem. 8, 1902 (1969).

258. R. D. Hall, J. Am. Chem. Soc. 29, 692 (1907).

259. M. C. Boswell and R. K. Iler, J. Am. Chem. Soc. 58, 924 (1936).

260. R. W. Cairns and E. Ott, J. Am. Chem. Soc. 56, 1094 (1934).

261. J. J. Lander, Acta Cryst. 4, 148 (1951).

262. J. J. Lander and L. A. Wooten, J. Am. Chem. Soc. 73, 2452 (1951).

263. B. E. Gushee, L. Katz, and R. Ward, J. Am. Chem. Soc. 79, 5601 (1957).

264. H. Bade, W. Bronjer, and W. Klemm, Bull. Soc. Chim. France, 1124 (1965).

265. L. P. Eddy and N.-G. Vannerberg, Acta Chem. Scand. 20, 2886 (1966).

266. P. Ray and B. Sarma, J. Ind. Chem. Soc. 25, 205 (1948).

267. P. Ray, A. Bhaduri, and B. Sarma, J. Ind. Chem. Soc. 25, 51 (1948).

268. H. C. M. Flynn, Jr., and G. D. Stucky, Inorg. Chem. 8, 332 (1969).

269. H. C. M. Flynn, Jr. and M. T. Pope, J. Am. Chem. Soc. 92, 85 (1970).

270. R. Bongon, Compt. Rend. 267C, 681 (1968).

271. W. Klemm and E. Huss, Z. Anorg. Allgem. Chem. 258, 221 (1949).

272. L. Stein, J. M. Neil, and G. R. Alms, Inorg. Chem. 8, 2472 (1969).

273. M. J. Reisfeld, L. B. Asprey, and R. A. Reunemann, J. Mol. Spectry. 29, 109 (1969).

274. H. Henkel, R. Hoppe, and G. C. Allen, J. Inorg. Nuc. Chem. 31 3855 (1969).

275. G. C. Allen and K. D. Warren, Inorg. Chem. **8**, 753 (1969).

276. M. J. Reisfeld, J. Mol. Spectry. **29**, 120 (1969).

277. C. K. Jørgensen, Acta Chem. Scand. **12**, 1539 (1958).

278. C. K. Jørgensen, Mol. Phys. **6**, 43 (1963).

279. W. E. Wageman, M. J. Reisfeld, and E. Fukushima, Inorg. Chem. **8**, 750 (1969).

280. A. D. Westland, R. Hoppe, and S. S. I. Kaseno, Z. Anorg. Allgem. Chem. **338**, 319 (1965).

281. H. Bode and E. Voss, Z. Anorg. Allgem. Chem. **290**, 1 (1957).

282. J. B. Goodenough, D. G. Wickham, and W. J. Croft, J. Phys. Chem. Solids **5**, 107 (1958).

283. G. J. Toussaint and G. Voss, J. Appl. Cryst. **1**, 187 (1968).

284. L. D. Dyer, B. S. Borie, Jr., and G. P. Smith, J. Am. Chem. Soc. **76**, 1499 (1954).

285. W. C. Koehler and E. O. Wollan, J. Phys. Chem. Solids **2**, 100 (1957).

286. J. B. Goodenough and P. M. Raccah, J. Appl. Phys. **36**, 1031 (1965).

287. P. R. Heikes and W. D. Johnston, J. Chem. Phys. **26**, 583 (1957).

288. D. S. McClure, J. Chem. Phys. **36**, 2757 (1962).

289. S. Geschwind and J. P. Remeika, J. Appl. Phys. **33**, 370 (1962).

290. R. S. Rubins and W. Low, Proc. 1st Intern. Conf. Paramag. Res. Jerusalem, 1962, vol. 1, p. 51, Acad. Press, N. Y. (1963).

291. U. P. Hochli and K. A. Muller, Phys. Rev. Letters **12**, 730 (1964).

292. A. J. Gerritsen and E. S. Sabisky, Phys. Rev. **125**, 1853 (1962).

293. W. C. Holton, J. Schneider, and T. L. Estle, Phys. Rev. **133A**, 1638 (1964).

294. D. M. Hannon, Phys. Rev. **164**, 366 (1967).

295. W. Low and J. T. Suss, Phys. Letters (Netherlands), **11** 115 (1964).

296. H. Bode and E. Voss, Z. Anorg. Allgem. Chem. **290**, 1 (1957).

297. G. C. Allen and K. D. Warren, Inorg. Chem. **8**, 1895 (1969).

298. K. Wahl and W. Klemm, Z. Anorg. Allgem. Chem. **270**, 69 (1952).

299. R. Scholder and U. Voelskow, Z. Anorg. Allgem. Chem. **266**, 256 (1951).

300. P. K. Jaiswall and S. Chandra, Chim. Anal. **51**, 493 (1969) and references therein.

301.  W. E. Blumberg, J. Eisinjet, and S. Geschwind, Phys. Rev. 130,
       900 (1963).

302.  R. Hoppe and R. Homann, Naturwiss. 53, 501 (1966).

303.  H. Kon, J. Inorg. Nuc. Chem. 25, 931 (1963).

Chapter 3

MAGNETIC ANISOTROPY

S. Mitra[†]
Department of Inorganic Chemistry
University of Melbourne
Australia

[†]Present address:  Tata Institute of Fundamental Research, Colaba,
Bombay-5, India.

## I. INTRODUCTION

Study of single crystal magnetic susceptibility of paramagnetic and diamagnetic compounds has been very useful in revealing an accurate description of many physicochemical properties of solids. A single crystal, in
general, shows directional properties depending upon the inherent asymmetry in its physical and chemical nature, and the magnetic susceptibility
in different directions of the crystal is different. A measurement of the
individual susceptibilities may, therefore, be of great value in understanding
the detailed properties of solids.

The individual (or principal) susceptibilities of a single crystal can be
determined, in principle, by a variety of techniques generally based on the
Faraday method [1]. While these methods are quite unsuitable for crystals
of lower symmetry (monoclinic and triclinic) they are not sufficiently
accurate for crystals belonging to higher symmetry. The method of direct
measurement of magnetic anisotropy is, on the contrary, very accurate
and suitable for crystals belonging to any system. In this method one measures directly the difference between the two susceptibilities and, hence,
the name "anisotropy." The measurement of magnetic anisotropy is of
considerable interest in a large number of problems and its direct determination has great significance. By incorporating with it the measurement
of average susceptibility or one of the principal susceptibilities, the three
principal susceptibilities can also be determined.

Measurement of magnetic anisotropy is perhaps a much older method
than is generally realized. The first measurements, for instance, were
reported by Stenger [2] and by König [3] on quartz and calcite as early as
1883. It was however Krishnan and co-workers who developed the method
to a very great extent and applied it successfully to a wide spectrum of
solids ranging from transition metal compounds to aromatic hydrocarbons,
minerals, semiconductors, natural products, etc. In a series of papers
published between 1933 and 1939 [4-9], most of their work on diamagnetic
and paramagnetic anisotropies is described. Besides improving the earlier
method used by Stenger and by König, Krishnan and Banerji [6] suggested
a new method, which later became known as the angle-flip method. After
the work of Krishnan and his colleagues, the measurement of magnetic
anisotropy became much more widely known and used. In 1950, Stout and
Griffel [10] modified Krishnan's original angle-flip method and applied it
to a very interesting system of some magnetically concentrated compounds
[11,12]. Since then many improvements have been added to the technique
and the method at present is accurate and sensitive.

Information obtained from the measurement of magnetic anisotropy on
para- and diamagnetic crystals is in general different. In the case of
paramagnetic transition metal compounds, magnetic anisotropy gives
accurate information about the nature of ligand field, the electronic structure of the paramagnetic ion and, in some favorable cases, convincing

evidence of stereochemistry. Magnetic anisotropy of a paramagnetic compound depends primarily upon the asymmetric nature of the ligand field, and hence its measurement is useful in deciding the true symmetry of the ligand field. Most of the earlier work done on the paramagnetic anisotropies is directed towards this end. While the anisotropies of octahedrally and tetrahedrally coordinated cobalt(II) and nickel(II) compounds are strikingly different, there does not appear a similar marked difference in anisotropy of other ions in these coordinations. The anisotropy, in fact, depends so intimately upon the detailed electronic structure of the metal ion that a detailed correlation between the anisotropy and stereochemistry may not be possible except in some simple cases. Measurement of anisotropy has, however, proved to be of particular help in understanding the electronic structure of the compounds belonging to stereochemistries other than octahedral or tetrahedral [13-17].

Information obtained from the measurement of anisotropy on diamagnetic compounds are obviously different from those on paramagnetic ones. Here the effect of crystalline electric field, which is so very important in the paramagnetic compounds, is absent and the anisotropy exhibited by the crystals is a consequence of the asymmetry inherent in their physical and chemical structure. It is striking to find the great variety of diamagnetic crystals studied for different reasons. The interest in the large anisotropy exhibited by aromatic hydrocarbons has been the subject of extensive experimental and theoretical investigations. Recent demonstration of the correlation of NMR chemical shift with the diamagnetic anisotropy of these compounds and its eventual relationship with their semiconductivity has aroused fresh interest in this area. Further the earlier work of Pauling [18], Lonsdale [19], and others on the correlation of diamagnetic anisotropy with the chemical bonding has provided a continuing interest in this field over the past few decades, and has been instrumental in initiating the studies on such interesting compounds as phthalocyanines and metallocenes.

While a very large amount of work on the paramagnetic and diamagnetic anisotropies has been done over the past few decades, most of the work is scattered over a large number of physics texts and journals. Unfortunately no general review in this area has so far been written. Further, neither the basic experimental details and principles involved in different methods of measurement, nor the application of this technique of various chemical problems is clearly documented. The purpose of this article is, therefore, twofold: (1) to provide the necessary experimental details of the various methods of measuring magnetic anisotropy and examine their relative merits; and (2) to review the work done in this area on paramagnetic and diamagnetic compounds, and evaluate the use of this technique in understanding the physical and chemical properties of solids.

Among the paramagnetic transition metal compounds the work done so far is mainly confined to the first row compounds. Some studies have been recently made on second row compounds. No data exist on the compounds of third row metals. However a very systematic study has been made in the area of rare earth compounds and has been included here. The theory of ligand fields in transition metal compounds has been intentionally omitted here, since many standard textbooks and review articles are now available on this subject [20-27].

In the discussion of diamagnetic anisotropy we have included simple inorganic compounds, organometallic compounds, minerals, polymers, aromatic and aliphatic hydrocarbons, and some metals and alloys. Some natural products have also been included. While a full appreciation of this vast area of studies is almost impossible within the limits and scope of this chapter, we have discussed their salient features. Those compounds which are chemically more important have been discussed in detail. Discussion on the aromatic hydrocarbons had to be kept at a minimum. This is unfortunate since they are historically the first compounds to be studied systematically and also constitute an interesting area of investigation.

## II.  EXPERIMENTAL DETAILS

### A.  General Methods of Measuring
### Magnetic Anisotropy

If a magnetic dipole of strength M is placed in a uniform magnetic field of intensity H, it will experience a couple (G) that will try to turn M in the direction of H.  The couple can be represented in vector notation as, [28],

$$G = M \wedge H \tag{1}$$

The above relation holds good whether the dipole is a permanent one or whether it is induced by the field itself. If a magnetizable anisotropic crystal is thus placed in a uniform field, the induced M is not, in general, parallel to H, and the crystal will therefore experience a couple. On the other hand in an isotropic crystal, M is parallel to H and so the couple is zero (we shall neglect here and in our subsequent discussion the effect of shape anisotropy; this topic will be discussed in detail in Section II, E). Equation 1 can be written

$$G_{ij} = - M_i H_j + M_j H_i \tag{2}$$

where,

$$M_i = \chi_{ij} H_j,$$

and hence Eq. 2 becomes

$$G_{ij} = - \chi_{ij} H_k H_j + \chi_{jk} H_k H_i \tag{3}$$

If the axes are chosen as the principal axes of $\chi_{ij}$, the components of $G_{ij}$ simplify to:

$$\begin{vmatrix} O & (\chi_1 - \chi_2)H_1H_2 & (\chi_3 - \chi_1)H_3H_1 \\ (\chi_1 - \chi_2)H_1H_2 & O & \\ -(\chi_3 - \chi_1)H_3H_1 & (\chi_2 - \chi_3)H_2H_3 & O \end{vmatrix}$$

It is seen that the couple depends on the <u>difference</u> between the susceptibilities (or on the magnetic anisotropy) and not on their absolute magnitudes. Evidently any method of measuring magnetic anisotropy would, in principle, involve determination of this couple acting on the crystal in a magnetic field. A convenient method of measuring the couple is to suspend the crystal with a fiber and measure the couple against the torque of the fiber. Within the framework of this principle, various methods have been used for the measurement of anisotropy and the actual details of the measurement have been considerably refined. It is, however, of some interest to note that all the methods are still used. It appears desirable, therefore, at this stage to summarize briefly all the existing methods and discuss their relative advantages and disadvantages. It will be noticed that all the methods are similar to a certain extent.

### 1. Oscillation Method

Let us suppose that a crystal is suspended in a uniform magnetic field H with a fine quartz fiber attached to a torsion head. Let $\chi_{max}$ and $\chi_{min}$ be the maximum and minimum values, respectively, of the susceptibility of the crystal in the plane of the magnetic field. When the magnetic field is applied, the greater of the two susceptibilities ($\chi_{max}$) will tend to "set" along the field direction. By alternately applying the field and rotating the crystal with the torsion head, the crystal can be brought to the zero position. At this position no couple acts on the crystal and, hence, no motion of the crystal on the application of the field. This position is called the "setting position" or "zero position" of the crystal.

The crystal is now rotated in the absence of the magnetic field with $T_0$ being the period of its oscillation. If now the magnetic field H is applied along $\chi_{max}$, then on rotating the crystal through an angle $\delta\phi$ about the axis of the fiber, a restoring couple

$$[C + (\chi_{max} - \chi_{min}) H^2] \delta\phi$$

acts upon the crystal, instead of the couple $C\delta\phi$ in the absence of the field,

C being the torsion constant of the fiber. A new time period $T_1$ is there-fore observed. Hence,

$$cT_o^2 = [C + (\chi_{max} - \chi_{min}) H^2] T_1^2$$

$$\chi_{max} - \chi_{min} = \frac{T_o^2 - T_1^2}{T_1^2} \cdot \frac{C}{H^2}$$

And, therefore,

$$\Delta\chi = \frac{T_o^2 - T_1^2}{T_1^2} \cdot \frac{C}{H^2} \cdot \frac{M}{m} \tag{5}$$

M and m are the molecular weight and mass, respectively, of the crystal. To determine the magnetic anisotropy of a crystal by this method it is sufficient, therefore, to measure its period of oscillation in a uniform magnetic field when suspended along different axes, and to compare them with the corresponding periods of oscillation outside the field.

The above method of measuring magnetic anisotropy was first used by Stenger [3] and by König [4] for measurements on calcite and quartz. It was however Krishnan and his co-workers [5] who developed and used this method extensively in their earlier studies on the diamagnetic and paramagnetic crystals and thus obtained some very stimulating results [4-6]. The great success they had in using this simple and unsophisticated method is indeed most striking.

It was however soon realized that the above method was not suitable in general. For crystals of small mass and low anisotropy, the determination of $T_1$ could not be accurately done. Indeed it was shown later by Mookherji [29], Stout and Griffel [10] and Datta [30] that the earlier measurements of Krishnan et al. [5] on some octahedrally coordinated $Ni^{2+}$ compounds, which are very feebly anisotropic, were in error while those on highly anisotropic compounds were in good agreement. This limitation of the oscillation method was first realized by Krishnan and his co-workers. Nevertheless the great success they achieved by using this method in the case of aromatic hydrocarbons and some highly anisotropic iron group compounds is quite remarkable. The method is still used by some workers [31, 32].

## 2. Critical Couple Method

As we have discussed above, the oscillation method is not very suitable in general [5]. Krishnan and Banerji [6], therefore, proposed a new method termed either the critical-couple method or the angle-flip method. This is probably the most common method of measuring magnetic anisotropy. In this method also, the crystal is suspended in a uniform magnetic field H on a fine quartz fiber attached to a torsion head and brought to the setting position as in the earlier method. In this position the greater of the two susceptibilities in the plane of the magnetic field lies along the field direction. If the torsion head be now slowly rotated away from the equilibrium setting position by an angle $\alpha$ with the magnetic field switched on, the crystal will also rotate but by a much smaller angle, say $\delta$, the twist on the fiber being balanced by a restoring couple due to the field and the crystal anisotropy. The relation governing the motion is,

$$C(\alpha - \delta) = \frac{1}{2} \frac{m}{M} \cdot H^2 (\chi_{max} - \chi_{min}) \sin 2\delta \qquad (6)$$

where the notation is the same as earlier. The restoring couple is a maximum when $\delta = 45°$; this corresponds to a critical angle of deflection, $\alpha_c$, such that

$$C(\alpha_c - \pi/4) = \frac{1}{2} \cdot \frac{m}{M} \cdot H^2 (\chi_{max} - \chi_{min})$$

or

$$\Delta\chi = \frac{2C}{H^2} \cdot \frac{M}{m} \cdot (\alpha_c - \pi/4) \qquad (7)$$

Thereafter the restoring couple begins to decrease but the torsional couple continues to increase, and the two no longer balance. Thus as soon as the value of $\alpha_c$ is overreached the fiber rapidly untwists itself and the crystal spins round. This spinning round of the crystal is indeed, as Lonsdale [33] puts it, most "dramatic."

If after the above spinning, the crystal is allowed to come to rest and the torsion head is further rotated in the same direction as before, when $\alpha$ reaches a value $\alpha_c + \pi$, the crystal rotation $\delta$ would reach $\frac{5\pi}{4}$, which would again correspond to a maximum value of the couple due to the field, and the slightest farther rotation of the torsion head would allow the crystal to flip round as before. Again when $\alpha$ just exceeds $\alpha_c + 2\pi$, there will be still another flipping and so on. From any of these critical values of $\alpha$, the value $\alpha_c$ can be readily determined and hence the anisotropy calculated from Eq. 7.

It may be mentioned here that Eq. 7 is only an approximate one. When the crystal just reaches its unstable position in the field, the direction of the greater susceptibility ($\chi$max) does not strictly make an angle of $\pi/4$, but is really slightly greater. If the true angle be $(\pi/4 + \sigma_c)$, then it can be shown that Eq. 7 modifies to Eq. 9,

$$\Delta\chi = \frac{\alpha_c -\pi - \sigma_c}{\cos 2\sigma_c} \cdot \frac{2MC}{mH^2} \tag{8}$$

where,

$$\sin 2\sigma_c = \frac{M}{m} \cdot \frac{C}{H^2} \cdot \frac{1}{\Delta\chi} \tag{9}$$

An estimate of the error involved in using the appropriate Eq. 7 instead of Eq. 8 can be made. For example, if $\alpha_c = 2\pi$, the error is about $\frac{1}{2}\%$. However if $\alpha_c$ is less than $\pi$, the error is more than 1% and hence significant. A value of $\alpha_c = \pi$ is quite a normal value, and could be even less for small, feebly anisotropic crystals. Further, in a low temperature measurement when the anisotropy is found to increase 10 to 20 times as we pass from room temperature to liquid nitrogen or helium temperatures, the value of $\alpha_c$ is usually kept moderate ($\simeq 100°$) at room temperature in order to avoid an inconveniently large value at low temperatures. It is surprising that in spite of this limitation in the use of Eq. 7 most workers have ignored this correction.

The critical-couple method has been used most extensively by Krishnan and his co-workers and by most of the later workers. The method is found to be sufficiently accurate for the crystals of high, moderate and even low anisotropy. Indeed Krishnan and Banerji [8] had remarkable success in measuring the magnetic anisotropy of even S-state ions, though they had to make special arrangements to eliminate the shape anisotropy. The method is simple and very sensitive. However there are some disadvantages with the method and we would like to record them here.

i. For crystals of low anisotropies and small mass, $\alpha_c$ becomes small at moderate values of H, and an estimate of the correction term in Eq. 8 becomes uncertain.

ii. The final position, i.e., when the crystal just spins around is difficult to determine accurately.

iii. In a temperature variation measurement it is difficult to check the magnetic field and the temperature just at the time when the crystal starts spinning round. This is natural because of the "dynamic" nature of the experiment.

## 3. Null-Deflection Method

From Eq. 4 it is apparent that the couple acting on an anisotropic crystal in a plane containing $\chi_{max}$ and $\chi_{min}$ as the two principal susceptibilities is given by

$$C = \frac{m}{2M} \cdot H^2 (\chi_{max} - \chi_{min}) \sin 2\phi$$

where $\phi$ is the angle between $\chi_{max}$ and the magnetic field. The couple is therefore a maximum when $\phi = \pi/4$. In the "null" deflection method, this property is utilized for a null measurement. The crystal is suspended in a uniform field and the setting position is determined as usual. This position is then noted and the crystal is rotated (with the field off) through 45°, so that the $\chi_{max}$ is now at 45° to the field. This position is noted accurately with some optical arrangement. This field is then applied and the torsion head rotated so as to bring the crystal back to its 45° position. If $\alpha_c$ be the angle through which the torsion head is rotated from the 45° position of the crystal, the gm molecular magnetic anisotropy is then given by

$$\Delta\chi = \frac{2M}{m} \cdot \frac{C}{H^2} \cdot \alpha_c \qquad (10)$$

It is obvious that the present method avoids the difficulties of Krishnan's critical couple method. Because of the static nature of the measurement, the temperature and the magnetic field can be easily checked at the time of making the final measurement.

The method was first described by Stout and Griffel [10]. In the original design, they used beryllium-copper wire as the torsion fiber for suspending the crystal. However the metallic-fiber is not very suitable for crystals of small mass and/or moderate anisotropy. Later workers have, therefore, replaced the metallic fiber with the original quartz fiber which, though fragile, gives much more accurate results. The null-deflection method of Stout and Griffel with the above modification seems to be the best method of determining magnetic anisotropy.

Recently Pointeau and Poquet [34] have described a method in which the crystal is suspended by a quartz fiber in a uniform magnetic field in a vacuum and the torque is measured as the field direction is changed. The method is essentially the same as Krishnan's critical couple method except that in this method the magnet is rotated instead of the crystal. This is, however, rather inconvenient and certainly a less practical method.

We shall now comment on a few points regarding the determination of C and H. The determination of the torsion constant of the suspension fiber (C) has been dealt with at length by Datta [35], Gordon [36], and others. The usual practice is to measure the period of oscillation of a small cylinder or sphere (made of aluminum or brass) with a known moment of inertia. Gordon [36] has observed that it is necessary, for accurate determination of the time period, to keep the oscillating pendulum under high vacuum and shielded from static charges. He has also pointed out that the torsion constant of the fiber tends to decrease when kept under tension for a long period of time, suggesting that the fiber elongates by plastic flow. It appears desirable, therefore, that the quartz fiber be recalibrated periodically.

The accurate measurement and control of the magnetic field is very important since the magnetic field occurs as $H^2$ in the expressions used for the calculation of magnetic anisotropy. Various conventional methods are used for this purpose, the most common being a search coil. Some commercial gaussmeters available now are also very accurate. The proton magnetic resonance method would obviously be the most accurate one.

From what has been said above, it is obvious that an absolute measurement of magnetic anisotropy can be done by measuring C and H. However it is often more convenient to measure the anisotropy with respect to some "secondary standard" crystal. There is some confusion as to the choice of a suitable secondary standard for this purpose. Since the determination of C and H is not very difficult, earlier workers measured the absolute anisotropy directly. Later workers have used copper sulfate pentahydrate as a calibrant. Selwood [37] also preferred this crystal for this purpose. Datta [30] suggested $K_2Ni(SO_4)_2.6H_2O$ as a calibrant, but the anisotropy of this crystal is very low and therefore not very suitable. $NiSO_4.6H_2O$ has also been used but this crystal again has the same disadvantage. Figgis and Lewis [38] suggested potassium ferricyanide as a suitable calibrant. However the single crystals of this compound appear to grow in two polymorphic forms [39, 40] (monoclinic and orthorhombic) and are reported to show complicated internal twinning [41] and are thus unsuitable. Recently Gregson and Mitra [42] have pointed out that $CuSO_4.5H_2O$ is probably the best calibrant for this purpose. A suitable calibrant for a magnetic anisotropy measurement should have the following characteristics:

1. Large well-developed single crystals free from all imperfections and twinning can be easily grown.

2. The crystals must be amendable to easy and accurate mounting along a particular axis.

3. The crystal must always grow in the same habit and be stable under normal conditions.

4. The crystal should have medium anisotropy along the particular axis.

It has been pointed out by Gregson and Mitra [42] that copper sulfate pentahydrate appears to satisfy these criteria. Large well-developed single crystals of this compound can easily be grown and they are very stable under normal conditions. The crystals always grow in the same form, i.e., as an elongated prism with the long axis as the c axis of the crystal. The crystal shows moderate anisotropy [44,45] and with the c-axis vertical, the anisotropy obeys a Curie law very closely [45]. The magnetic anisotropy of this crystal with the c-axis vertical has been measured by many workers and the agreement is good [43, 44, 46, 30]. However some more accurate measurements are needed on this crystal with the aim of establishing a convenient calibrant.

Before closing this section it will be profitable to mention some of the magnetic anisotropy instruments which have been used for the measurement at room and low temperatures. Besides the brief description of their apparatus by Stenger [2] and by König [3], Krishnan et al. [4-7] first described the details of their apparatus and the experimental techniques. The first low temperature apparatus for this purpose was also described by Krishnan et al. [9]. They used petroleum ether as the cryostatic bath and liquid air as the refrigerant. Since petroleum ether becomes very viscous below about -140°C, the cryostat was not very successful at lower temperatures. Bose [47] described a gas-flow type cryostat using liquid air as the refrigerant which was very successful in this temperature range.

Stout and Griffel [10] described their apparatus which operated successfully between room and 14°K. Brandt [48] has described an apparatus for the measurement of anisotropy of metals below 1.0°K. Schoffa et al. [49] have also recently described an instrument for the measurement of magnetic anisotropy at low pressure and temperatures. Zijlstra [50] has outlined a device for the rapid measurement of magnetic anisotropy at high temperatures.

Gordon [36], Rogers [51], Datta [30,35], and Fleischmann and Turner [52] have suggested some improvements in the critical couple method for greater accuracy.

Recently Van Mal [53] has developed a method of measuring magnetic anisotropy of single crystals by the Hartshorn inductance bridge. It consists of a low temperature sample holder which can hold a sample as wide as the inner diameter of the solenoid magnet and which can be rotated into any position. A pickup coil which rotates along with the sample is employed. The sensitivity is found to be good.

Guha Thakurta and Mukhopadhyay [54] have described an apparatus, similar to that of Bose [47], which works down to about 65°K. The accuracy of measurement has been claimed to be very high.

Neogy et al. [55] have recently described a magnetic anisotropy balance which uses an electrodynamic method for balancing the couple experienced by a crystal in the magnetic field. This arrangement eliminates handling of the balance during operation. The apparatus has been found to be satisfactory for low temperature measurements.

## B. Crystal Symmetry and the Magnetic Tensors

We have discussed in the previous section the various general methods of measuring the magnetic anisotropy of a single crystal. One is usually concerned with the measurement of principal crystalline anisotropies of a crystal belonging to a certain crystal system. Since the orientation of the crystalline magnetic tensors in a crystal is governed by the macroscopic symmetry of the crystal, the procedure and the details of the measurement of principal crystalline anisotropies depend upon the symmetry inherent in the crystal. It is, therefore, desirable to examine the relationship between the symmetry of a crystal and its magnetic tensors before we proceed to the description of the methods used for the measurement on crystals belonging to different crystal systems.

An ideal crystal is defined to be a body in which the atoms are arranged in a lattice in such a way that the atomic arrangement appears the same when viewed from all latice points. Most crystals possess symmetry in addition to the repetition expressed by the crystal lattice. This symmetry is best described by analyzing it into symmetry elements. There are many different symmetry elements, such as, center of symmetry, mirror plane, glide plane, etc., and each of these elements represents certain symmetry operation. All the symmetry operations are however not independent. For example, a singlefold inversion axis is the same as a center of symmetry, a twofold inversion axis is equivalent to a plane of symmetry normal to the axis and so on.

In the study of the physical properties of crystals we are not concerned with the relative positions of the symmetry elements, but only with their orientation. Hence the possible macroscopic symmetry elements in a crystal reduces to the following:

1.  Center of symmetry
2.  Mirror plane
3.  1-, 2-, 3-, 4-, or 6-fold rotation axes
4.  1-, 2-, 3-, 4-, or 6-fold inversion axes.

In a combination of these elements we may regard all the members as passing through a single point; the possible combinations of macroscopic symmetry elements are accordingly called a point-group.

It is known that crystals are divided into 32 crystal classes according to the point group symmetry they possess. The 32 crystal classes are conveniently grouped into seven crystal systems. Each of these seven systems possesses characteristic symmetry elements which determine the magnetic symmetry. We shall now examine these crystal systems and see how the magnetic tensors are oriented (Fig. 1).

## 1. Triclinic

$$a \neq b \neq c \quad ; \quad a \neq \beta \neq \gamma$$

The crystals belonging to this system possess no symmetry other than a onefold axis of rotation or inversion. Hence there is no relationship between the crystal axes and the magnetic axes, and the magnetic suscepti- bility tensors are randomly oriented in the crystal (Fig. 1).

## 2. Monoclinic

$$a \neq b \neq c \quad ; \quad \alpha = \gamma = 90° \neq \beta$$

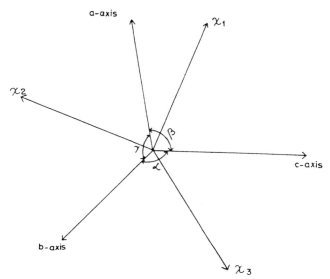

FIG. 1. Orientation of crystalline magnetic tensors in triclinic system.

This crystal possesses a single twofold axis such that a rotation of 180° will bring the system to self-coincidence. The twofold axis is taken as the b-axis of the crystal and is perpendicular to the (010) plane. Because of this symmetry, one of the principal crystalline susceptibilities (say $\chi_3$) is along the b-axis of the crystal and the other two principal susceptibilities ($\chi_1$ and $\chi_2$) lie in the (010) plane, such that $\chi_1$ and $\chi_2$ make angles $\psi$ and $\phi$ with the c- and a-axes, respectively of the crystal (Fig. 2). The relationship between $\phi$, $\psi$, and $\beta$ is,

$$\beta = \pi/2 + \psi + \phi \tag{11}$$

## 3. Orthorhombic

$$a \neq b \neq c \quad ; \quad \alpha = \beta = \gamma = 90°$$

This crystal possesses three mutually perpendicular twofold axes and these are parallel to the crystal axes, a,b,c. Because of this higher symmetry, the three principal crystal susceptibilities lie along the crystal axes, and are usually denoted by $\chi_a$, $\chi_b$, and $\chi_c$ (see Fig. 3).

## 4. Uniaxial

The magnetic axes of tetragonal, trigonal, and hexagonal crystals (also called the uniaxial crystals) are all similarly oriented; so they are grouped together.

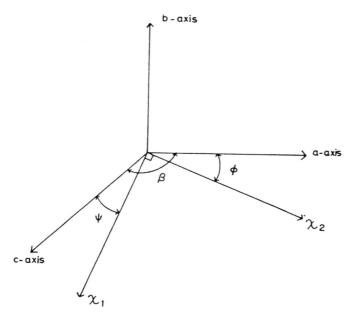

Fig. 2. Orientation of crystalline magnetic tensors in monoclinic system.

**a. Tetragonal.** a=b≠c ; $\alpha=\beta=\gamma=90°$ This crystal possesses a 4-fold rotation or inversion axis. A rotation of 90° about that axis brings the structure to self-coincidence. This axis is taken parallel to the c-axis of the crystal.

**b. Trigonal.** a=b≠c ; $\alpha=\beta=90°$, $\gamma=120°$ This crystal possesses a single threefold axis (rotation or inversion) parallel to its c-axis.

**c. Hexagonal.** a=b≠c ; $\alpha=\beta=90°$, $\gamma=120°$ This crystal has a single sixfold axis parallel to the c-axis.

In the above three crystal systems, the crystalline principal susceptibilities lie along the crystal axis, and $\chi_a = \chi_b$. $\chi_c$ is the susceptibility along the crystal symmetry axis (i.e. the c-axis) and so it is designated as $\chi_{\parallel}$ Accordingly $\chi_a = \chi_b = \chi_{\perp}$.

## 5. Cubic

$$a=b=c \quad ; \quad \alpha=\beta=\gamma=90°$$

This crystal possesses 4 threefold axes parallel to the body diagonal of the cube and 6 twofold axes parallel to the cube face diagonal. Because of

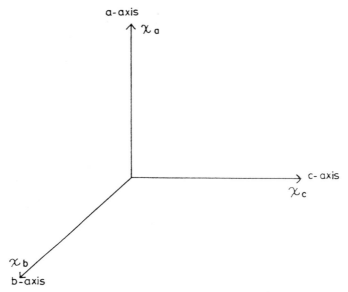

FIG. 3. Orientation of crystalline magnetic tensors in orthorhombic system.

this high symmetry, $\chi_1 = \chi_2 = \chi_3$, and the crystal shows no magnetic anisotropy (except the apparent anisotropy due to its shape).

## C. Measurement of the Principal Crystalline Anisotropies

### 1. Triclinic System

As we have mentioned earlier, there is no relationship between the crystal axes and the magnetic axes because of the lack of the necessary symmetry in the crystal, so the magnetic axes ($\chi_1, \chi_2, \chi_3$) are arbitrarily oriented in the crystal. It is, therefore, a difficult problem to determine the principal crystal anisotropies (or the susceptibilities) of triclinic crystals. Some methods have been suggested and used from time to time; however, none of the methods are very convenient or satisfactory.

The first attempt in this direction was made by Krishnan and Mookherji [43-45]. In a series of papers, they worked out an approximate method for the determination of the principal crystalline anisotropies of a triclinic crystal and applied it very successfully to the triclinic $CuSO_4.5H_2O$. The great success of their work lies in determining the orientation of the principal susceptibilities $(\chi_1, \chi_2, \chi_3)$ from the magnetic anisotropy data alone. Because of the importance of this work, we shall discuss briefly the method adopted by these workers.

The magnetic anisotropies of $CuSO_4.5H_2O$ crystals were measured in eight different orientations. The experimental data were then plotted on a stereographic projection. For any given suspension of the crystal, there is one plane which is horizontal. In the stereographic projection, the different crystal suspensions for which magnetic measurements were made, were numbered, and the poles of the corresponding horizontal planes were denoted by the same number (see Fig. 3a). From each of these poles a great circle was drawn to represent the direction of a maximum suscepti- bility. As will be seen from the figure, all these great circles pass very close to a certain point M which is marked by a small circle with a central dot. This indicates that the crystal has an axis of approximate magnetic symmetry, namely along the normal to the plane which has M as its pole; this axis was found to make angles of 156°, 65°, and 52° with the a, b, c axes respectively; and that the susceptibility of the crystal along this approximate axis is less than that along the perpendicular directions; denoting the two susceptibilities by $\chi_{||}$ and $\chi_{\perp}$, $\chi_{||}$ is less than $\chi_{\perp}$. Assuming the above conclusions of uniaxial symmetry Krishnan and Mookherji showed that the anisotropy $\Delta\chi$ for a given plane should be given by the relation

$$\Delta\chi = (\chi_{\perp} - \chi_{||}) \sin^2\theta \qquad (12)$$

where $\theta$ is the angle between the normal to the plane and the axis of sym- metry; in other words $\Delta\chi/\sin^2\theta$) should be constant. This was exactly what they observed (see Table 1) and the value of $(\chi_{\perp} - \chi_{||})$ calculated in this manner was found to be $300 \times 10^{-6}$ at 30°C.

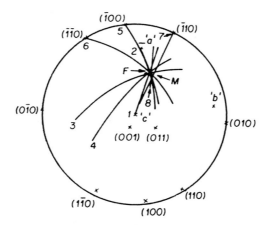

FIG. 3a. Stereographic projection.

TABLE 1

| Mode of suspension | $\Delta\chi$ | $\theta$ | $\Delta\chi\sin^2\theta$ |
|---|---|---|---|
| c-axis vertical | 183 | 52° | 295 |
| a-axis vertical | 55 | 24-5° | 320 |
| ($\overline{1}$00) and ($\overline{1}$11) vertical | 261 | 86° | 262 |
| ($\overline{1}\overline{1}$0) and ($\overline{1}$11) vertical | 263 | 79° | 273 |
| (100) horizontal | 163 | 45° | 326 |
| ($\overline{1}$10) horizontal | 113 | 38° | 298 |
| ($\overline{1}$11) horizontal | 26 | 16° | 342 |
| (110) horizontal | 220 | 61° | 288 |

In a later paper, Krishnan and Mookherji modified their approach slightly and were able to determine the principal crystalline anisotropies more accurately:

$$\chi_3 - \chi_1 = 280$$
$$\chi_3 - \chi_2 = 10,$$

and the orientation of the principal crystalline susceptibilities relative to crystallographic axes. This was a remarkable result which points out the use of this method in structural investigations of paramagnetic crystals.

In spite of the great success of Krishnan and Mookherji in determining the crystal susceptibilities of $CuSO_4.5H_2O$, their method is obviously approximate and very tedious. Mathur [56] tried to improve upon this method, but again the improvements still only lead to approximate solution.

The first attempt to determine exactly and directly the principal suscepti-bilities of triclinic crystals was made by Ghose and Bagchi [57]. We shall discuss the method outlined by them.

The equation of the ellipsoid of crystal susceptibility referred to an orthogonal co-ordinate system x, y, z, is

$$\chi_{11}x^2 + \chi_{22}y^2 + \chi_{33}z^2 + 2\chi_{12}xy + 2\chi_{23}yz + 2\chi_{13}xz = 1 \qquad (13)$$

where $\chi_{11}$ etc. are the six independent components of susceptibility tensor referred to the x, y, z system, and are given by

$$\chi_{11} = \frac{1}{\eta} \Sigma \ (K_1\alpha_1^2 + K_2\alpha_2^2 + K_3\alpha_3^2)$$

$$\chi_{22} = \frac{1}{\eta} \Sigma \ (K_1\beta_1^2 + K_2\beta_2^2 + K_3\beta_3^2)$$

$$\chi_{33} = \frac{1}{\eta} \Sigma \ (K_1\gamma_1^2 + K_2\gamma_2^2 + K_3\gamma_3^2)$$

$$\chi_{12} = \frac{1}{\eta} \Sigma \ (K_1\alpha_1\beta_1 + K_2\alpha_2\beta_2 + K_3\alpha_3\beta_3) \qquad (14)$$

$$\chi_{13} = \frac{1}{\eta} \Sigma \ (K_1\alpha_1\gamma_1 + K_2\alpha_2\gamma_2 + K_3\alpha_3\gamma_3)$$

$$\chi_{23} = \frac{1}{\eta} \Sigma \ (K_1\beta_1\gamma_1 + K_2\beta_2\gamma_2 + K_3\beta_3\gamma_3)$$

the summation being taken over the $\eta$ inequivalent molecules in the unit cell, and the $\alpha$'s, $\beta$'s, and $\gamma$'s are the direction cosines of the principal molecular susceptibilities ($K_1$, $K_2$, and $K_3$) with respect to x, y, z coordinate system; i.e.,

|       | x | y | z |
|-------|-----------|-----------|-----------|
| $K_1$ | $\alpha_1$ | $\beta_1$ | $\gamma_1$ |
| $K_2$ | $\alpha_2$ | $\beta_2$ | $\gamma_2$ |
| $K_3$ | $\alpha_3$ | $\beta_3$ | $\gamma_3$ |

$$(15)$$

If the values of the above six tensor component (e.g. $\chi_{11}$, $\chi_{22}$, etc.) are determined, the three principal crystalline susceptibilities are given by the three roots of $\chi$ in the determinantal equation

$$\begin{vmatrix} \chi_{11}-\chi & \chi_{12} & \chi_{13} \\ \chi_{12} & \chi_{22}-\chi & \chi_{23} \\ \chi_{13} & \chi_{23} & \chi_{33}-\chi \end{vmatrix} = 0 \qquad (16)$$

The direction cosines of any one of the principal crystalline susceptibilities, say $\chi_i$, relative to x, y, z axes respectively are given by

$$f_i = [\chi_{12}\chi_{23} - \chi_{13}(\chi_{22} - \chi_i)]Q$$

$$g_i = [\chi_{12}\chi_{13} - \chi_{23}(\chi_{11} - \chi_i)]Q \qquad (17)$$

$$h_i = [(\chi_{11} - \chi_i)(\chi_{22} - \chi_i) - \chi_{12}^2]Q$$

where Q can be determined from the relation

$$f_i^2 + g_i^2 + h_i^2 = 1 \qquad (18)$$

To find $\chi_{11}$, $\chi_{22}$, etc., measurements of $(\chi_{max} + \chi_{min})$ in six nonparallel planes of the crystal were suggested. The $(\chi_{max} + \chi_{min})$ is actually the result of two independent measurements,—the anisotropy which gives $(\chi_{max} - \chi_{min})$, and the maximum susceptibility $(\chi_{max})$.

The value of $(\chi_{max} + \chi_{min})$ in a plane whose normal has the direction cosines $\xi$, $\eta$, $\zeta$ is given by the expression:

$$\chi_{max} + \chi_{min} = \chi_{11}(1-\xi^2) + \chi_{22}(1-\eta^2) + \chi_{33}(1-\zeta^2)$$

$$- 2\chi_{12}\xi\eta - 2\chi_{13}\xi\zeta - 2\chi_{13}\eta\zeta \qquad (19)$$

Hence working with six planes, one obtains six linear equations of the type of Eq. 19 and so the six unknowns $\chi_{11}$, $\chi_{22}$, $\chi_{33}$, $\chi_{13}$, $\chi_{12}$, $\chi_{23}$ can then be determined.

The method is obviously quite direct and involves no approximation. It is, however, unfortunate that no experimental verification of the method appears to have been published.

Following the method of Ghose and Bagchi, Neogy [58] showed that the three principal susceptibilities of a triclinic crystal can be uniquely determined from the measured values of $\chi_{max}$ and $(\chi_{max} - \chi_{min})$ in any three mutually perpendicular planes. The method thus reduces the number of measurements from six to three. However, to resolve the ambiguity in sign in the off-diagonal elements $\chi_{12}$, $\chi_{13}$ and $\chi_{23}$, he later suggested [59] an additional three sets of measurements. The method thus offers no real advantage.

Ghose [60] has recently discussed the covariant matrix in the determination of the principal susceptibilities of a crystal. He points out that for any triclinic crystal, five observations suffice to determine the absolute values of the principal susceptibilities and principal axes. In a mathematical treatment of susceptibility tensors he shows that the maximum susceptibility, M, and any error, ΔM, in its measurement cancel out. Based on this treatment, he outlines a method for measurements on triclinic crystals giving greater accuracy. In a subsequent paper [61] he has shown that the variation of anisotropy in a set of planes perpendicular to any given plane can be expressed in terms of just five coefficients which are analogous to Fourier coefficients and are explicit functions of the elements of the susceptibility matrix. These coefficients, and from them the principal anisotropies and their direction cosines, and can be calculated from just five measurements of anisotropy. The greatest advantage of his method is that, unlike some earlier methods, no measurement of $\chi_{max}$ is necessary. This is significant, since the measurement of $\chi_{max}$ involves a different kind of apparatus and is supposed to be much less accurate than anisotropy measurement. It is, however, again unfortunate that none of his methods have been experimentally verified.

## 2. Monoclinic System

In monoclinic crystals, one of the principal crystalline susceptibilities (say $\chi_3$) coincides with the b-axis of the crystal. $\chi_1$ and $\chi_2$ lie in the (010) plane, making angles $\psi$ and $\phi$ with the c- and a-axes, respectively, of the crystal (see Fig. 2). Evidently if the crystal be suspended with the b-axis vertical, $(\chi_1-\chi_2)$ is directly measured:

$$(\Delta\chi)_{\text{b-axis vert.}} = (\chi_1 - \chi_2) \tag{20}$$

Here $\chi_1$ is by definition taken to be greater than $\chi_2$.

If the crystal is suspended with a-axis vertical, the anisotropy in the horizontal plane can be related to the principal anisotropy in the following manner. The orientation of the magnetic and crystallographic axes with a-axis vertical is shown in Fig. 4. If the crystal sets with b-axis parallel to the magnetic field, then the anisotropy with a-axis vertical is given by

$$(\Delta\chi)_a = \chi_3 - (\chi_2 \text{Sin}^2\phi + \chi_1 \text{Cos}^2\phi)$$

$$= \chi_3 - \chi_2 \text{Sin}^2\phi - \chi_1 + \chi_1 \text{Sin}^2\phi$$

$$= [(\chi_1-\chi_2)\text{Sin}^2\phi - (\chi_1-\chi_3)] \tag{21}$$

Similarly if the b-axis for this suspension sets perpendicular to the magnetic field,

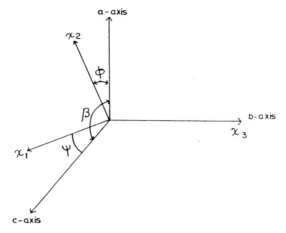

FIG. 4.  Orientation of principal crystalline susceptibilities in monoclinic system when suspended with a-axis vertical.

$$(\Delta\chi)_a = (\chi_2 Sin^2\phi + \chi_1 Cos^2\phi) - \chi_3$$

$$= - [ (\chi_1 - \chi_2)Sin^2\phi - (\chi_1 - \chi_3)] \tag{22}$$

Combining both Eqs. 21 and 22, we can write,

$$(\Delta\chi)_a = \pm [ (\chi_1 - \chi_2)Sin^2\phi - (\chi_1 - \chi_3)] \tag{23}$$

where the plus and minus signs are chosen as the b-axis sets parallel and perpendicular to the magnetic field.

If the crystal is suspended with the c-axis vertical, the orientation of the magnetic and crystallographic axes is shown in Fig. 5. Here again if the b-axis sets parallel to the magnetic field direction, the anisotropy with the c-axis vertical is given by

$$(\Delta\chi)_c = \chi_3 - [\chi_1 Sin^2\psi + \chi_2 Cos^2\psi]$$

$$= [ (\chi_1 - \chi_2)Cos^2\psi - (\chi_1 - \chi_2)] \tag{24}$$

Likewise if the b-axis sets perpendicular to the magnetic field,

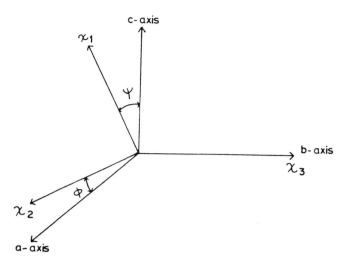

FIG. 5. Orientation of principal crystalline susceptibilities in a monoclinic system when suspended with c-axis vertical.

$$(\Delta\chi)_c = -[(\chi_1-\chi_2)\cos^2\psi - (\chi_1-\chi_3)] \qquad (25)$$

The expressions with other crystallographic planes horizontal can be similarly deduced. For instance, with (001) plane horizontal,

$$(\Delta\chi)_{[001]\text{p.h.}} = \pm[(\chi_1-\chi_2)\cos^2\phi - (\chi_1-\chi_3)] \qquad (26)$$

where the choice of sign is same as above.

Thus measurements with the b-axis vertical and any other axes vertical or any plane horizontal, will give $(\chi_1-\chi_2)$ and $(\chi_1-\chi_3)$. By measuring the average susceptibility or any one of the principal susceptibilities (usually $\chi_1$), the three principal crystalline susceptibilities can be readily deduced.

To measure $\psi$, the following procedure has been suggested by Datta [35]. A good single crystal of $NiSO_4.6H_2O$ (tetragonal system) which usually grows with a large reflecting (001) face, is mounted with the tetrad axis horizontal and the torsion head adjusted till the crystal is under no torsion in the magnetic field. In this position the (001) face lies exactly parallel to the magnetic field. The lamp and scale arrangement is so set up that a bright image of the filament of the lamp is observed on a scale on reflection from the (001) face of the crystal. The orientation of any other crystal, suspended in the place of this crystal can then be determined with respect to the field keeping the position of the lamp and scale fixed. A monoclinic crystal is suspended with the b-axis vertical and the setting position is determined as described before. At this position, $\chi_1$ lies along the field. The angle between $\chi_1$ and a- or c-axes can then easily be determined by rotating the crystal so that either the a- or c-axis is just along the field when again the bright reflection from the face of the crystal containing the above axes is obtained in the telescope. For a crystal with a well-developed (001) plane, the angle between $\chi_1$ and the a-axis (i.e. $90 - \phi$) is directly obtained. If the c-axis is well-developed, $\psi$ can be directly measured. The sign of $\phi$ and $\psi$ is determined by noting the position of any other (h0l) plane with respect to the $\chi_1$ direction. The convention for the sign of $\phi$ or $\psi$ is that they are taken to be positive if they lie within the monoclinic angle $\beta$. By measuring one of the angles, the other can be calculated from Eq. 11.

### 3. Orthorhombic System

In orthorhombic crystals, the principal crystalline susceptibilities lie along the crystallographic axes (see Fig. 3). Hence by suspending the crystal with any of the crystal axes vertical, the principal anisotropy of the horizintal plane can be directly measured.

The following three suspensions are used:

(i) a-axis vertical

$$(\Delta\chi)_a = (\chi_b \sim \chi_c) \tag{27}$$

(ii) b-axis vertical

$$(\Delta\chi)_b = (\chi_a \sim \chi_c) \tag{28}$$

(iii) c-axis vertical

$$(\Delta\chi)_c = (\chi_a \sim \chi_b) \tag{29}$$

The sign of the anisotropies are determined by noting the crystal axis lying along the magnetic field at the setting position.

### 4. Uniaxial System

(Tetragonal, Trigonal and Hexagonal)

For uniaxial crystals, the crystal is suspended with its symmetry axis (c-axis) horizontal, and the principal anisotropy is directly measured,

$$(\Delta\chi) = (\chi_{||} - \chi_{\perp}) \tag{30}$$

The sign of the anisotropy is determined by noting whether the c-axis lies along or normal to the magnetic field at the setting position.

We would like here to make some comments concerning the low temperature measurements on different crystal systems.

Since in the triclinic system none of the crystal susceptibilities are fixed by the crystal symmetry, a small variation in their orientation with temperature is feasible. This change is evidently reflected in a change in the setting position of the crystal with temperature. Hence it is desirable to determine the setting position at every temperature at which a measurement is made.

In the monoclinic crystal, $\chi_1$ and $\chi_2$ are not fixed by the symmetry of the crystal. Hence a change in their orientation with temperature is likely and it is therefore desirable to check the setting position at each temperature with the b-axis vertical. Since the expressions for the calculation of $(\chi_1-\chi_3)$ or $(\chi_2-\chi_3)$ involves $\phi$ or $\psi$, the value of $\phi$ or $\psi$ must be determined at every temperature. To determine if $\phi$ or $\psi$ is increasing or decreasing, the position of the crystal axes should be noted before placing the crystal suspension system in the cryostat. Any change in the setting position with temperature can then be easily identified as an increase or decrease in $\phi$ $\psi$ by the direction in which the torsion head is rotated. It is important to note here that the change in the setting-position is only possible for the b-axis vertical suspension. No such change is expected if the crystal is suspended with any other axis vertical or horizontal, since in this case the b-axis (or the $\chi_3$) lies in the horizontal plane.

In the orthorhombic and uniaxial crystals, no change in the setting position with temperature is expected as the principal susceptibilities are fixed by the crystal symmetry.

## D. Calculation of Molecular Anisotropy

We have discussed in the previous section the methods of measuring the principal crystalline anisotropies for crystals belonging to different systems. The principal molecular anisotropy is, in general, different and is connected to the crystal anisotropies by certain mathematical relationships which are again different for different crystal systems. Since the molecular anisotropy is of primary importance, it is necessary at this stage to examine the relationship between the molecular and crystal anisotropies in different crystal systems. The classic work in this direction was done by Lonsdale and Krishnan [62] who worked out the exact mathematical relationships. We shall in the following paragraphs examine these relationships and point out the simplified expressions. Further a detailed account of recent calculations for determining molecular anisotropy of triclinic crystals will be examined, since they are not included in the Lonsdale and Krishnan paper. Our approach will be mainly that with paramagnetic crystals in mind but the relations deduced can be very readily used for diamagnetic anisotropies as well.

## 1. Triclinic System

We have noted earlier the method adopted by Krishnan and Mookherji [43, 44] for the determination of principal crystalline anisotropies of triclinic crystals. Having determined the principal crystal anisotropy of $CuSO_4.5H_2O$ (which has two inequivalent molecules in the unit cell), they pointed out that the molecular anisotropy can be calculated in the present case by the approximate formula,

$$K_{||} - K_\perp = 2(\chi_{||} - \chi_\perp) \qquad (31)$$

It is important to note that the above method is not general and is probably true only for the specific case they considered. Further it involves the determination of crystal anisotropy in different planes, which is obviously quite a tedious process.

A convenient and more direct method for finding the molecular anisotropy of triclinic crystals was developed by Ghose and Mitra [63]. They pointed out that molecular anisotropy can be calculated directly without having recourse to determining the crystal anisotropies first. Further they showed that the axial molecular anisotropy $(K_{||} - K_\perp)$ can be determined from measurements in a single plane. We shall briefly discuss their method.

The direction cosines of the principal molecular susceptibilities ($K_1$, $K_2$, and $K_3$) with respect to an orthogonal coordinate system (x,y,z) is expressed in the usual matrix form

|  | x | y | z |
|---|---|---|---|
| $K_1$ | $\alpha_1$ | $\beta_1$ | $\gamma_1$ |
| $K_2$ | $\alpha_2$ | $\beta_2$ | $\gamma_2$ |
| $K_3$ | $\alpha_3$ | $\beta_3$ | $\gamma_3$ |

In the most general case, consider there are n (n>1) such molecules in the unit cell, which are all differently oriented. The crystal susceptibility ellipsoid in the (x,y,z) system can be written as

$$\chi_{11}x^2 + \chi_{22}y^2 + \chi_{33}z^2 + 2(\chi_{12}xy + \chi_{13}xz + \chi_{23}xz) = 1$$

where $\chi_{11}$, $\chi_{22}$ etc. have the same meaning as in Eq. 14. Let the working plane for the magnetic measurement of the crystal be any crystallographic plane whose Miller indices are (h,k,l), and let $\xi$, $\eta$, $\zeta$ be the direction cosines of the normal to this plane relative to the x,y,z axes. If $\chi_{max}$ and $\chi_{min}$ be the maximum and minimum susceptibility in the above plane, then

$$\chi_{max} + \chi_{min} = \chi_{11} + \chi_{22} + \chi_{33} - (\chi_{11}\xi^2 + \chi_{22}\eta^2 + \chi_{33}\zeta^2$$

$$+ 2\chi_{12}\xi\eta + 2\chi_{13}\xi\zeta + 2\chi_{23}\eta\zeta)$$

$$= K_1 [ 1 - \frac{1}{\eta} \sum_\eta \{ (\alpha_1^2\xi^2 + \beta_1^2\eta^2 + \gamma_1^2\zeta^2) + 2(\alpha_1\beta_1\xi\eta$$

$$+ \alpha_1\gamma_1\xi\zeta + \beta_1\gamma_1\eta\zeta) \} ]$$

$$+ K_2 [ 1 - \frac{1}{\eta} \sum_\eta \{ (\alpha_2^2\xi^2 + \beta_2^2\eta^2 + \gamma_2^2\zeta^2) + 2(\alpha_2\beta_2\xi\eta$$

$$+ \alpha_2\gamma_2\xi\zeta + \beta_2\gamma_2\eta\zeta) \} ]$$

$$+ K_3 [ 1 - \frac{1}{\eta} \sum_\eta \{ (\alpha_3^2\xi^2 + \beta_3^2\eta^2 + \gamma_3^2\zeta^2) + 2(\alpha_3\beta_3\xi\eta$$

$$+ \alpha_3\gamma_3\xi\zeta + \beta_3\gamma_3\eta\zeta) \} ] \qquad (32)$$

It is apparent that $K_1$, $K_2$, $K_3$ can be uniquely determined by measuring $(\chi_{max} + \chi_{min})$ in two different planes and by using the average suscepti- bility data. Equation 32 can, however, be greatly simplified by a suitable choice of the plane of measurement. For instance, with the c-axis vertical,

$$\xi = 1 \quad , \quad \eta = \zeta = 0$$

and then Eq. 32 reduces to

$$\chi_{max} + \chi_{min} = K_1[ 1 - \frac{1}{\eta}\Sigma\alpha_1^2 ] + K_2[ 1 - \frac{1}{\eta}\Sigma\alpha_2^2 ] + K_3[ 1 - \frac{1}{\eta}\Sigma\alpha_3^2 ] \quad (33)$$

With the (010) plane horizontal: $\zeta = 1$, $\xi = \eta = 0$, and so Eq. 32 reduces to,

$$\chi_{max} + \chi_{min} = K_1[ 1 - \frac{1}{\eta}\Sigma\gamma_1^2 ] + K_2[ 1 - \frac{1}{\eta}\Sigma\gamma_2^2 ] + K_3[ 1 - \frac{1}{\eta}\Sigma\gamma_3^2 ] \quad (34)$$

The above expressions (Eqs. 33 and 34) are further simplified if we consider a uniaxial symmetry of the molecule, i.e., $K_1 = K_2 = K_\perp$ and $K_3 = K_{||}$. With the c-axis vertical,

$$\chi_{max} + \chi_{min} = K_{||}(1 - \frac{1}{\eta}\Sigma\alpha_3^2) + K(1 + \frac{1}{\eta}\Sigma\alpha_3^2) \qquad (35)$$

The measurements in only one plane and the average susceptibility can give the values of $K_{||}$ and $K_\perp$. Equations 35 can be simplified further if there is only one molecule in the unit cell:

$$\chi_{max} + \chi_{min} = K_{||}(1 - \alpha_3^2) + K(1 + \alpha_3^2) \qquad (36)$$

It is important to note that $(\chi_{max} + \chi_{min})$ consists of two independent measurements,—an anisotropy measurement which gives $(\chi_{max} - \chi_{min})$, and $\chi_{max} \cdot \chi_{max}$ can be measured by a Faraday or Curie balance using a flexible suspension [64].

The above method of Ghose and Mitra, although showing for the first time that molecular anisotropies of a triclinic crystal can be determined "directly" without any approximations, has certain disadvantages. First, it requires two sets of instruments, one for the $(\chi_{max} - \chi_{min})$ and another for the $\chi_{max}$. Second, for crystals of low anisotropy, $(\chi_{max} - \chi_{min})$ is usually much smaller in comparison to $\chi_{max}$, and so large errors are possible in the calculations of molecular susceptibilities.

Recently Ghosh [65] has improved the above method and has shown that the molecular anisotropy can be obtained from the measurement of magnetic anisotropy alone. Since this method appears to be the most convenient among the existing ones, we shall examine it here briefly.

For the orthogonal coordinate system x,y,z, if $\chi_{max}$ and $\Delta\chi$ be the observed maximum susceptibility and the anisotropy in x-y plane and $\theta$ be the angle which the direction of maximum susceptibility in this plane makes with the x axis, then the susceptibility tensor matrix can be written as,

$$\chi_{11} = \chi_{max} - \Delta\chi \cdot \sin^2\theta \qquad (37)$$

$$\chi_{22} = \chi_{max} - \Delta\chi \cdot \cos^2\theta \qquad (38)$$

$$\chi_{12} = \Delta\chi \cdot \sin\theta \cos\theta \qquad (39)$$

Eliminating $\chi_{max}$ between Eqs. 37 and 38, we get,

$$(\chi_{11} - \chi_{22}) = \Delta\chi \ . \ Cos2\theta \tag{40}$$

Substituting the values of $\chi_{11}$, $\chi_{22}$, and $\chi_{12}$ from Eq. 14 in Eqs. 39 and 40, we get

$$(K_2-K_1)\Sigma\alpha_2\beta_2 + (K_3-K_1)\Sigma\alpha_3\beta_3 = \frac{\eta.\Delta\chi}{2} \ . \ Sin2\theta \tag{41}$$

$$(K_2-K_1)\Sigma(\alpha_2^2-\beta_2^2) + (K_3-K_1)\Sigma(\alpha_3^2-\beta_3^2) = \eta\Delta\chi \ . \ Cos2\theta \tag{42}$$

Eliminating $\theta$ between Eqs. 41 and 42, we get

$$\{(K_2-K_1)\Sigma(\alpha_2^2-\beta_2^2) + (K_3-K_1)\Sigma(\alpha_3^2-\beta_3^2)\}^2$$

$$+ 4\{(K_2-K_1)\Sigma\alpha_2\beta_2 + (K_3-K_1)\Sigma\alpha_3\beta_3\}^2 = \eta^2(\Delta\chi)^2 \tag{43}$$

The summations are for n ions. Here $\Delta\chi$ is the anisotropy in the (x-y) plane of the crystal. Similarly for the anisotropy in the x-z plane of the crystal ($\Delta\chi'$) we have,

$$\{(K_2-K_1)\Sigma(\alpha_2^2-\gamma_2^2) + (K_3-K_1)\Sigma(\alpha_3^2-\gamma_3^2)\}^2$$

$$+ 4\{(K_2-K_1)\Sigma\alpha_2\gamma_2 + (K_3-K_1)\Sigma\alpha_3\gamma_3\}^2$$

$$= \eta^2 \ . \ (\Delta\chi')^2 \tag{44}$$

Thus the values of $(K_2-K_1)$ and $(K_3-K_1)$ can be obtained from Eqs. 43 and 44. But there will be, in general, four alternative sets of solutions of which only one is correct. To resolve this ambiguity, Ghose suggests noting the approximate value of $\theta$ in one of the positions for measuring the anisotropy of the crystal and calculate the approximate values of the principal molecular anisotropies by Eqs. 41 and 42. Of the four alternative sets of solutions previously obtained, the one which is closest to the approximate solution will be correct.

If a uniaxial symmetry of the molecule is assumed, then from Eq. 43,

$$K_{||}-K_\perp = \frac{\pm \eta \ . \ \Delta\chi}{\{(\Sigma\alpha_3^2-\Sigma\beta_3^2)^2 + 4(\Sigma\alpha_3\beta_3)^2\}^{1/2}} \tag{45}$$

Thus the principal molecular anisotropy can be calculated from a <u>single measurement of anisotropy</u> in any conveniently chosen plane. The correct sign in Eq. 45 is determined by noting the sign of $\Sigma \alpha_3 \beta_3$ and the direction of the maximum susceptibility in the magnetic field.

## 2. Monoclinic System

The exact mathematical relationships between the crystalline and molecular susceptibilities for monoclinic system were worked out by Lonsdale and Krishnan [62]. In the following we shall discuss the expressions deduced by them and point out their simplified forms and advantages. The notation used by us below is the same as earlier; however it differs slightly from that of Lonsdale and Krishnan.

The maximum symmetry in the monoclinic system is that of reflection plus inversion. Let us consider that there are two inequivalent molecules in the unit cell. This is quite a common occurrence in the monoclinic crystals. Let (a,b,c′) be an orthogonal set of axes, consisting of the crystal axis a, the rotation axis b, and the c′ normal to these two directions in the (010) plane. Relative to these axes the direction cosines of $K_1$, $K_2$, and $K_3$ of any one molecule is given by

$$
\begin{array}{c|ccc}
 & a & b & c' \\
\hline
K_1 & \alpha_1 & \beta_1 & \gamma_1 \\
\\
K_2 & \alpha_2 & \beta_2 & \gamma_2 \\
\\
K_3 & \alpha_3 & \beta_3 & \gamma_3 \\
\end{array}
\tag{46}
$$

The direction cosines of the principal molecular axes of the other molecule is simply given by changing the sign of $\beta$'s in Eq. 46.

The general equation of ellipsoid in the monoclinic system [66] is written as,

$$
K_{aa}x^2 + K_{bb}y^2 + K_{c'c'}z^2 + 2K_{ac'}xz = 1 \tag{47}
$$

where,

$$
K_{aa} = K_1\alpha_1^2 + K_2\alpha_2^2 + K_3\alpha_3^2
$$

$$
K_{bb} = K_1\beta_1^2 + K_2\beta_2^2 + K_3\beta_3^2 \tag{48}
$$

$$K_{c'c'} = K_1\gamma_1^2 + K_2\gamma_2^2 + K_3\gamma_3^2$$

$$K_{ac'} = K_1\alpha_1\gamma_1 + K_2\alpha_2\gamma_2 + K_3\alpha_3\gamma_3$$

The component $K_{bb}$ is along the b-axis and so,

$$\chi_3 = K_{bb} = K_1\beta_1^2 + K_2\beta_2^2 + K_3\beta_3^2 \tag{49}$$

and $K_{aa}$, $K_{c'c'}$ and $K_{ac'}$ lie in the (010) plane.

Now the direction cosines of $\chi_1$, $\chi_2$ and $\chi_3$ with respect to (a,b,c') are (see Fig. 2):

|          | a            | b   | c'            |
|----------|--------------|-----|---------------|
| $\chi_1$ | $-\sin\phi$  | 0   | $\cos\phi$    |
| $\chi_2$ | $-\cos\phi$  | 0   | $-\sin\phi$   |
| $\chi_3$ | O            | 1   | O             |

$$(50)$$

After suitable transformation of $K_{aa}$ etc. from the (a,b,c') system to (x, y, z) system, one obtains,

$$\chi_1 = K_{aa} + K_{ac'}\ \tan(90-\phi) \tag{51}$$

$$\chi_2 = K_{c'c'} - K_{ac'}\ \tan(90-\phi) \tag{52}$$

Hence,

$$\tan(90-\phi) = \frac{\chi_1 - K_{aa}}{K_{ac'}} = \frac{K_{c'c'} - \chi_2}{K_{ac'}} \tag{53}$$

$$\chi_1 = \frac{1}{2}[K_{aa} + K_{c'c'} \pm \{(K_{aa} - K_{c'c'})^2 + 4K_{ac'}^2\}^{\frac{1}{2}}] \tag{54}$$

$$\chi_2 = \frac{1}{2}[K_{aa} + K_{c'c'} \mp \{(K_{aa} - K_{c'c'})^2 + 4K_{ac'}^2\}^{\frac{1}{2}}] \tag{55}$$

the choice of sign is made so that $\chi_1 > \chi_2$.

Thus $K_1$, $K_2$, $K_3$ can be uniquely determined for a monoclinic crystal from Eqs. 49, 54, and 55, if the $\chi_i$'s and $\alpha$'s are known. It may also be pointed out that Eq. 53 gives $\phi$, which can also be experimentally measured, thus offering a convenient check on the measurement.

Now, in general, the two molecular susceptibilities in the symmetric plane of the molecule are expected to be very nearly the same. Hence, $K_1 = K_2 = K_\perp$ and $K_3 = K_{||}$. Substituting these simplifications in Eqs. 49, 54, and 55, we get for $K_\perp > K_{||}$,

$$\chi_1 = K_\perp$$

$$\chi_2 = K_{||} + (K_\perp - K_{||})\beta_3^2 \qquad (56)$$

$$\chi_3 = K_\perp - (K_\perp - K_{||})\beta_3^2$$

Equations 56 can be combined together and written as

$$K_\perp - K_{||} = (\chi_1 - \chi_2) + (\chi_1 - \chi_2) \qquad (57)$$

Similarly for $K_{||} > K_\perp$, we get

$$\chi_1 = K_{||} - (K_{||} - K_\perp)\beta_3^2$$

$$\chi_2 = K_\perp \qquad (58)$$

$$\chi_3 = K_\perp + (K_{||} - K_\perp)\beta_3^2$$

Equations 58 can be likewise combined and written as

$$K_{||} - K_\perp = 2(\chi_1 - \chi_2) - (\chi_1 - \chi_3) \qquad (59)$$

It is seen from Eqs. 57 and 59 that the molecular anisotropy can be calculated directly from the measured values of crystal anisotropies and no detailed X-ray structural data are required for this purpose. This is a great advantage since the assumption of uniaxial symmetry is nearly true in most cases. However, the use of these approximate expressions (Eqs. 57 and 59) requires a prior knowledge of the sign of molecular anisotropy $(K_{||} - K_\perp)$, which is often available from other experiments in favorable cases. On the other hand, Eqs. 49, 54, and 55 yield $K_1$, $K_2$, and $K_3$ uniquely but require detailed structural data. Further the accuracy of the $K_i$'s depends very much upon the structural parameters. It is often observed that, because of the complicated nature of the equations envolved in calculations of $K_1$, $K_2$, $K_3$, the results are very sensitive to changes (sometimes within the accuracy of their measurement) in the $\chi_i$'s or $\alpha$'s and small changes in them make a singificant change in the calculated values of $K_i$'s, and so difference in the calculated values of $K_1$ and $K_2$ becomes questionable. An alternate method of calculating $K_1$, $K_2$, and $K_3$ from the structural parameter is set out below. If the $\alpha''$'s, $\beta''$'s, and $\gamma''$'s are the direction cosines of $K_1$, $K_2$, and $K_3$ with respect to the $\chi_1$, $\chi_2$, and $\chi_3$ axes, then

$$\chi_1 = K_1 \alpha'^2_1 + K_2 \alpha'^2_2 + K_3 \alpha'^2_3$$

$$\chi_2 = K_1 \beta'^2_1 + K_2 \beta'^2_2 + K_3 \beta'^2_3 \qquad (60)$$

$$\chi_3 = K_1 \gamma'^2_1 + K_2 \gamma'^2_2 + K_3 \gamma'^2_3$$

The X-ray results usually give the direction cosines of $K_1$, $K_2$, and $K_3$ with respect to a, b, c' axes. The necessary transformation can, therefore, be readily effected by Eq. 50. The advantage of this method over that of Lonsdale and Krishnan is that unlike that of Lonsdale and Krishnan it does not require solution of a quadratic equation, which appears to make the calculation of Lonsdale and Krishnan so very sensitive.

In the case of Eq. 56 it can be easily shown that for ($K_\perp > K_{||}$), the angle between the $K_{||}$ and the b-axes of the crystal ($\delta$) is given by

$$\mathrm{Cos} 2\delta = \frac{\chi_2 - \chi_3}{K_\perp - K_{||}} \qquad (61)$$

[In the notation followed earlier, $\delta = \mathrm{Cos}^{-1} \beta_3$].
Similarly for $K_{||} > K_\perp$,

$$\mathrm{Cos} 2\delta = - \frac{\chi_1 - \chi_3}{K_{||} - K_\perp} \qquad (62)$$

Equations 61 and 62 offer the determination of $\delta$ only from the magnetic anisotropy measurements. This angle can also be obtained from the X-ray structural data, and thus offers a convenient check on the measurement. Again within the approximation of uniaxial symmetry, Eq. 53 can be simply written as

$$\tan(90-\phi) = -\frac{\alpha_3}{\gamma_3} \tag{63}$$

where $\alpha_3$, $\gamma_3$ are the same as in Eq. 46. Since $\phi$ can be directly measured (see Section II,C,2), Eq. 63 is important. Thus combining Eqs. 61, 62 and 63 one can, in favorable cases, predict the orientation of the symmetry axis (and hence of a molecular plane) with respect to a, b, and c' axes. Unfortunately this result is not general and will be discussed later.

As a special case, if the unit cell contains only one molecule (or all the molecules in the unit cell are "magnetically equivalent"), then the crystal and molecular susceptibilities become equivalent.

### 3. Orthorhombic System

In the orthorhombic system, the crystalline principal susceptibilities lie along the crystal axes and are connected with the molecular susceptibilities by the following relations:

$$\chi_a = K_1 \sum_n \alpha_1^2 + K_2 \sum_\eta \alpha_2^2 + K_3 \sum_\eta \alpha_3^2$$

$$\chi_b = K_1 \sum_n \beta_1^2 + K_2 \sum_\eta \beta_2^2 + K_3 \sum_\eta \beta_3^2 \tag{64}$$

$$\chi_c = K_1 \sum_n \gamma_1^2 + K_2 \sum_\eta \gamma_2^2 + K_2 \sum_\eta \gamma_3^2$$

where $\alpha_i$, $\beta_i$, $\gamma_i$'s are the direction cosines of the principal molecular susceptibilities ($K_1$, $K_2$, $K_3$) relative to crystal axes (a, b, c). The summation is taken over all independently oriented molecules (not connected by symmetry operation inherent in the space group) in the unit cell. Obviously, by knowing the angular parameters, the $K_i$'s can easily be calculated from the measured values of $\chi_a$, $\chi_b$, $\chi_c$. The above relations can be however simplified if we assume uniaxial symmetry of the molecule i.e. $K_1 = K_2 = K_\perp$, $K_3 = K_{||}$, and that all the molecules in the unit cell are connected by symmetry operations inherent in the space group. Hence Eq. 64 becomes,

$$\chi_a = (K_{||}-K_\perp)\alpha_3^2 + K_\perp$$

$$\chi_b = (K_{||}-K_\perp)\beta_3^2 + K_\perp$$

$$\chi_c = (K_{||}-K_\perp)\gamma_3^2 + K_\perp$$

Hence,

$$K_{||} - K_\perp = \frac{\chi_a - \chi_b}{\alpha_3^2 - \beta_3^2}$$

$$= \frac{\chi_b - \chi_c}{\beta_3^2 - \gamma_3^2} \tag{65}$$

$$= (\chi_a - \chi_c)/\alpha_3^2 - \gamma_3^2$$

where $\alpha_3$, $\beta_3$, $\gamma_3$ are the direction cosines of the symmetry axis ($K_{||}$-axis) with respect to the crystal axes. It is clear from Eq. 65 that the sign and magnitude of the molecular anisotropy can be determined independently from each of the three measurements of crystal anisotropy. This offers a very convenient check on the accuracy of the individual measurements.

Lasheen and Tadros [67] have recently examined the errors involved in the calculations of molecular susceptibilities by Eq. 64. They point out that when the ratios of the crystal magnetic susceptibilities are nearly equal to the corresponding ratios of the squares of the direction cosines of one of the molecular principal axes, the uncertainties in the relative molecular susceptibility along the other two molecular principal axes will be relatively large.

As a special case for the orthorhombic system, when the unit cell contains only one molecule (or all the molecules are magnetically equivalent), the crystalline and molecular susceptibilities become identical.

## 4. Uniaxial System

We shall consider here the crystals belonging to the trigonal, tetragonal and hexagonal systems. In these crystal systems,

$$\chi_a = \chi_b = \chi_\perp \quad \text{and} \quad \chi_c = \chi_{||}$$

Now, from Eq. 64,

$$\chi_{||} = \chi_c = K_1 \gamma_1^2 + K_2 \gamma_2^2 + K_3 \gamma_3^2$$

$$\frac{\chi_a + \chi_b}{2} = \chi_a = \chi_b = \chi_\perp = K_1 \frac{\alpha_1^2 + \beta_1^2}{2} + K_2 \frac{\alpha_2^2 + \beta_2^2}{2} + K_3 \frac{\alpha_3^2 + \beta_3^2}{2} \tag{66}$$

If we assume uniaxial symmetry of the molecule ($K_1 = K_2 = K_\perp$, $K_3 = K_{||}$), then

$$K_{||} - K_\perp = (\chi_{||} - \chi_\perp) / \left( \frac{3\gamma_3^2 - 1}{2} \right) \tag{67}$$

The above relation (Eq. 67) is sometimes also expressed as:

$$K_{||} - K_\perp = (\chi_{||} - \chi_\perp) / (\frac{3}{2} - \frac{3}{2} \sin^2 \delta - \frac{1}{2})$$

$$\text{(where } \cos \delta = \gamma_3 \text{)}$$

or,

$$K_{||} - K_\perp = (\chi_{||} - \chi_\perp) / (1 - \frac{3}{2} \sin^2 \delta) \tag{68}$$

If the unit cell contains one molecule (or all the molecules in the unit cell are magnetically equivalent), $\delta = 0$, and hence $(K_{||} - K_\perp) = (\chi_{||} - \chi_\perp)$.

### E. Shape Anisotropy

When a crystal is placed in a magnetic field, it will experience a couple being a combination of several factors tending to rotate the crystal into a preferred orientation. In the case of a paramagnetic crystal, the couple is made of following combinations: paramagnetic anisotropy, anisotropy of its diamagnetism, and shape anisotropy. In the case of a diamagnetic crystal also, the couple acting on it is partly due to its shape. In the following paragraphs we shall discuss the origin, magnitude and ways of eliminating the shape anisotropy.

If we consider a cubic crystal where by definition the crystal susceptibility is isotropic, the crystal will still experience a couple when placed in a magnetic field. This arises from two factors:

a. If the magnetic field $H_{ex}$ in which the crystal is placed is ideally homogeneous before the crystal is brought in, the magnetic field inside the crystal is changed to $H_{eff}$ where

$$H_{ex} - H_{eff} = NI \tag{69}$$

and I is the intensity of magnetization and N the demagnetizing factor. This shows that the apparent susceptibility of the crystal is anisotropic if the demagnetizing factor is anisotropic.

b. Actual magnetic fields deviate slightly from complete homogeneity as does the induced magnetization inside the sample when placed in the magnetic field. This in effect induces a couple to act on the crystal tending to rotate it into the strongest part of the field.

There are several methods by which the shape effect can either be completely eliminated or at best minimized. Obviously a field as homogeneous as possible should be used in order to minimize the effect of (b) above. No general theoretical treatment of this effect seems to have been made, although it can be shown that the shape anisotropy is proportional to the square of the mean susceptibility of the crystal [28]. Hence it is negligibly small for almost all cases except perhaps in the cases where the paramagnetic anisotropies are extremely small [7].

One of the earliest methods of eliminating shape anisotropy was to suspend the crystal in a liquid bath of the same volume susceptibility as the mean susceptibility of the crystal. This procedure was used by Krishnan and Banerji [7] in their studies on some paramagnetic S-state compounds. The method is clearly unsuitable for crystals soluble in the liquids used or for those with very high susceptibility. More often the crystal is ground into a cylinder and the anisotropy measured with the cylinder axis perpendicular to the plane of the magnetic field. However in this case the crystal can easily become contaminated, and further difficulties may be encountered if the crystal is hygroscopic, efflorescent or chemically unstable. It would then seem best to use the crystal in its natural shape and allow for the shape anisotropy by calculation or by empirical relationship between known shapes and anisotropies.

The shape anisotropy can be calculated theoretically from Eq. 69 if the demagnetizing factor N is known. The value of N can only be calculated for certain geometrical shapes where the assumption of uniform magnetization is valid. Tables and graphs of N for oblate and prolate ellipsoids of revolution have been presented by Stoner [68] and Osborne [69] and applied to the calculation of shape anisotropy by Uyeda et al. [70], and by Rhodes and Rowlands [71]. The demagnetizing factors for thin circular discs have been calculated by Snoek [72].

The only systematic experimental investigation on shape anisotropy is by Majumdar [73] whose aim was to try and obtain shape anisotropy data on cubic crystals and thus to apply shape anisotropy corrections to non-cubic crystals. Cubic aluminum potassium, chromium potassium and iron ammonium alums as well as single crystals of potassium bromide were cut into rectangular shapes with differing length (l), breadth (b), and height (h) and the anisotropy (assumed to be only due to the asymmetry in shape) measured.

Majumdar found that the shape anisotropy depended not only upon the ratio of the dimensions in the horizontal plane (l and b) but also on the height (h), since the couple acting on different sections in the horizontal plane of a crystal of a given l and b varies with the height h and the observed value is an average. So in the investigations two dimensions were held constant while the third was varied. The variation of $(\Delta\chi/\chi)$ with the height for ferric alum and KBr is illustrated in Fig. 6. Similarly variation of $(\Delta\chi/\chi)$ with l/b for KBr and ferric alum is shown in Fig. 7. It is observed that $(\Delta\chi/\chi)$ tends to become zero for l/b = 1 (see Fig. 7). Even for the (l/b) = 4, $\Delta\chi/\chi$ is much less than 1% and hence not significant.

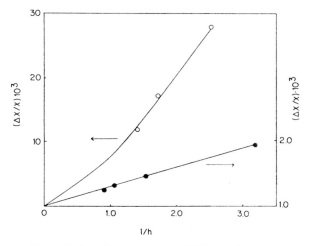

FIG. 6. Variation of $(\Delta\chi/\chi)$ with height (1/h) for ferric alum (filled circles) and KBr (hollow circles).

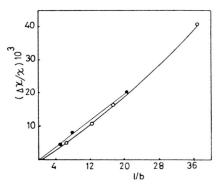

FIG. 7. Variation of $(\Delta\chi/\chi)$ with l/b for ferric alum (filled circles) and KBr (hollow circles).

### F.  Anisotropy Due to Diamagnetism
### in Paramagnetic Crystals

We have mentioned in the previous subsection that a part of the couple experienced by a paramagnetic crystal is due to the anisotropy of its diamagnetism. Hence a correction is necessary in the observed anisotropy of a paramagnetic crystal to obtain its "true" anisotropy. Theoretically the diamagnetic anisotropy can be eliminated since it should be independent of temperature, whereas the paramagnetic anisotropy varies with temperature. But practically this method of separating diamagnetic anisotropy is neither convenient nor accurate, since the exact nature of temperature variation of anisotropy is not known. Krishnan and Banerji [7] have, therefore, suggested a convenient method for its determination in monoclinic crystals.

While measuring the anisotropy of a series of diamagnetic magnesium and zinc Tutton salts, Krishnan and Banerji [7] made the following observations:

a. The magnetic anisotropy of these diamagnetic Tutton salts are very much smaller than those of the paramagnetic ones. In fact the correction in their case was only significant for the manganous Tutton salts which being in S-state, show extremely small anisotropy.

b. The diamagnetic anisotropy shows very little variation from crystal to crystal. In the Tutton salts studied, these variations were observed to occur only when the monovalent atom is changed, or when the $SO_4$ group is replaced by $SeO_4$; when, on the other hand, the divalent atom (say Mg) is replaced by Zn, there is practically no variation in the anisotropy. It may, therefore, be presumed that usually when the diamagnetic anions (Zn or Mg) are replaced by paramagnetic ions (e.g. Cu, Ni, etc.) the anisotropy of the diamagnetism of the crystal remains the same.

With the above assumptions, Krishnan and Banerji [7] deduced the relevant mathematical relationships for correcting the observed paramagnetic anisotropies in the monoclinic system.

Let OA in Fig. 8 be the direction of $\chi_1$ axis as actually observed, OD that of the diamagnetic $\chi_1$-axis, and OS that of the paramagnetic $\chi_1$ axis. Let their relative inclinations be as marked in the figure. If $(\chi_1-\chi_2)_o$ be the observed anisotropy in the (010) plane, $(\chi_1-\chi_2)_d$ the diamagnetic part, and $(\chi_1-\chi_2)_p$ the paramagnetic part, then

$$\tan 2\sigma = P/Q \qquad (69a)$$

where,

$$P = (\chi_1-\chi_2)\sin 2\sigma \qquad (70)$$

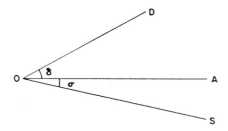

FIG. 8. Orientation of diamagnetic and paramagnetic axes in a paramagnetic crystal.

$$Q = (\chi_1 - \chi_2)_o - (\chi_1 - \chi_2)\cos 2\sigma \tag{71}$$

Thus $\sigma$, and, therefore, the directions of the paramagnetic axes are known. The "true" anisotropy is given by

$$(\chi_1 - \chi_2)_p = (P^2 + Q^2)^{\frac{1}{2}} \tag{72}$$

and

$$(\chi_1 - \chi_3)_p = \frac{1}{2}[(\chi_1 - \chi_2)_p - (\chi_1 - \chi_2)_o + (\chi_1 - \chi_2)_d]$$
$$+ [(\chi_1 - \chi_2)_o - (\chi_1 - \chi_3)_d] \tag{73}$$

Thus by measuring the anisotropy of an isomorphous diamagnetic crystal, one can easily "correct" the observed anisotropy for the diamagnetism. In the case of crystals belonging to higher symmetry, the correction can be applied directly.

The correction for the diamagnetic anisotripy is usually, as mentioned above, very small. For example, the anisotropy of Mg or Zn Tutton salts is of the order of 1 x $10^{-6}$ $cm^3$/mole, and hence negligibly small for most compounds. In fact the correction is normally significant only for the S-state and some $Cr^{3+}$ compounds. A striking case where the diamagnetic anisotropy has very important contribution is the metal phthalocyanines. Because of a large delocalized $\pi$-system the metal-free phthalocyanine shows very high anisotropy. We shall discuss its significance later on.

### III. SURVEY OF EXPERIMENTAL RESULTS

#### A. Crystal Structure

Particulars will be given in this subsection about the crystal structure of a number of compounds which will be the subject of our subsequent discussion.

## 1. Tutton Salts

The Tutton salts form an extensive isomorphous series with the general formula $M'_2M''(SO_4)_2.6H_2O$ where $M'$ is a monovalent cation (e.g., K, Rb, $NH_4$ etc.), $M''$ is a divalent cation (e.g., Mn, Cu, Zn, etc.), and S = S or Se. The crystals belong to the monoclinic space group, and generally grow with well-developed (001) plane. The crystal structure of one member of the series, $(NH_4)_2Mg(SO_4)_2.6H_2O$ was investigated in detail by Hoffmann [74] about forty years ago. Recently the detailed crystal structures of $Ni(NH_4SO_4)_2.6H_2O$ and $Cu(NH_4SO_4)_2.6H_2O$ have been studied [75, 76]. There are two molecules in the unit cell, and each divalent cation is surrounded by a distorted octahedron of six water molecules. Four of these lie very nearly in a square with the divalent cation approximately at its center, the other two water molecules lie above and below this plane. In the copper ammonium sulfate the octahedron is elongated perpendicular to this square plane of the water molecules; in the nickel ammonium sulfate it is compressed in this direction.

## 2. Fluosilicates and Chlorostannates

The general formula for the fluosilicates and chlorostannates is $M''SxX_6.6H_2O$, where $M''$ is a divalent cation (e.g. Cu, Ni, Zn, etc.). For fluosilicates Sx = Si and X = F, and for chlorostannates Sx = Sn and X = Cl. The crystals belong to the trigonal system and grow in hexagonal pillars parallel to the trigonal axis. Pauling [77] studied the crystal structure of $NiSnCl_6.6H_2O$ and observed that all the divalent fluosilicates are isostructural with the chlorostannate. The unit cell of the crystal was shown to contain only one molecule. Recent neutron diffraction studies [78] on $FeSiF_6.6H_2O$ has confirmed the basic structural findings of Pauling. The crystals have been shown to have a distorted CsCl structure, the $[M''(H_2O)_6]$ and $[SiF_6]$ (or $SnCl_6$) octahedra being stacked in compact column with their trigonal axes parallel to that of the crystal. Hence the crystalline and magnetic ellipsoids are identical, and

$$\chi_{||} = K_{||} \quad , \quad \chi_{\perp} = K_{\perp}$$

## 3. Ethylsulfates

The general formula of the ethylsulfate is $M''' (C_2H_5SO_4)_3.9H_2O$, where $M'''$ is a trivalent rare-earty ion. The crystals belong to the hexagonal system and grow in hexagonal prisms parallel to the sixfold axis. The detailed structure has been investigated by Ketelaar [79]. The rare earth ion has been shown to be at the center of a triangle of water molecules, and in parallel planes above and below there are two more triangles of water molecules, but rotated through $\pi/3$ with respect to the middle one; the ethylsulfate radicals lie in the middle plane, between the water molecules. The unit cell of the crystal contains two such molecules, but they have been shown to be completely equivalent, each having trigonal symmetry about the axis parallel to the crystal hexagonal axis, i.e., c-axis of the crystal. Hence the crystal and molecular susceptibilities become identical:

## 4. Octahydrated Sulfates

The general formula of octahydrated sulfates of rare earths is $M'''_2(SO_4)_3.8H_2O$, where $M'''$ is a trivalent rare earth ion. The crystals belong to monoclinic system and form an isomorphous series except when $M''' = Ce^{3+}$. Zachariasen [80] reported the space group of $Sm_2(SO_4)_3.8H_2O$ as $C_{2h}^6$. The unit cell contains eight paramagnetic ions. Four out of these eight molecules in the unit cell are oriented parallel to the other four and shifted to a glide plane. The four molecules in any set are related amongst themselves by a twofold axis b and a plane of reflection (010). Thus there are two magnetically inequivalent molecules per unit cell. This is in agreement with the ESR studies on $Gd_2(SO_4)_3.8H_2O$ [81].

## 5. Phthalocyanines

Transition metal phthalocyanines form an extensive series of isomorphous compounds with divalent metal atoms. The crystals are known to show three polymorphic forms ($\alpha$, $\beta$ and $\gamma$), of which the $\beta$-form is believed to be the most stable. The crystal structure of $\beta$-phthalocyanines was resolved by Robertson [82] some thirty years ago. Detailed structural data have been reported for metal-free and nickel phthalocyanines, and those of Co(II), Cu(II), Mn(II), Fe(II), etc. have been shown to be completely isostructural with them. The single crystals of metal-free and metal phthalocyanines belong to the monoclinic space group ($P2_1/a$) with two inequivalent molecules in the unit cell. The molecules have been shown to be perfectly planar; a typical molecular geometry of Ni phthalocyanine has been shown in Fig. 9. The planar molecules are stacked in the lattice so that they occur at intervals of about 4.7 Å along the b-axis of the crystal, and the two aza-methine N-atoms of each molecule lie exactly above or below the metal atom (in the case of metal phthalocyanines) which belongs to the nearest neighbors. These axially located N-atoms lie at about 3.38 Å and complete a grossly elongated "octahedron" about each metal atom since the equatorial distance between metal and nitrogen atoms is about 1.83 Å (see Fig. 10). The planes of the phthalocyanine molecules are found to be inclined to b-axis at an angle of $45°$.

## B.  Paramagnetic Transition Metal Compounds

Among the paramagnetic transition metal compounds studied to date, the

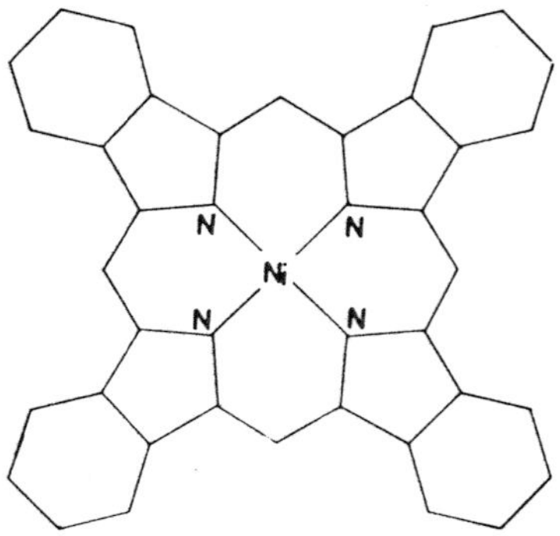

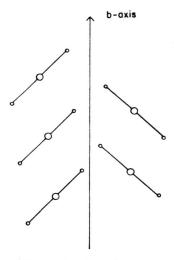

FIG. 9.  Molecular geometry of nickel phthalocyanine.

b-axis

FIG. 10.  Orientation of the molecular planes in metal phthalocyanines. The bigger and the smaller circles represent, respectively, the metal and N atoms.

study of magnetic anisotropy is confined to the compounds of first row metals, except the single example of $Ru(acac)_3$ which belongs to the $4d^5$ configuration.  Our discussion below will therefore be with reference to the first row transition metal compounds.  The $Ru(acac)_3$ will be discussed with the other compounds of $d^5$ configuration.  All the data discussed below are expressed in $10^{-6}$ cm$^3$/mole unless otherwise mentioned.

## 1. $d^1$ ($Ti^{3+}$, $V^{4+}$)

Although the compounds of $d^1$ configuration are expected to be interesting as they have only one d electron, there is a great dearth of experimental data on this configuration. The compounds of trivalent titanium are usually very sensitive to air and hence are rather unsuitable for this type of work. However the compounds of $V^{4+}$ are stable at ordinary temperatures, and are suitable for magnetic studies. Recently Gregson and Mitra [83] have studied the magnetic anisotropy of vanadyl acetylacetonate between room and liquid nitrogen temperatures.

Single crystals of vanadyl acetylacetone, $VO(acac)_2$, belong to the triclinic system with one inequivalent molecule in the unit cell. The co-ordination around the $V^{4+}$ is approximately square pyramidal [84]. The metal atom is coordinated to the four oxygen atoms from the two acetylacetonate group in the basal plane, and the fifth oxygen atom above the basal plane completes the pyramidal configuration. The $V^{4+}$ ion is slightly above the plane of the acetylacetonate bridges such that the V = O bond distance, about 1.56A, is shorter than the corresponding distance in the basal plane. The structure further reveals that the four-fold axis (along the fifth oxygen atom) is perpendicular to the b-axis of the crystal. This makes a very convenient simplification of the situation, since, although it belongs to the triclinic crystal system, one can directly measure the molecular anisotropy. At $300°K$, Gregson and Mitra report

$$K_\perp - K_{||} = 76 \times 10^{-6} \text{ cm}^3/\text{mole}$$

$$\text{and } \overline{\chi} = 1247 \qquad (\overline{\mu} = 1.73 \text{ BM}).$$

The anisotropy is observed to be very small, which indicates that the ground term is an orbital singlet and is far removed from the excited terms. The electronic structure deduced from the magnetic data is shown in Fig. 11. The position of the $d_z{}^2$ orbital cannot be ascertained from the magnetic data. Under such condition the expression for the principal susceptibilities can be written as

$$K_{||} = \frac{N\beta^2}{4kT} \cdot g_{||}^2 + \frac{8N\beta^2 \kappa_{||}^2}{\Delta E_{||}}$$

$$K_\perp = \frac{N\beta^2}{4kT} \cdot g_\perp^2 + \frac{2N\beta^2 \kappa_\perp^2}{\Delta E_\perp}$$

The temperature variation of $K_{||}$ and $K_\perp$ are given in Fig. 12. Both are

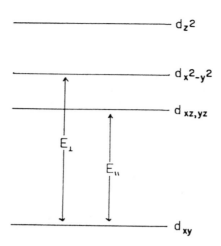

FIG. 11. Probable ordering of energy levels in vanadyl acetylacetonate (not to scale).

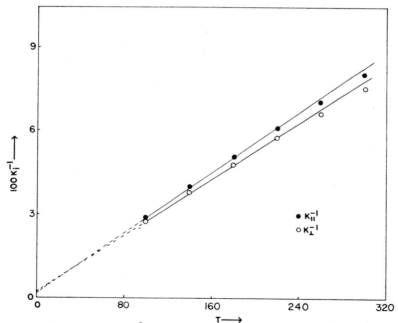

FIG. 12. Variation of $K_{||}^{-1}$ and $K_{\perp}^{-1}$ of $VO(acac)_2$ with temperature.

observed to obey Curie law very closely. The position of the excited states deduced from the magnetic measurements agrees very well with the electronic spectra.

## 2. $d^2$ ($V^{3+}$)

In this configuration the data exist only for trivalent vanadium compounds. Potassium trisoxalatovanadate (III) trihydrate and vanadium doped in corundum have been studied over a wide range of temperature. In both these cases, the $V^{3+}$ ion is in a slightly distorted octahedral field.

The theory of magnetic susceptibility for the $V^{3+}$ ion in an approximate octahedral ligand field has been examined by Van den Handel and Seigert [85], Figgis et al. [86], Chakravarty [87], Brumage et al. [88] and Bose et al. [89]. The ground state of a free $V^{3+}$ ion is $^3F$. Under an octahedral ligand field, the ground term $^3F$ splits up into a degenerate orbital triplet $^3T_1$ lying lowest, with another triplet $^3T_2$ and a singlet $^3A_2$ lying successively above it. Under an axial component of the predominant octahedral field, the triplets split into a singlet and a doublet. The further splitting and intermixing of ground levels due to spin-orbit coupling is shown in Fig. 13. The excited free ion term is also shown in the figure. Since the

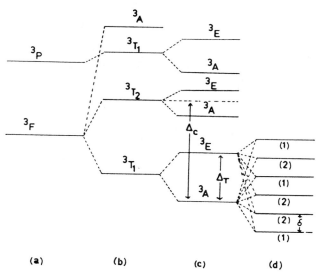

FIG. 13. Stark pattern for $V^{3+}$ ion in a slightly distorted octahedral ligand field (not to scale); (a) Free-ion, (b) octahedral field, (c) Axial (trigonal field), and (d) Spin orbit coupling. The numbers in brackets represent the total degeneracy.

spin-orbit coupling parameter is here quite small ($\lambda = 102$ cm$^{-1}$ for free ion), the anisotropy will depend very largely on the extent of the splitting of $^3T_1$ level under axial field.

The magnetic anisotropy of potassium trisoxalatovandate trihydrate has been studied by Bose et al.[89] between 300° and 100°K. The crystal belongs to the monoclinic space group $C_{2h}^5$ [90] with two enantiomorphous pairs of molecules, each pair differing from the other by a rotation of 180° about the b'-axis of the crystal. Each $V^{3+}$ ion is surrounded by an approximate octahedron of six oxygen atoms belonging to three oxalate groups arranged in the so-called propeller model, the octahedron being elongated along the propeller axis. The symmetry of the crystal field around $V^{3+}$ ion is presumed to be trigonal. The detailed data on the crystal anisotropies of this compound do not appear to have been published; the values of molecular anisotropy at four temperatures have only been reported by Bose et al. (see Table 1a). The room temperature anisotropy is very low, $\Delta K/\overline{K}$ being about 1.5%, and it increases to about 7.5% at 100°K. The principal susceptibilities ($K_{||}$, $K_\perp$) obey Curie law very closely ($\theta_{||} \simeq 2°$ and $\theta_\perp \simeq 6°$). These facts suggest that the ground term under the axial field must be the singlet and that $\Delta_T$ should be much larger. The expressions for principal susceptibilities as deduced by Brumage et al.[88] for the trivalent vanadium under the trigonal field (and $\delta << \kappa T$) are as below:

$$K_{||} = \frac{2}{3} NM_o^2 \{ g_{||}^2 [\kappa(T+\delta/3\kappa)]^{-1} + 6\alpha^{12}/\Delta_c$$

$$+ 4(\alpha+2)^2 \alpha^2 \lambda^2 / \Delta_T^2 \}$$

$$K_\perp = \frac{2}{3} NM_o^2 \{ g_\perp^2 [\kappa(T-\delta/6\kappa)]^{-1} + 3\alpha^{12}/\Delta_c$$

$$+ 3\alpha^2 [1-4(4\alpha^2+\alpha-2)\lambda^2/3\Delta_T^2] \} \qquad (75)$$

TABLE 1a

| Temperature (°K) | $(K_{||} - K_\perp)$ | $K_{||}$ | $K_\perp$ |
|---|---|---|---|
| 300 | 44.22 | 3298 | 3254 |
| 240 | 162.82 | 4108 | 3946 |
| 200 | 283.1 | 4953 | 4670 |
| 100 | 711.0 | 9764 | 9053 |

where $\alpha$, $\alpha^1$ are the Lande factors; $\Delta_c$, $\Delta_T$ and $\delta$ are the same as in Fig. 13. It will be seen from Eq. 75 that both $K_{||}$ and $K_{\perp}$ are expected to follow Curie Weiss law. Because the $\Delta_T$ is much smaller than $\Delta_c$, Eq. 75 also predicts that the high frequency component in $K_{\perp}$ should be larger than the corresponding term in $K_{||}$. This is borne out by the data on $K_3V(C_2O_4)_3$ $.3H_2O$. The high frequency terms associated with $K_{||}$ and $K$ (as obtained by the graphical extrapolation method by the present author) are $200 \times 10^{-6}$ and $350 \times 10^{-6}$, respectively.

Following parameters have been deduced by Bose et al. from their data:

$$\alpha = 1.158, \; \alpha^1 = 1.39, \; \kappa_{||} = 0.855, \; \kappa_{\perp} = 0.68$$

$$\lambda_{||} = 57 \quad \lambda = 63 \quad \delta = 7.0 \text{ cm}^{-1}.$$

The value of $\Delta_T$ has been found to vary with temperature from $930 \text{ cm}^{-1}$ at $300°$ to $720 \text{ cm}^{-1}$ at $100°K$.

Brumage et al. [88] have studied the magnetic anisotropy of $V^{3+}$ doped in corundum over the temperature range of $4$-$500°K$. The magnetic susceptibilities follow a Curie-Weiss law between $77$-$500°K$. At temperatures below $30°K$, the magnetic anisotropy becomes very large and deviations from Curie-Weiss law is observed for both $K_{||}$ and $K_{\perp}$. They observe that an excellent fit between Eq. 75 and their data is obtained over the whole range when

$$\alpha = 1.29 \quad \alpha^1 = 1.52 \quad \delta = 8.4 \text{ cm}^{-1}$$

$$\lambda = 95 \text{ cm}^{-1} \text{ and } \Delta_T \simeq 1100 \text{ cm}^{-1}.$$

It is interesting that $\Delta_T$ for the trisoxalate and the $V^{3+}$ doped in corundum is so similar. Since the packing in these two crystals is not expected to be the same, it seems likely that the effective axial field arises from the Jahn-Teller effect phenomenon rather than from exigencies of lattice packing. This may be the reason why the $\Delta_T$ in the vanadium ammonium alum is also so similar [86], because the effect of lattice packing in the alum and the above crystals is expected to be different.

## 3. $d^3$ ($Cr^{3+}$)

In this configuration also, the magnetic anisotropy data is very meagre. The only crystals studied are trisoxalates of trivalent chromium. The $Cr^{3+}$ ion in all these cases has an approximate octahedral (with a small trigonal distortion) environment.

The free ion ground state of the $Cr^{3+}$ ion is $^4F$. Under an octahedral field it is split up into a singlet ($^4A_2$) and two orbital triplets ($^4T_1$ and $^4T_2$), with the singlet lying some $17000 \text{ cm}^{-1}$ below the first higher state. The detailed splitting of the ground state under axial and spin-orbit coupling perturbations is shown in Fig. 14. The spin-orbit coupling parameter for the free-ion is +87 $cm^{-1}$. It is therefore evident that the $Cr^{3+}$ ion will behave as if it is in S-state. Thus one would expect these crystals to be very feebly anisotropic and the anisotropy obeying very closely $\frac{1}{T^2}$ law.

Krishnan et al. [9] studied the magnetic anisotropy of $(NH_4)_3Cr(C_2O_4)_3$ $.3H_2O$. Their room temperature (30°C) results are given below,

$$\chi_1 - \chi_2 = 4.3$$
$$\bar{\chi} = 5440 \qquad \chi_3 - \chi_2 = 14.6.$$
$$\chi_1 - \chi_3 = -10.3$$

It is observed that the maximum anisotropy, namely $\chi_3 - \chi_2 = 14.6$ is only $\frac{1}{4}\%$ of the mean susceptibility, and is indeed of the same order of magnitude as the anisotropy exhibited by S-state ions (see later). The observed anisotropy is nearly equal to the diamagnetic anisotropy of analogous crystal $K_3Al(C_2O_4)_3.3H_2O$,

$$\chi_1 - \chi_2 = 6.9 \quad , \quad \chi_1 - \chi_3 = 13.1$$

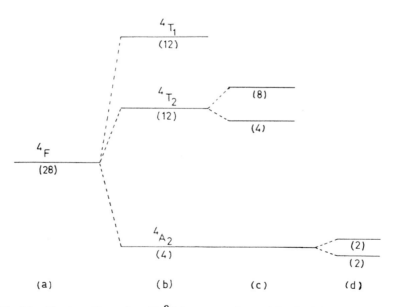

FIG. 14. Stark pattern for $Cr^{3+}$ in a pseudo-octahedral ligand field: (a) Free-ion, (b) Octahedral field, (c) Axial field, (d) Spin-orbit coupling. The numbers in brackets represent the total degeneracy.

Krishnan et al. have also studied the temperature variation of its magnetic anisotropy. With b-axis vertical they find that the value of $\psi$ changes by about 78° as the temperature is lowered from 30° to -183°C, and the value of anisotropy in the horizontal plane increases by 1.9 times. With c-axis vertical the anisotropy in the horizontal plane increases by 12 times in this temperature range. No satisfactory explanation has been given for this unusually large variation in $\psi$ and the small thermal variation in $(\chi_1 - \chi_2)$.

Recently Saha [92] has reported the single crystal studies on $K_3Cr(C_2O_4)_3$ .$3H_2O$ which belongs to monoclinic system at room temperature. Anisotropies are observed to be very feeble, and to increase by about 10 times between room temperature and 90°K, thus in good agreement with the $\frac{1}{T^2}$ law. He further observes a very marked change in the setting angle with temperature when the crystal is suspended with a-axis vertical, thus indicating some sort of phase transition in the crystal, destroying its monoclinic symmetry. It is of interest that there is no marked effect of this transition on the temperature variation of anisotropy or average susceptibility (see Fig. 15).

### 4. $d^4$

No magnetic anisotropy data appears to have been published for $d^4$ configuration.

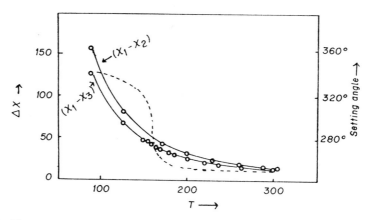

FIG. 15. Variation of principal crystalline anisotropies and the setting angle for $K_3Cr(C_4O_3)_3.3H_2O$ with temperature (reference 92). The left hand ordinate is for principal anisotropies, while the right hand one is for the setting angle (shown by broken curve).

## 5. $d^5$ ($Fe^{3+}$, $Mn^{2+}$, $Ru^{3+}$)

i. High-spin Complexes. A huge amount of data exist for high-spin $3d^5$ configuration. The earlier studies by Krishnan and Banerji were mainly on manganous Tutton salts and some ferric trisoxalates. In both these cases the co-ordination around the metal ion is approximately octahedral. The free ion ground term is $^6S_{5/2}$ and is far below the higher terms. Since the ground state has no orbital moment, the cubic and the axial ligand field has no effect on it. The spin-orbit coupling however removes partially its 6-fold spin degeneracy into three Kramers doublets. Because of this removal of spin degeneracy, the S-state ions are expected to show a very small, but finite, anisotropy which otherwise would have vanished. Further the anisotropies are expected to obey a $\frac{1}{T^2}$ law of temperature dependence and the principal susceptibilities Curie law. These theoretical predictions have been verified by the experiments of Rabi [93], Jackson [94], Krishnan and Banerji [7], Guha [95], and Krishnan et al. [9]. The room temperature magnetic anisotropy of manganous Tutton salts and some ferric compounds are summarized in Table 2. The anisotropies

TABLE 2

| Crystals | $\chi_1-\chi_2$ | $\chi_1-\chi_3$ | $\Psi$ | Ref. |
|---|---|---|---|---|
| $R_2Mn(SO_4)_2.6H_2O$ | | | | |
| R = $NH_4$ | 8.7 | 6.0 | $-16°$ | 7 |
| | 40 | $-60$ | $+14°$ | 93 |
| = Rb | 8.9 | 6.2 | $-12°$ | 7 |
| = Cs | 7.0 | 5.0 | $-12.5°$ | 7 |
| = Tl | 8.9 | 7.2 | $-16.5°$ | 7 |
| $K_2Mn(SO_4)_2.4H_2O$ | 4.3 | 1.5 | $-61°$ | 7 |
| (does not belong to the Tutton series) | | | | |
| $R_2Mn(SeO_4)_2.6H_2O$ | | | | |
| R = $NH_4$ | 8.9 | 6.3 | $-20°$ | 7 |
| = Rb | 9.0 | 7.3 | $-14°$ | 7 |
| = Tl | 10.2 | 8.5 | $-14°$ | 7 |
| $K_3Fe(C_2O_4)_3.3H_2O$ | 28 | 77 | $-1°$ | 7 |
| $Na_3Fe(C_2O_4)_3.5H_2O$ | 31 | 24 | $-2°$ | 7 |

are observed to be extremely small, $\Delta\chi/\bar{\chi}$ about 0.07%. The measurement of these small anisotropies usually poses many problems which we have discussed earlier.

Krishnan and Banerji [7] also measured the anisotropy of some isomorphous zinc and magnesium Tutton salts. A few typical values are given in the Table 3. The anisotropies are of the same order as the S-state ions.

Krishnan et al. [9] studied the temperature variation of anisotropy of manganous ammonium sulfate hexahydrate. For low temperature measurements the crystals were cut in the form of cylinders.

The low temperature data on this crystal are tabulated in Table 4. The corrected anisotropies refer to the values after diamagnetic correction. It is seen from the Table that both $(\chi_1-\chi_2)$ and $(\chi_1-\chi_3)$ obey very closely $\frac{1}{T^2}$ law.

We have discussed earlier the splitting of the $^6S$ term in the crystal field and spin-orbit coupling, the effect of the latter being to split the $^6S$ term into three Kramers doublets. The separation between these Kramers doublets plays a vital role in the production of extremely low temperature by adiabatic demagnetization, and in some very low temperature phenomenon such as specific heat, entropy etc. For example, manganous ammonium sulfate has been used in the classic experiments of Kurti and Simon [96] for the production of temperatures of the order of 0.1°K. A prior knowledge of the separation of these spin states is very helpful for the choice of a

TABLE 3

| Crystals | $\chi_1-\chi_2$ | $\chi_1-\chi_3$ | $\Psi$ | Ref. |
|---|---|---|---|---|
| $R_2Zn(SO_4)_2.6H_2O$ | | | | |
| R = $NH_4$ | 1.2 | 0.8 | +6° | 7 |
| = Rb | 0.9 | 0.6 | +23.5° | '' |
| $R_2Mg(SO_4)_2.6H_2O$ | | | | |
| R = $NH_4$ | 1.1 | 0.6 | -5° | '' |
| = Rb | 0.9 | 0.7 | +13° | '' |
| = K | 1.2 | 0.7 | -2° | '' |
| $MgNa_2(SO_4)_2.4H_2O$ (Bloedite) | 0.5 | 0.7 | +66° | '' |
| $K_3Al(C_2O_4)_3.3H_2O$ | 6.9 | 13.1 | +19° | '' |
| $Na_3Al(C_2O_4)_3.5H_2O$ | 4.9 | 8.5 | +9° | '' |

TABLE 4

| Temp. | $\Psi$ | $(\chi_1-\chi_2)$ | $(\chi_1-\chi_2)_{corr}$ | $(\chi_1-\chi_3)$ | $(\chi_1-\chi_3)_{corr}$ |
|-------|--------|-------------------|--------------------------|-------------------|--------------------------|
| 303°K | -16 | 8.7 | 7.7 | 6.0 | 5.4 |
| 194.3 | -16 | 21.0 | 20.0 | 13.0 | 12.4 |
| 90 | -16 | 93.0 | 92.0 | 49.0 | 48.4 |

particular compound for such experiments. Krishnan and Banerji calculated these splittings from their data on the single crystals of a number of manganous Tutton salts and had very remarkable success in predicting these separations and some other low temperature properties. Because of the great importance of their work, we shall outline below the method of their calculation.

For the S-state ions, the three principal susceptibilities are given by the expression [7],

$$\chi_1 = C(\frac{1}{T} + \frac{2\gamma}{T^2} + \ldots) \tag{76}$$

$$\chi_2 = C(\frac{1}{T} - \frac{\gamma+s}{T^2} + \ldots) \tag{77}$$

$$\chi_3 = C(\frac{1}{T} - \frac{\gamma-s}{T^2} + \ldots) \tag{78}$$

$$C = N\mu^2/3\kappa \quad , \quad \mu = g\beta[\, S(S+1)]^{\frac{1}{2}}$$

where $\mu$ is the magnetic moment of the ion; N is the Avagadro number, $\kappa$ the Boltzmann constant, g the spectroscopic splitting factor ( = 2), and $\beta$ the Bohr magneton. Also,

$$\gamma = (12a + 8b)/35\kappa \quad , \quad s = (4\sqrt{10}.c + 12\sqrt{2}.e)/35\kappa$$

where a, b, c are the constants of the crystal field splitting, which are of course very small in comparison with kT at room temperature. From Eqs. 76, 77 and 78 we obtain for the Stark anisotropies,

$$\frac{\chi_1 - \chi_2}{\overline{\chi}} = \frac{3\gamma + s}{T} \tag{79}$$

$$\frac{\chi_1 - \chi_3}{\overline{\chi}} = \frac{3\gamma - s}{T} \tag{80}$$

For manganous ammonium sulfate at 30°C,

$$\frac{\chi_1 - \chi_2}{\overline{\chi}} = \frac{7 \cdot 8}{13,830} \quad , \quad \frac{\chi_1 - \chi_3}{\overline{\chi}} = \frac{5 \cdot 5}{13,800}$$

whence $\qquad \gamma = 0.049°\qquad S = 0.025°$

To obtain a closer estimate of the separation of the spin-levels, Krishnan and Banerji made some simplifying assumptions. They assumed that the spin-orbit coupling removes the total degeneracy of the ground state and that the six Stark levels so produced are equally spaced at energy interval of $\kappa\theta_s$, where $\theta_s$ is a temperature characteristic of the crystalline splitting. The effect of crystalline electric field in splitting $^6S$ level would then be considered to be equivalent to that of a magnetic field $H_s$ given by the relation,

$$g\beta H_s = \kappa\theta_s.$$

So, if this equivalent magnetic field $H_s$ is determined $\theta_s$ is known.

Since $\gamma$ and $S$ are very small ($\sim 0.5°$) in comparison with the room temperature $T$ (=303°K) the expression (76-78) can be rewritten in the form

$$\chi_1 = C/T - \Theta_1 \quad , \quad \chi_2 = C/T - \Theta_2 \quad , \quad \chi_3 = C/T - \Theta_3$$

where $\qquad \Theta_1 = 2\gamma \quad , \quad \Theta_2 = -(\gamma+s) \quad , \quad \Theta_3 = -(\gamma-s)$ $\tag{81}$

and $\qquad \Theta_1 + \Theta_2 + \Theta_3 = 0$

The new Eq. 81 are of the Weiss type and in the usual treatment the Curie temperature $\Theta_i$, occurring in them, which in the present crystals really arise from the crystalline electric fields, would be considered as due to certain internal magnetic fields. The Weiss field is proportional to the intensity of magnetization whereas $H_s$ is constant. Thus Weiss field is the magnetic equivalent of the crystal field so far as its influence on the magnetization is concerned, whereas $H_s$ is the appropriate magnetic equivalent of the crystalline field as regards its effect on the splitting of the $^6S$ levels. The two fields would then be identical only if the splitting of $^6S$ state by the crystalline field were actually into six equally spaced levels, which as we have mentioned earlier is not so; the actual splitting is into three doubly degenerate Kramers spin doublets.

This twofold degeneracy does not differentiate change in the signs of the spin moments, and hence the crystalline splitting does not produce any magnetization by itself. If however by the application of an external magnetic field the degeneracy is removed, the crystalline splitting will exert an influence on the magnetization which will increase continually with the increase in the latter; when saturation is reached and the spin moments are all aligned along the same direction, the contribution to the magnetization field as arising from crystalline splitting would have reached its maximum value, which is naturally $H_S$. Thus

$$H_S = 3\kappa \Theta / \mu$$

Applying this formula to crystals there are three Curie temperatures to be considered. For example, for $(NH_4)_2Mn(SO_4)_2.6H_2O$ the $\Theta_i$ are:

$$\Theta_1 = -0.10° \qquad \Theta_2 = -0.07° \qquad \Theta_3 = -0.03°.$$

In dealing with phenomenon that occur at temperatures not much lower than $0.1°K$, the field corresponding to $\Theta_1$ would be the predominant field, and we may put,

$$H_S = 3\kappa \Theta_1 / \mu$$

which gives,

$$\theta_S = 3g\beta \Theta_1 / \mu$$

Thus $\theta_S$ can be easily calculated. We tabulate below (Table 5) the values of $\theta_S$ for different crystals of $Mn^{2+}$ compounds as calculated by Krishnan and Banerji.

It may be remarked here that the values of $\theta_S$ calculated by the above approximate method agree very well with the experimental observations. For example, for $Mn(NH_4SO_4)_2.6H_2O$, Kurti and Simon find $\theta_S = 0.11°K$ which agrees well with the above calculated value of $0.10°K$.

From the known characteristic temperature $\theta_S$, the entropy-temperature and specific heat-temperatures curves of these crystals in the neighborhood of $0.1°K$ can be easily predicted. For example, entropy can be calculated directly from the Stark pattern assumed;

$$S = R \left[ T. \frac{d \log_e Q_{(2J+1),\theta_S}}{dT} + \log_e Q_{(2J+1), \theta_S} \right] \qquad (82)$$

TABLE 5

| Crystal | Characteristic temperatures ($\theta_s$) |
|---------|------------------------------------------|
| $MnSO_4.R_2SO_4.6H_2O$ | |
| R = $NH_4$ | 0.10°K |
| Rb | 0.10 |
| Cs | 0.09 |
| Tl | 0.10 |
| $MnSeO_4.R_2SeO_4.6H_2O$ | |
| R = $NH_4$ | 0.10 |
| Rb | 0.10 |
| Tl | 0.10 |
| $MnSO_4.K_2SO_4.4H_2O$ | 0.04 |

where

$$Q_{(2J+1), \theta_s} = 1 + e^{-\theta s|T} + e^{-2\theta s/T} + \ldots + e^{-2J\theta s/T}$$

Thus the entropy-temperature curve at these low temperatures can be easily plotted from the known values of $\theta_s$.

Mixed Manganous Tutton Salts. Krishnan and Banerji [7] have also studied the effect of magnetic dilution on the magnetic anisotropy of the manganous ammonium sulfate hexahydrate. The results of their measurements are shown in Table 5. A small change in the magnetic anisotropy upon dilution is observed. It indicates that a part of the Stark anisotropy is perhaps due to exchange interaction even in these highly diluted crystals. Unfortunately the detailed data are not available on the crystal.

Guha [95] and Jackson [94] studied the principal susceptibilities of some $Mn^{2+}$ and $Fe^{3+}$ crystals over a temperature range. The following crystals were studied by them.

$MnSO_4.4H_2O$. This crystal belongs to monoclinic system and its susceptibility along b-axis ($\chi_3$) was measured by Jackson. $\theta = 2°$

$Fe(acac)_3$. The temperature variation of its susceptibility along a- and c-axes was measured by Guha and Jackson. $\theta$ along both directions is found to be ~ 4°.

$K_3Fe(C_2O_4)_3 \cdot 3H_2O$. Guha reported the temperature variation of $\chi_1$ which obeys Curie law with $\theta = 0$. Jackson measured the susceptibility along c-axis of the crystal and found $\theta = 0$.

$MnF_2$. Manganous fluoride is a typical example of antiferromagnetic substance. The magnetic anisotropy of this crystal was studied by Stout and Griffel [11] between 12° and 295°K. The magnetic anisotropy study on this crystal and some similar compounds by Stout and his co-workers forms a very important series of work on antiferromagnetic compounds and reveals some fundamental aspects of their behavior. $MnF_2$ has the tetragonal rutile crystal structure with each $Mn^{2+}$ ion surrounded by two nearest neighbors at a distance of 3.3103 Å in the direction of the c-axis, eight neighbors at a distance of 3.8229 Å in the (111) directions and four at a distance of 4.8734 Å in the (001) plane.

At 295.7°K, $\chi_{||} - \chi_{\perp} = 13.3 \times 10^{-6}$ which is only about 0.1% of the mean susceptibility. This low anisotropy at room temperature is consistent with the $^6S$ ground state of $Mn^{2+}$ ion, and is of the same order as that observed for manganous Tutton salts. Further at high temperatures $\chi_{||} > \chi_{\perp}$, and hence the states corresponding to the alignment of the spin in the direction of the c axis will be energetically more favored than others. As temperature is lowered, the anisotropy rises reaching a maximum at about 120°K, and then falls rapidly. At about 77°K, the anisotropy passes through zero and as the temperature is lowered it becomes negative $(\chi_{\perp} > \chi_{||})$ and then increases very rapidly such that at 11.9°K, $\chi_{\perp} - \chi_{||} = 24,400 \times 10^{-6}$ (see Fig. 16). The temperature variation of principal susceptibilities are shown in Fig. 17. Below the Curie temperature ($\simeq 70°K$) when the long range ordering begins to set in, $\chi_{||}$ decreases very rapidly and appears to reach zero at 0°K, indicating that the spins are now all lined up in the direction of c axis, whereas $\chi_{\perp}$ increases very slightly.

The exchange forces cannot in themselves produce any anisotropy. However, once the magnetic ions are coupled together by the exchange forces in ordered groups the energy needed to turn the spins with respect to the crystal lattice is the energy of a few tenths of a wave number for each individual ion times the effective number in a group. Below the Curie temperature the size of this group rapidly increases. With this addition, Van Vleck's theory [97] of antiferromagnetism is in excellent agreement with the results of Griffel and Stout on $MnF_2$.

(ii) Intermediate Spin. Ferric monohalodialkyldithiocarbamates. Recently Martin and co-workers [98, 99] have discovered a unique series of iron (III) dithiocarbamates where the iron (III) atom has almost a square pyramidal configuration with $S = 3/2$. The X-ray structural data on one member of this series, chlorodiethyldithiocarbamate iron (III), by Hoskin et al. [98] show that the crystal belongs to the monoclinic system with two inequivalent molecules in the unit cell. The $Fe^{3+}$ ion is observed to be slightly above

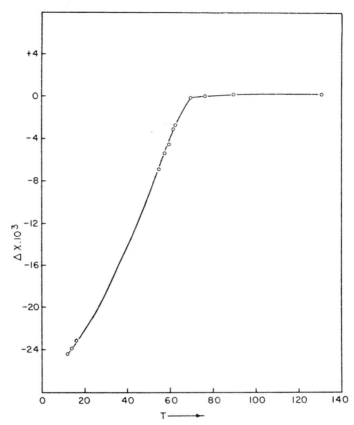

FIG. 16. Variation of magnetic anisotropy of $MnF_2$ with temperature (°K) (Ref. 11).

the basal plane with the Fe-Cl bond distance of about 2.26 Å. The molecular geometry of this crystal is shown in Fig. 18. The magnetic anisotropy of some of these compounds has been recently studied by Martin and Mitra [100]. At room temperature the anisotropies are about 10-12% of the average susceptibility. This, together with the mean moment of these compounds $\bar{\mu} = 3.98$, suggests that the ground state of these compounds is probably $^4A_1$ deriving from a configuration like $(b_2)^2(e)^2(a_1)^1$. It can be easily shown that an alternate ground state $^4E$ is not compatible with the sign of the anisotropy. The principal susceptibilities obey a Curie-Weiss law (Fig. 19).

It is interesting that the ESR measurements on these crystals show, unlike the magnetic anisotropy, considerable anisotropy in g-values:

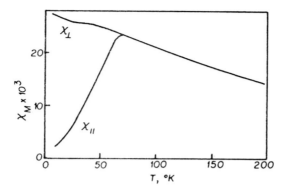

FIG. 17.  Temperature variation of $\chi_{||}$ and $\chi_{\perp}$ of $MnF_2$ (Ref. 11).

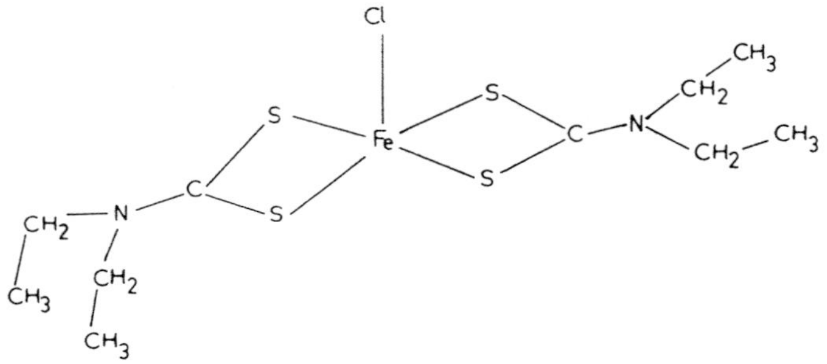

FIG. 18.  Molecular geometry of chloroferric diethyldithiocarbamate (Ref. 98).

$$g_{||} \simeq 2.8 \quad , \quad g_{\perp} \simeq 4.5$$

The reason for this discrepancy is that ESR is being observed only from $^4A_1$ state, while the magnetic anisotropy is an average effect of this and many other excited terms. Spin-orbit coupling splits the $^4A_1$ term into $M_S = \pm 3/2$, and $M_S = \pm\frac{1}{2}$ levels, of which only $M_S = \pm\frac{1}{2}$ is resonating (see Fig. 20). There is perhaps some mixing in of the excited $M_S = \pm 3/2$ level as, which causes the departure of the g-values from the ideal values of

$$g_{||} = 2.0 \quad , \quad g_{\perp} = 4.0$$

A point can be mentioned here that in the chlorodiethyl derivative, $M_S = \pm 3/2$ lies below the $M_S = \pm\frac{1}{2}$, whereas in the bromodiethyl derivative this is just the reverse.

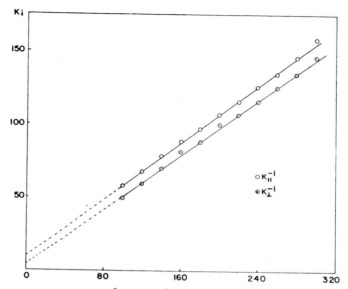

FIG. 19. Variation of $K_{\parallel}^{-1}$ and $K_{\perp}^{-1}$ of ferric diethyldithiocarbamate with temperature (°K).

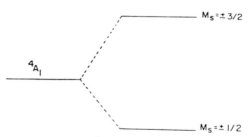

FIG. 20. Splitting of $^4A_1$ term by spin-orbit coupling.

Manganous Phthalocyanine. The crystal structure of this compound has been discussed earlier (see sub-section III A). It has recently been shown that the $Mn^{2+}$ ion in this compound has spin S = $3/2$. At room temperature the magnetic anisotropies are:

$$(\chi_1 - \chi_2) = 671 \quad , \quad (\chi_1 - \chi_3) = 466.$$

The anisotropy of the crystal is close to that of halo-iron(III) dithiocarbamates. This fact, together with the value of average moment suggest that the compound provides an example of intermediate spin situation (S = $3/2$). The electronic structure deduced from the magnetic data is shown in Fig. 21 [14]. The ground state is deduced to be $^4A_{1g}$ (Table 6).

The temperature variation of $K_{||}$ and $K_{\perp}$ down to 90°K (Fig. 22) and of the average moment down to 1.5°K (Fig. 23) establishes that the Mn atoms in this compound interact ferromagnetically. The deduced electronic struc- ture of the molecule explains successfully the observed ferromagnetism. The coplanar MnPc molecules are stacked in the lattice in a manner such that not only are their molecular planes parallel but also two diagonally opposed aza-methine N-atoms of each molecule lie exactly above or below the metal atom which belongs to the nearest neighbors. These axially located N-atoms at 3.38 Å complete a grossly elongated "octahedron" about each metal atom. Although linear chains of Mn atoms run parallel to the b-axis with an intermetallic separation of 4.75 Å, this distance is considered to be too great for direct overlap of manganese 3d orbitals to contribute significantly to the magnetic properties. However the availability of 90° Mn-N-Mn pathways suggests a mechanism for ferromagnetic inter- action via superexchange [102-104]. For example, while an axially situated $a_{1g}$ orbital centered on any Mn ion is nonorthogonal to the $2p_z$ orbitals of the neighboring aza-methine N atoms (and hence the delocalized $\pi$-systems of the nearest neighboring Pc molecules), these $\pi$-systems themselves are orthogonal to $a_{1g}$ orbitals centered on the nearest neighboring Mn atoms. Ferromagnetic coupling along the chain, therefore, results. Likewise, the alternate superexchange pathway which involves eg orbitals centered on Mn is served by the orthogonality of the aza-methine N-atoms with the orbitals centered on the neighboring Mn atoms [14].

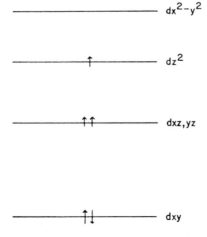

FIG. 21.  Electronic structure of manganous phthalocyanine (Ref. 14).

TABLE 6

| A = Mn | | |
|---|---|---|
| $(NH_4)_2(SO_4)_2.6H_2O$ | | |
| B = Mg | C = Zn | |

| Composition of Crystal | $\psi$ | $(\chi_1 - \chi_2)_C$ |
|---|---|---|
| A + nB | | |
| n = 0 | $-17.5°$ | 7.7 |
| 0.36 | $-20.0°$ | 6.9 |
| 1.75 | $-24.5°$ | 5.9 |
| 5.12 | $-27.5°$ | 5.4 |
| A + nC | | |
| n = 0 | $-19°$ | 7.8 |
| 1.05 | $-24°$ | 6.6 |
| 4.62 | $-36°$ | 5.3 |

(iii)  Low Spin.  Magnetic anisotropy of two low-spin $d^5$ compounds have been measured.  In both the cases the molecule has an approximate octahedral environment around the metal atom.  The electronic configuration of the compounds is, therefore, $t_{2g}^5$.  There is one unpaired electron in the $t_{2g}$ shell, and hence the ground term is $^2T_{2g}$.  The excited state $^2E$ is far removed and is therefore usually neglected in the calculation.  In the axial field, the ground term splits into an orbital singlet and a doublet, which are further split up into three orbital singlet by the spin-orbit coupling perturbation.  The ground term in the octahedral field being an orbital triplet, the crystals are expected to be highly anisotropic, and the temperature dependence of principal susceptibilities of low spin $d^5$ compounds have been done by Howard [105] and Kamimura [106].  Bose et al. [107] have deduced the expressions of principal susceptibilities for $^2T_{2g}$ term with the specific example of $Ti^{3+}$ compounds ($d^1$); however it can be readily changed to the low-spin $d^5$ cases [108].

$K_3Fe(CN)_6$.  The morphology of this crystal has been shown to be pseudo-orthorhombic [109].  The crystal structure has been reported by Baskhatov [110], and by Baskhatov and Zhdanov [111] and the unit cell is shown to be monoclinic with monoclinic angle $\beta = 90°$.  The iron atom is co-ordinated to six $(CN)^-$ groups forming an approximate octahedron.  The magnetic

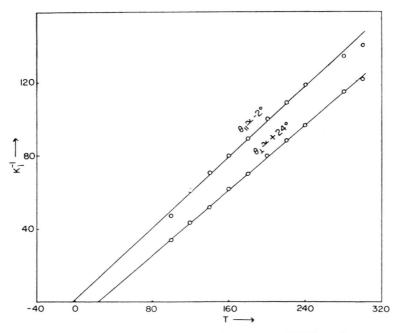

FIG. 22. Variation of principal molecular susceptibilities of manganous phthalocyanine with temperature (°K).

anisotropy and principal susceptibilities of this crystal have been studied over a wide range of temperatures by Jackson [112] and Guha [95].

Krishnan and Mookherji [8] have reported the room temperature data on this crystal. At 303°K

$$\chi_1 - \chi_2 = 451, \quad \chi_1 - \chi_3 = 123, \quad \bar{\chi} = 1940.$$

The maximum anisotropy of the crystal is thus about 21% of the average susceptibility. This is in contrast to the nearly isotropic nature of the high spin ferric compounds, but is consistent with the orbital triplet, as discussed above, being the ground term in the octahedral field. The electron spin resonance data on this crystal [113] shows that the symmetry of the ligand field is strongly rhombic and the ground triplet is split up into three orbital singlets with an over-all separation of about a few hundred cm$^{-1}$.

Guha [95] has reported the anisotropy data down to about 80°K. Jackson's measurements on principal susceptibilities extends down to about 14°K. The data of these two workers differ appreciably at room and low temperatures (see Fig. 24). The ESR data [113] have been shown to be consistent

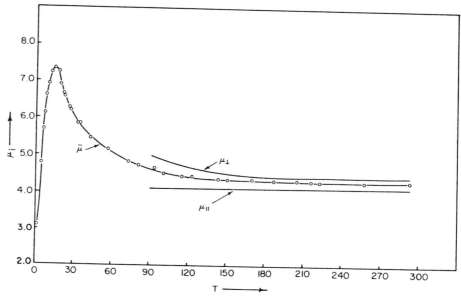

FIG. 23.  Temperature variation (°K) of average and principal moment of manganous phthalocyanine. Because of the magnetic field–dependence of the average susceptibility below about 14°K, the values of magnetic moment below this temperature were calculated for the highest value of magnetic field [14].

with the principal susceptibilities of Jackson. However the later average susceptibility measurement of Figis [114] agrees with data of Guha. The temperature of anisotropy is shown in Fig. 25. $(\chi_1 - \chi_3)$ is observed to rise to maximum at about 120°K and falls off on further lowering of temperature. It is interesting to note that a phase transition has been reported in this crystal at about this temperature by some Russian workers [115].

Ruthenium Acetylacetonate $(4d^5)$. The $Ru(acac)_3$ crystals belong to the monoclinic system with two inequivalent molecules in the unit cell [90]. The symmetry of the molecule is distorted octahedral.

Gregson and Mitra [108] have measured its magnetic anisotropy between 300-80°K. At 300°K,

$$\chi_1 - \chi_2 = 307 \quad , \quad \chi_3 - \chi_2 = 654 \quad , \quad K_{||} - K_\perp = 961.$$

The anisotropy of the molecule is thus very large, about 70% of the average susceptibility. This is consistent with the expected behavior of 2nd row transition metal compounds where the spin-orbit coupling is very large. The temperature variation of the principal moments is shown in Fig. 26.

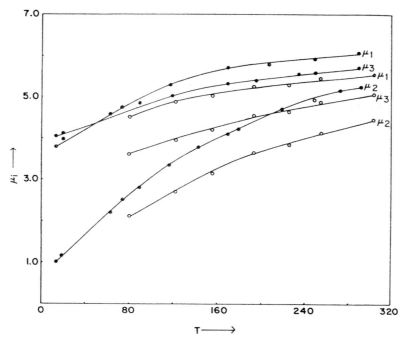

FIG. 24. Temperature variation ($^\circ$K) of principal crystalline moments of $K_3Fe(CN)_6$. Solid circles are the experimental data of Jackson [94, 112], while the open circles represent Guha's data [95].

The data over the entire range of temperature can be satisfactorily explained within the limits of a $t_{2g}^5$ configuration with the following parameters:

$$\Delta = -600 \text{ cm}^{-1}, \quad \lambda = -1000 \text{ cm}^{-1} \text{ and } k = 0.90,$$

where $\Delta$ is the axial field splitting and k is the orbital reduction parameter. An interesting point is that if $|\Delta|$ is increased, the principal moments separate out in such a way that the average moment remains nearly the same. Thus the average moment in this case is extraordinarily insensitive to $\Delta$. This explains the large discrepancy in $\Delta$ deduced by Figgis et al. [116] from the average moment and by Gregson and Mitra from the anisotropy measurements.

## 6. $d^6$ ($Fe^{2+}$)

In this configuration the anisotropy data exists for a number of high spin

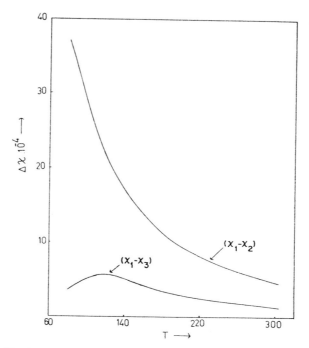

FIG. 25. Variation of principal crystalline anisotropies of $K_3Fe(CN)_6$ with temperature (°K) (Ref. 95).

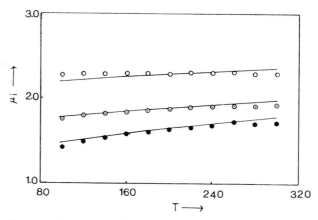

Fig. 26. Temperature-variation of principal molecular moments of ruthenium trisacetylacetonate. The circles are the experimental data: 0, $\mu_{||}$; 0, $\mu_{\perp}$; and 0, $\overline{\mu}$. The solid curves are the calculated ones using $\lambda = -120$ cm$^{-1}$, $\Delta = -500$ cm$^{-1}$, and k = 0.9,

octahedrally coordinated $Fe^{2+}$ ions. Ferrous phthalocyanine is the only crystal which has been studied outside the octahedral coordination.

In the octahedral ligand field, the free-ion $^5D$ ground-term of the Fe (II) ion is split into a triplet ($^5T_2$) and a doublet $^5E_2$, the triplet lying some 10,000 $cm^{-1}$ below the doublet. The free-ion spin-orbit coupling is about -103 $cm^{-1}$ and, hence, the effect of the excited $^5E$ level can be neglected. The tetragonal component of the field splits the ground $^5T_2$ state into a singlet and doublet, which are further split and inter-mixed by the spin-orbit coupling perturbation (see Fig. 27). In the trigonal axial field the singlet $^5B$ lies lowest and the ordering of the fine structure spin levels is also changed.

The ground term in the octahedral ligand field being an orbital triplet, one expects a very large anisotropy for the Fe(II) compounds. Temperature variation of magnetic anisotropy, particularly principal susceptibilities are also expected to be interesting.

Theoretical calculation of magnetic susceptibility and anisotropy of $^5T_2$ term has been carried out by many workers. Bose et al. [117] used a crystal field approach in $D_{4h}$ symmetry. Palumbo [118], Bose and Rai [119] and Eicher [120] have carried out similar calculations in $D_{3h}$ symmetry. Recently König and Chakravarty [121] have done very extensive calculations of magnetic susceptibility and anisotropy under $D_{4h}$ and $D_{3h}$ symmetry, and have applied it to existing experimental data.

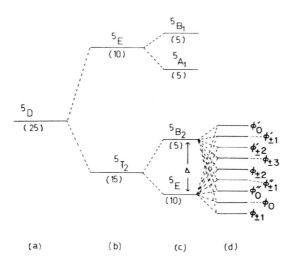

FIG. 27. Stark-pattern for $Fe^{2+}$ ion: (a) Free-ion, (b) Octahedral field, (c) Axial-field (Tetragonal), (d) Spin-orbit coupling. The values in parentheses represent the total degeneracy.

<u>Tutton Salts:</u> A large number of crystals belonging to Tutton series has been studied both at room and low temperatures. The room temperature anisotropy data is summarized in Table 7. The anisotropies are observed to be, as expected, very large, the maximum crystal anisotropy being about 19% of the average susceptibility. An interesting point is the variation in anisotropy among these isomorphous crystals, which indicates the sensitive nature of anisotropy to small changes in the packing of the lattice caused presumably by the cations.

Bose [122] and Guha Thakurta and Mukhopadhyay [54] have measured the anisotropy of $(NH_4)_2Fe(SO_4)_2.6H_2O$ and $K_2Fe(SO_4)_2.6H_2O$ down to about 70°K. Data on these two crystals have been explained on theoretical models based on crystal field and molecular orbital approximations. We shall discuss them below in some more detail.

(a) $(NH_4)_2Fe(SO_4)_2.6H_2O$:- The variation of its principal molecular moments with temperature is shown in Fig. 28. Bose et al. [117] observe that the data over the temperature range can be explained with $\lambda = -80$ cm$^{-1}$ and $\Delta$ varying from 650 cm$^{-1}$ and 400 cm$^{-1}$ between 300° and 90°K. König and Chakravarty [121], however, observed that with $\lambda = -100$ cm$^{-1}$, a satisfactory fit was observed with $\Delta \simeq 1100$ and the orbital reduction parameter (k) varying between 0.8 and 0.6.

(b) $K_2Fe(SO_4)_2.6H_2O$:- The temperature variation of principal moments of this crystal is similar to that of the ammonium derivative, though the

TABLE 7

| Crystal | Temp. (°K) | $\chi_1-\chi_2$ | $\chi_1-\chi_3$ | $\Psi°$ | Ref. |
|---|---|---|---|---|---|
| $Fe(SO_4).R_2(SO_4).6H_2O$ | | | | | |
| R = K | 303 | 1841 | -314 | 58.8 | 5 |
| | 300 | 1520 | -320 | 55 | 93 |
| | 300 | 1696 | -279 | 57.9 | 30 |
| NH$_4$ | 303 | 2582 | 213 | 54.3 | 5 |
| | 300 | 2179 | 180.1 | 53.4 | 30 |
| Rb | 303 | 2270 | 150 | -27.3 | 8 |
| Cs | 303 | 2540 | 570 | -23.9 | 8 |
| Tl | 303 | 2220 | 30 | -26.2 | 8 |
| $Fe(SeO_4).Tl(SeO_4).6H_2O$ | 303 | 2230 | 360 | -28.5 | 8 |

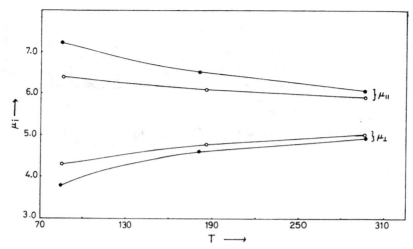

FIG. 28. Temperature variation of principal molecular moments of ferrous ammonium sulfate (●) and ferrous potassium sulfate (0). The data are taken from Ref. 122.

anisotropy of this crystal is smaller than the ammonium one. König and Chakravarty [121] find that $\lambda = -100$ cm$^{-1}$, $\Delta = 500$ cm$^{-1}$, and k = 0.8 gives good fit at room temperature. However, the data at lower temperatures were fitted by changing the values of $\Delta$ and k.

The values of k deduced by König and Chakravarty in the above two compounds are very low at lower temperatures. For example, a value of k = 0.6 or 0.5 for these ionic compounds does not seem very realistic if the conventional significance of this parameter is assumed.

Diluted Tutton Salts:- Joglekar [123] studied the effect of magnetic dilution on the anisotropy of $(NH_4)_2Fe(SO_4)_2 \cdot 6H_2O$. As we have remarked earlier, magnetic anisotropy is very sensitive to any small change in the long-range packing of the lattice, and the measurement of magnetic anisotropy of the diluted crystals is expected to be interesting. Joglekar's results are summarized in Table 8. The effect on the anisotropy of the same diluent is observed to be small. However, there is a large effect on the anisotropy when the host lattice is changed from Zn to Mg, and probably reflects the small difference in the atomic radii of these host atoms. Joglekar's study is, unfortunately, limited to a very small range of dilution. It would be interesting to study the effect of more extensive substitution on the anisotropy, particularly using the isomorphous Zn compound as zinc has almost the same atomic radius as the Fe atom.

Ferrous Fluosilicate Hexahydrate:- Because of the special crystallographic features (see Section III A), this crystal has been studied over a very wide

TABLE 8

| Crystal[a] | $\chi_1 - \chi_2$ | $\chi_1 - \chi_3$ | $\Psi$ |
|---|---|---|---|
| A + 0.0  B$_1$ | 2380 | 190 | +53° |
| A + 1.02 B$_1$ | 2340 | 110 | 52° |
| A + 1.83 B$_1$ | 2350 | 110 | 52° |
| A + 0.89 B$_2$ | 2650 | 210 | 53° |
| A + 2.74 B$_2$ | 2560 | 200 | 53° |

[a]A = $Fe(NH_4)_2(SO_4)_2.6H_2O$; B$_1$ = $Zn(NH_4)_2(SO_4)_2.6H_2O$; and B$_2$ = $Mg(NH_4)_2(SO_4)_2.6H_2O$.

range of temperature. Majumdar and Datta [124] have reported the anisotropy data between 300° and 90°K. Jackson [125] has measured its principal susceptibilities down to about 1.5°K. At room temperature the crystal shows high anisotropy, about 25% of the average susceptibility, and is of the same order of magnitude as in Tutton salts. It may be mentioned here that neutron diffraction studies [78] on this crystal shows that the water octahedron is elongated along the [111] direction, while the magnetic measurements show that $K_\perp > K_{||}$. A simple point charge model will predict $K_{||} > K_\perp$.

The temperature variation of principal moments is interesting (Fig. 29). It is observed that $\mu_{||}$ almost approaches zero as the temperature decreases, while $\mu_\perp$ first increases slightly with decreasing temperature and reaches a value of about 5.8 BM at about 20°K, and then decreases sharply to 4.5 BM at about 4°K. These behaviors of the principal moments have been very satisfactorily explained by many workers on the crystal field and molecular orbital approximations, assuming $^5B_2$ level lying below $^5E$ in the axial part of the Stark pattern (Fig. 27). Under such circumstances the spin-orbit coupling leaves a nonmagnetic level ($\phi_o$) lowest in the manifold of spin levels. The separation of this level ($\phi_o$) from the next excited "magnetic" level ($\phi_{\pm 1}$) has been calculated by various workers and is deduced to be about 10 cm$^{-1}$. König and Chakravarty have deduced $\Delta = -760$ cm$^{-1}$ at 77.3°K. This is in good agreement with the value resulting from Mossbauer spectroscopy [126].

Ferrous Phthalocyanine:- The electronic structure of ferrous phthalocyanine has been the subject of long discussion. The variation of its average and principal magnetic moments down to helium temperature is shown in Fig. 30, and it has been recently explained on the assumption of a large zero-field splitting of the ground spin triplet deriving from a spin

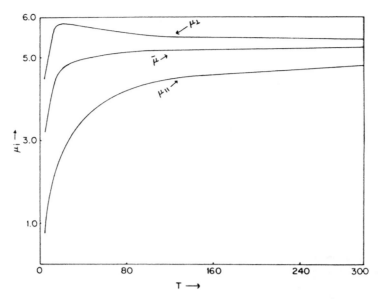

FIG. 29.  Temperature variation of average and principal moments (Ref. 124 and 125) of $FeSiF_6 \cdot 6H_2O$.

configuration of S=1(126a, 127).  This interpretation is consistent with the magnetic anisotropy of the crystal of this compound measured recently by Martin and Mitra [127].  Preliminary calculations suggest that an orbitally non-degenerate spin quintate may be the ground term deriving from a S=1 configuration [127].  A large zero-field splitting is deduced under such situation.  The magnetization curve obtained at 1.5°K supports such conclusion.

$FeF_2$:  This compound has rutile-type structure like $MnF_2$.  The magnetic anisotropy of its single crystals was measured by Stout and Matarrese [12] between room temperature and about 14°K.  At room temperature the crystal shows high anisotropy, amounting to about 20% of the average susceptibility, and is of the same order as the anisotropy exhibited by magnetically dilute high spin ferrous compounds.  Between room temperature and about 100°K, the anisotropy rises with decreasing temperature, reaching a maximum at about 95°K, and then drops rapidly and becomes negative and assumes a very large value in the hydrogen temperature range (see Fig. 31).  The temperature variation of anisotropy of $FeF_2$ is thus very similar to $MnF_2$.  The variation of its principal susceptibilities with temperature is shown in Fig. 32.  It indicates that the spins are aligned antiparallel along the c-axis of the crystal.  Recently Stout et al. [128] have extended the measurement on this crystal down to about 1°K.  Between 1° - 4°K, they find $\chi_{||} = 0.56 \pm .01) \times 10^{-3}$ and $\chi = 16.47 \times 10^{-3}$ cm$^3$ mole$^{-1}$.

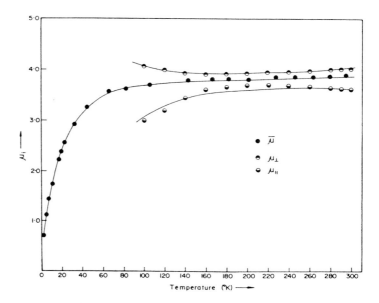

FIG. 30. Temperature variation of average and principal magnetic moments of ferrous phthalocyanine solid curve is calculated with $g_{||} = g_{\perp}$ = 2.74 and zero-field splitting parameter (D) = 64 cm$^{-1}$

from the value of $\chi_{||}$, they deduce the value of effective spin-orbit coupling parameter to be $\lambda =$ -116±25 cm$^{-1}$, which indicates almost an ionic metal-ligand bond in this compound.

A point of interest is that the anisotropy at low temperatures varies quartically with temperature, in contrast to Kubo's spin wave theory [129] which predicts that the anisotropy should vary as $T^2$. The discrepancy is perhaps due to the large anisotropy produced by the ligand field which was neglected by Kubo.

## 7. d$^7$ (Co$^{2+}$)

In this configuration a very large amount of data exists for divalent cobalt compounds with different stereochemistries. In particular the octahedrally and tetrahedrally coordinated compounds have been most extensively studied. Recently some square planar compounds have also been studied.

(a) Octahedrally Co-ordinated Compounds. The free ion ground term of

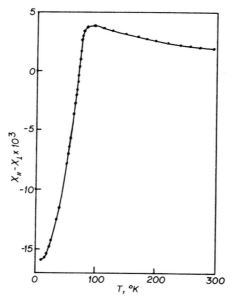

FIG. 31. Temperature variation of magnetic anisotropy of $FeF_2$ (Ref. 12).

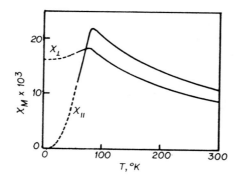

FIG. 32. Temperature variation of principal susceptibilities of $FeF_2$ (Ref. 12).

the cobaltous ion is $^4F$. In the octahedral ligand field, this term is split into two triplets ($^4T_1$ and $^4T_2$) and a singlet ($^4A_1$) in the increasing order of energy. The axial field splits the ground triplet ($^4T_1$) into a doublet and a singlet separated by an energy of the order of $10^2$ $cm^{-1}$. The further splitting and intermixing by spin-orbit coupling is shown in Fig. 32a.

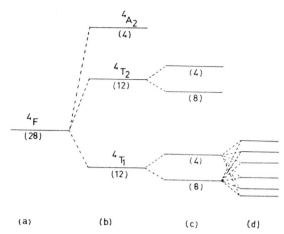

FIG. 32a. Stark pattern of cobaltous ion under a distorted octahedral ligand field: (a) Free-ion, (b) Octahedral field, (c) Axial field, (d) Spin-orbit coupling. The values in parentheses represent total degeneracy.

Since the ground state in the octahedral field is an orbital triplet, a large anisotropy is expected. The temperature dependence of the principal moments is expected to be interesting in revealing the nature of the ligand field.

Tutton Salts: The single crystals of the compounds belonging to the Tutton series have been very extensively studied by various workers. The room temperature data on these crystals are summarized in Table 9. The anisotropies, as expected, are observed to be very high. An interesting point is the large discrepancy in the data on $(NH_4)_2Co(SO_4)_2.6H_2O$ of Jackson [130] and all other later workers. As pointed out by Krishnan et al. [5], the discrepancy was due to the method employed by Jackson, which is not very suitable for monoclinic crystals. The wrong sign of $\psi^\circ$ observed by Jackson for the $K_2Co(SO_4)_2.6H_2O$ is also due to this reason.

As in the ferrous Tutton salts, the anisotropies are very sensitive to small changes in the packing of the crystal lattice. This is clear from the large variation in anisotropy in this isomorphous series due to the different cations.

The temperature variation of anisotropies and the principal susceptibilities of a number of compounds of this series has been measured. Jackson [130] reported the data on the $Co(NH_4SO_4)_2.6H_2O$ crystals down to about 14°K. Unfortunately his data are in error. Bose [131] has measured the anisotropies of potassium and ammonium sulfate derivatives and the cobalt ammonium fluoroberylate between room temperature and about 80°K. Guha [95] has reported very detailed magnetic anisotropy and principal

TABLE 9

| Crystal | Temp. (°K) | $\chi_1 - \chi_2$ | $\chi_1 - \chi_3$ | $\Psi°$ | Ref. |
|---|---|---|---|---|---|
| $Co(SO_4).R_2(SO_4).6H_2O$ | | | | | |
| R = K | 303 | 2532 | 1832 | -15.5 | 5 |
| | 300 | 2370 | 1670 | +20.2 | 130 |
| | 300 | 2432 | 1772 | -15.6 | 30 |
| $NH_4$ | 303 | 3023 | 1541 | -43.2 | 5 |
| | 300 | 1530 | 240 | -31.18 | 130 |
| | 300 | 3042 | 1543 | -43.3 | 30 |
| | 300 | 3200 | 1490 | -13 | 93 |
| Rb | 303 | 2440 | 1220 | -24.5 | 8 |
| Cs | 303 | 2600 | 500 | -44.9 | 8 |
| Tl | 303 | 2010 | 960 | -37.8 | 8 |
| $Co(SeO_4).R_2(SeO_4).6H_2O$ | | | | | |
| R = K | 303 | 3130 | 3090 | -13.6 | 8 |
| $NH_4$ | 303 | 3120 | 1980 | -44.4 | 8 |
| Rb | 303 | 2900 | 2250 | -20.0 | 8 |
| Tl | 303 | 2450 | 1950 | -31.5 | 8 |
| $Co(NH_4)_2(BeF_4)_2.6H_2O$ | 303 | 2930 | 1550 | -38.1 | 8 |

susceptibility data on a number of crystals in the liquid nitrogen range. Guha Thakurta and Mukhopadhyay [54] have recently remeasured the anisotropy of the cobalt ammonium and potassium sulfates in the liquid nitrogen range. The temperature variation of principal moments for a typical case is shown in Fig. 33.

Detailed theory for the magnetic susceptibility and anisotropy of the octahedrally coordinated cobalt compounds was worked out by Schlapp and Penney [132], Uryû [133] and by Bose et al. [134]. With a purely octahedral field, Schlapp and Penney find that the principal moments are identical, and they fall with the fall of temperature, slowly at first and quickly later. With small rhombic field the effective moments become naturally different and they decrease as the temperature is lowered, two of them in the same

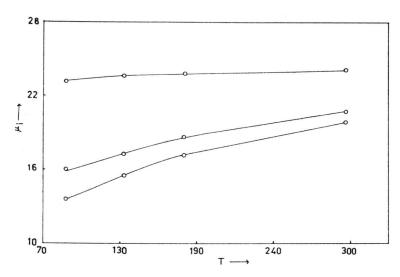

FIG. 33. Temperature variation of principal crystalline moments of $Rb_2Co(SeO_4)_2.6H_2O$ (Ref. 95) as $\mu_1$, $\mu_3$, and $\mu_2$.

way as with a purely cubic field, while the third having the largest value decreases slowly. If the rhombic field is very strong, the principal moments further separate out, and the greater moment now increases with the fall in temperature while the smaller one decreases. Thus the temperature variation of principal moments offers a convenient way to understand the nature and strength of the ligand field in these compounds. The temperature variation of the principal moments of the Tutton salts (see Fig. 33) indicates that the rhombic field in these compounds is smaller. The striking case of strong rhombic field will be discussed later.

Bose et al. [134] have discussed the temperature variation of anisotropy of some compounds. They deduce a small variation in the tetragonal field splitting with temperature.

Cobalt Acetate Tetrahydrate:- The crystal belongs to the monoclinic system with two molecules in the unit cell [135]. Each cobalt ion is surrounded octahedrally by four water molecules and two oxygens of acetate groups. The cobalt-water distances are 2.11 and 2.06 A and the cobalt oxygen 2.12 A. The magnetic anisotropy of this crystal has been reported by Guha [136] and by Mookherji and Mathur [137]. At 300°K, Mookherji and Mathur report,

$$\chi_1 - \chi_2 = 5450, \quad \chi_1 - \chi_3 = 248, \quad \overline{\chi} = 13,480,$$

thus indicating that the anisotropy of this compound is much greater than that of the Tutton series compounds. The temperature variation of its principal moments is shown in Fig. 34. It is observed that while $\mu_{||}$ decreases with the decrease in temperature, $\mu_{\perp}$ increases. Guha [136] has explained this behavior very successfully in terms of the theory of Schlapp and Penney [132] which indicates, as mentioned earlier, such a behavior for a strong rhombic crystal field. The large anisotropy of the crystal and its molecular geometry are also consistent with this conclusion.

$CoSO_4 \cdot 7H_2O$: The crystal belongs to the monoclinic system. The structure is assumed to be approximately octahedral with six water molecules surrounding the cobalt ion. The room temperature data were reported by Krishnan et al. [6]. At 30°C, they report,

$$\chi_1 - \chi_2 = 1928 \quad , \quad \chi_1 - \chi_3 = 1094, \quad \psi = -54.4°.$$

Guha [95] studied the temperature dependence of its principal anisotropies. The variation of its principal crystal moments are shown in Fig. 35. The

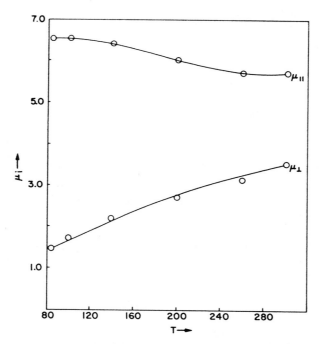

FIG. 34. Temperature variation (°K) of principal molecular moments of cobalt acetate tetrahydrate (Ref. 136 and 137)

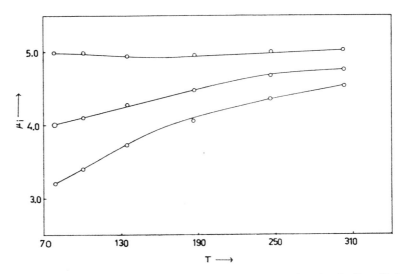

FIG. 35. Variation of principal crystalline moments of cobalt sulfate heptahydrate with temperature (°K) (Ref. 95) as $\mu_1$, $\mu_3$, and $\mu_2$.

trend of their temperature variation suggests that the rhombic field in this crystal is probably larger than that in the Tutton salts.

$CoSeO_4.6H_2O$: This crystal also belongs to the monoclinic system, and is supposed to have an approximate octahedral coordination of six water molecules. At 30°C, Krishnan and Mookherji [8] report,

$$\chi_1 - \chi_2 = 3950, \quad \chi_1 - \chi_3 = 2390 \quad \psi = 60.5°.$$

The temperature variation of its principal moments is similar to $CoSO_4$ .$7H_2O$. This indicates that the nature and strength of the ligand field in these two compounds is probably very similar.

Cobalt Fluosilicate Hexahydrate: Majumdar and Data [138] have done very detailed studies on this crystal. The crystal is highly anisotropic. The temperature variation of its anisotropy is interesting (see Fig. 36). A reversible phase transformation was observed in the crystal at about 246°K. Below the transition temperature, a large anisotropy develops in the symmetry plane, indicating a lowering in the symmetry of the crystal. Low temperature X-ray studies support this observation [139, 140].

It is interesting to note that no such phase transformation is observed in the diluted compound (Fig. 36).

The symmetry of the ligand field in this crystal at room temperature is

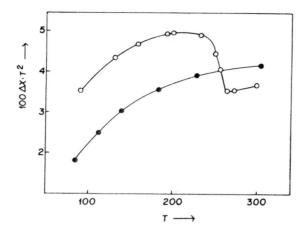

FIG. 36. Temperature variation (°K) of magnetic anisotropy of concentrated (open circles) and diluted (solid circles) cobalt fluosilicate hexahydrate (Ref. 138).

trigonal. Bose et al. [141] have deduced explicit expressions for the principal susceptibilities of the cobalt ion under a trigional ligand field.

Cobalt Hydrofluoride Hexahydrate: Dutta Roy and Ghosh [142] have studied the magnetic properties of this crystal between room temperature and 90°K. The behavior of this crystal is very similar to $CoSiF_6 \cdot 6H_2O$. A reversible phase transition at about 246°K was observed. This is accompanied by violent changes in the crystal and an anisotropy appears in the symmetry plane.

(b) Tetrahedrally Co-ordinated Compounds. As is well known, the cubic part of the Stark pattern in a tetrahedrally coordinated ion is inverted with respect to that of the corresponding octahedral compounds. Accordingly the $^4F$ free ion ground term of the $Co^{2+}$ ion is split in the tetrahedral ligand field in such a way that an orbitally non-degenerate $^4A_2$ state lies lowest. The next excited state is an orbital triplet ($^4T_2$) lying some 3000 cm$^{-1}$ above it. The axial part of the ligand field has no effect on the ground state, but it splits the excited $^4T_2$ state into a singlet and a doublet to the extent of a few hundred cm$^{-1}$. The spin orbit coupling partially lifts the spin degeneracy of the ground state, and the detailed splitting is shown in Fig. 37. The ground term for the tetrahedral Co(II) ion being an orbital singlet, its magnetic properties is expected to be very different from those of octahedral ones [143, 144]. The magnetic anisotropy is expected to be fairly small (as contrasted to the highly anisotropic octahedral ones) and the average moment closer to the spin only value. However, the details of the magnetic behavior may not obey such a simple relationship because

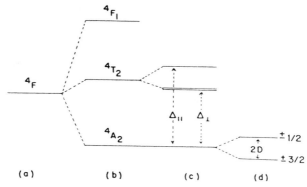

FIG. 37. Stark pattern of cobaltous ion in a distorted tetrahedral ligand field: (a) Free-ion, (b) Tetrahedral field, (c) Axial field, (d) Spin-orbit coupling.

of the much reduced overall ligand field splitting and the nature of chemical bonding.

Magnetic anisotropies of a few tetrahedral cobaltous compounds have been studied by different workers. $Cs_3CoCl_5$ has been studied in great details. $Cs_2CoCl_4$ and $K_2Co(NCS)_4$ have also been investigated over a temperature range. We shall now discuss them individually.

$Cs_3CoCl_5$: The single crystals of this compound belong to a tetragonal system with four magnetically equivalent molecules in the unit cell [145]. Hence the molecular and crystalline magnetic ellipsoids are identical, i.e., $K_{||} = \chi_{||}$ and $K_{\perp} = \chi_{\perp}$.

The magnetic anisotropy of this crystal was first reported by Krishnan and Mookherji [146, 8]. At 30°C, they find,

$$\chi_{||} - \chi_{\perp} = 650$$

which is only 7% of the average susceptibility as against about 30% of the octahedral ones. The difference between their average moments is not so marked. This indicates the usefulness of this technique in deciding the stereochemistry in favorable cases.

Bose et al. [147, 148] and Figgis et al. [149] have independently reported the data at low temperatures. The ligand field parameters deduced by Bose et al. [148] are in good agreement with the ESR [150] and find structure optical spectra [151]. The metal-ligand bond is observed to be fairly ionic. van Stapele et al. [152] have reported the measurement of its principal susceptibilities down to helium temperatures. The variation of its principal susceptibilities shown in Fig. 38.

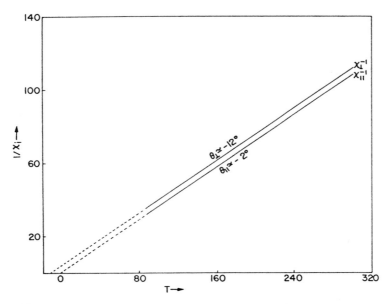

FIG. 38. Variation of principal susceptibilities of $Cs_3CoCl_5$ with temperature (°K) (Ref. 147).

$Cs_2CoCl_4$: This crystal belongs to orthorhombic system with two inequivalent molecules in the unit cell [153]. The magnetic anisotropy of this crystal was measured by Krishnan and Mookherji [154] and by Figgis et al. [149]. But the data of these two groups of workers appear to be in error. Mitra [155] has recently reported the correct data. At 20°C,

$$\chi_b - \chi_c = 375, \quad \chi_a - \chi_c = 268, \quad \chi_b - \chi_a = 115.$$

The molecular anisotropy is calculated to be,

$$K_{||} - K_{\perp} = 645,$$

which is similar to that observed for $Cs_3CoCl_5$.

$K_2Co(CNS)_4 \cdot 4H_2O$: This crystal also belongs to the orthorhombic system. The crystal anisotropy was reported by Figgis et al. [149]. The molecular anisotropy calculated by them appears to be very small. For example, at 280°K, they find

$$K_{||} - K_{\perp} = 118,$$

whereas for $Cs_3CoCl_5$ and $Cs_2CoCl_4$, the corresponding value lies between 640 and 650. It is difficult to understand the reason for such a large reduction in the anisotropy.

(c) Planar Compounds. Cobalt phthalocyanine which has almost a square planar stereochemistry (see Section III, A) has been studied over a temperature range [15]. The crystal is highly anisotropic and has one unpaired spin. The temperature variation of principal moments is shown in Fig. 39. It is observed that $\mu_\perp > \mu_{||}$, and that while $\mu_{||}$ is almost temperature-independent, $\mu_\perp$ decreases with the fall of temperature. It has been shown that these features are in good agreement with the unpaired electron being in the $d_{z^2}$ orbital. This indicates the usefulness of the technique in deciding the ground state of such systems.

## 8. $d^8$ ($Ni^{2+}$)

In this configuration the data exist mainly for the octahedrally co-ordinated Ni(II) compounds. Some data has also been reported for square planar compounds, but, as expected, they are all diamagnetic and hence will be discussed in the sub-section dealing with diamagnetic anisotropies.

The free ion ground state of the $Ni^{2+}$ ion is $^3F$. In the octahedral ligand field, this level splits up into a singlet ($^3A_{2g}$) and two triplets ($^3T_{2g}$ and $^3T_{1g}$) in the ascending order of energy. The splitting of the ground and excited states due to the axial field and spin-orbit coupling is shown in Fig. 40. The ground state in the octahedral field being an orbital singlet, the crystals are expected to be feebly anisotropic, and the principal susceptibilities obeying Curie law very closely.

Tutton Salts: The anisotropy of the crystals belonging to the Tutton series

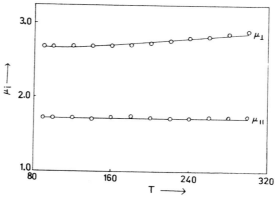

FIG. 39. Variation of principal molecular moments of cobalt phthalocyanine with temperature ($^\circ K$) (Ref. 15).

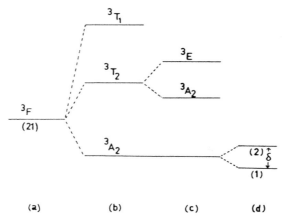

FIG. 40. Stark pattern of Ni(II) ion in an approximate octahedral ligand field: (a) Free ion, (b) Octahedral field, (c) Axial field, (d) Spin-orbit coupling.

have been very extensively studied by a large number of workers. The room temperature data on the crystals are summarized in Table 10. The anisotropies are fairly small, about 7-10% of the average susceptibility. An interesting point is that the anisotropy is smallest when the cation is $(NH_4)^+$.

Temperature variation of anisotropy of all the compounds listed in Table 10 have been measured by Guha [95] and Bose et al. [156]. The principal susceptibilities obey Curie law, and its temperature variation is shown for a typical case in Fig. 41. Guha has successfully fitted the anisotropy data with the theory of Schlapp and Penney [132], and obtained a satisfactory agreement with temperature-independent parameters. Bose et al. [156] and Bose and Chatterji [157] however deduce the variation of ligand field with temperature.

$\underline{NiSO_4.7H_2O}$: This crystal belongs to orthorhombic system, and the Ni(II) ion has an approximate octahedral symmetry with six water molecules. The symmetry of the crystal field is approximately tetragonal. The magnetic anisotropy of this crystal has been measured by Krishnan et al. [5], Mookherji [158] and Bose et al. [156]. The agreement is not bad. At 300°K, Bose et al. calculated,

$$K_\perp - K_{||} = 255.3,$$

which is of the same order as observed in Tutton salts. Temperature variation of its anisotropy down to about 90°K has been reported by Bose et al. [156].

TABLE 10

| Crystal | $\chi_1-\chi_2$ | $\chi_1-\chi_3$ | $\Psi°$ | Ref.[a] |
|---|---|---|---|---|
| $Ni(SO_4).R_2(SO_4).6H_2O$ | | | | |
| R = K | 158 | 155 | -12.5 | 5 |
| | 160 | 180 | -17.0 | 93 |
| | 131.2 | 130.5 | -11.4 | 30, 156 |
| NH$_4$ | 110 | 106 | -13.9 | 5 |
| | 60 | 40 | -7 | 93 |
| | 93.8 | 91.5 | -13.4 | 30, 156 |
| Rb | 144 | 137 | -11.0 | 8 |
| | 143.3 | 136.8 | -11.4 | 30, 156 |
| Cs | 134 | 127 | -10.7 | 8 |
| | 132.2 | 126.1 | -10.1 | 30, 156 |
| Tl | 114 | 108 | -20.9 | 8 |
| | 113.7 | 107.4 | -10.1 | 30, 156 |
| $Ni(SeO_4).R_2(SeO_4).6H_2O$ | | | | |
| R = K | 146 | 146 | -13.0 | 8 |
| | 127.3 | 107.2 | -16.5 | 30, 156 |
| NH$_4$ | 116 | 96 | -27.9 | 8 |
| | 92.6 | 79.6 | -22.1 | 30, 156 |
| Rb | 157 | 147 | -12.9 | 8 |
| | 136.5 | 117.0 | -12.7 | 30, 156 |
| Cs | 164 | 135 | -17.6 | 8 |
| | 131.5 | 110.7 | -11.3 | 30, 156 |
| Tl | 123 | 113 | -17.1 | 8 |
| | 122.5 | 116.5 | -17.1 | 30, 156 |
| $NiR_2(BeF_4)_2.6H_2O$ | | | | |
| R = K | 132.1 | 128.5 | -13.4 | 30, 156 |
| NH$_4$ | 107 | 106 | -14.3 | 8 |
| | 105.8 | 101.1 | -12.5 | 30, 156 |
| Rb | 122.3 | 118.1 | -10.8 | 30, 156 |
| Cs | 117.3 | 115.2 | -9.1 | 30, 156 |
| Tl | 121.0 | 117.9 | -11.1 | 30, 156 |

[a]Data of Refs. 30 and 156 refer to 300°K and that of Refs. 5 and 8 to 303°K.

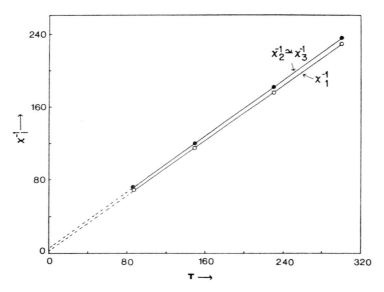

FIG. 41. Temperature variation (°K) of principal susceptibilities of $K_2Ni(SeO_4)_2.6H_2O$ (Ref. 95).

$NiSO_4.6H_2O$ and $NiSeO_4.6H_2O$:- These two crystals are isostructural and belong to the tetragonal system [159]. The Ni(II) ion is octahedrally coordinated to six water molecules and the symmetry of the ligand field is tetragonal. The magnetic anisotropy of the sulfate derivative has been studied by a number of workers. Except the original data of Krishnan et al. [5], the data of later workers [158, 156, 10] agree very closely (see Table 11). The temperature variation of its anisotropy has been studied by Krishnan et al. [9] between room temperature and about 140°K, by Bose et al. [156] down to 90°K, and by Stout and Griffel [10] between 293° and about 14°K. The data of Stout and Griffel over this wide range of temperature can be represented by,

$$(\chi_\perp - \chi_{||})T^2 = 1.686 + 1.502 \times 10^{-2}\,T + 13.7 \times 10^{-6}\,T^2.$$

Magnetic anisotropy of $NiSeO_4.6H_2O$ has been measured by Krishnan and Mookherji [8] at room temperature. Bose et al. [156] report the data at low temperatures. The behavior is very similar to the sulfate one.

It is interesting to note that the molecular anisotropy of the hexahydrated nickel sulfate and selenate is about 40% greater than that of the $NiSO_4.7H_2O$ or nickel Tutton salts. Since all these crystals have a very similar microsymmetry, this large difference is rather surprising.

TABLE 11

| Crystal | Temperature (°K) | $\chi_\perp - \chi_{||}$ | Ref. |
|---------|------------------|--------------------------|------|
| NiSO$_4$. 6H$_2$O | 303 | 109 | 5 |
| | 300 | 83 | 158 |
| | 293.7 | 84.4 | 10 |
| | 300 | 80.8 | 30 |
| NiSeO$_4$. 6H$_2$O | 303 | 90 | 5 |
| | 300 | 86.6 | 30 |

Hexahydrated Nickel Fluosilicate and Chlorostannate:- As mentioned earlier (see Section III, A), the Ni(II) has an approximate octahedral coordination. The symmetry of the ligand field is trigonal.

Datta [30] has reported the data on NiSiF$_6$.6H$_2$O at room temperature. Bhattacharya and Majumdar [160] have reported the data on both these crystals down to 90°K. At room temperature, the crystals are very feebly anisotropic, about 0.3% of the mean susceptibility. The principal susceptibilities obey Curie law very closely. The zero field splitting is estimated to be very small, about 0.50 cm$^{-1}$.

## 9. d$^9$ (Cu$^{2+}$)

In this configuration a huge amount of data has been reported on the divalent copper compounds having different stereochemistries. We shall discuss them below.

(a) Octahedrally Coordinated Compounds. In the octahedral ligand field the free ion ground term of the divalent copper ion $^2$D is split into an orbital triplet and an orbital doublet, the doublet lying some 12,000 cm$^{-1}$ below the triplet. The axial component (tetragonal) of the main octahedral field splits the ground doublet into two Kramers doublets to the extent of a few 10$^3$ cm$^{-1}$ (see Fig. 42). The spin orbit coupling is, however, fairly strong ($\lambda_{\text{free ion}}$ = -829 cm$^{-1}$). The anisotropy is, therefore, expected to be large. An interesting situation arises when the axial field is trigonal which cannot remove the degeneracy of the ground $^2$E state. Such a situation obtains, for example, in CuSiF$_6$.6H$_2$O, and has been discussed in great detail by Öpik and Pryce [161].

Tutton Salts. The single crystals belonging to this series have again been

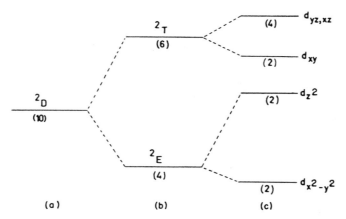

FIG. 42. Stark pattern of Cu(II) ion in a distorted octahedral ligand field: (a) Free-ion, (b) Octahedral field, (c) Axial (tetragonal) field.

studied very extensively by a large number of workers. The room temperature data are collected in Table 12. The anisotropies are observed to be large, about 30% of the average susceptibility. Here again a large difference in the anisotropy is observed among these isomorphous crystals.

The temperature variation of the magnetic anisotropy of all the compounds listed in Table 12 was studied down to about 80°K by a number of workers [162-164, 95]. Bose et al. [162] have done the most extensive investigation on these compounds. Among all these compounds, the temperature variation of anisotropy of copper ammonium sulfate is somewhat anomalous and is shown in Fig. 43, along with that of copper potassium sulfate which is a typical representation of other compounds of this series. The explanation of this anomalous behavior of the copper ammonium sulfate may lie in its structure which reveals an abnormal hydrogen bonding in this compound [76].

Bose et al. [162] observe that while fitting their molecular anisotropy data, calculated on the assumption of uniaxial symmetry, over the temperature range, a small but significant $1/T^2$ term persists. Since each orbital level in the Stark pattern has only a twofold Kramers degeneracy, a $1/T^2$ term is not ordinarily expected in these magnetically dilute compounds. They attribute this discrepancy to a continuous thermal variation of the ligand field.

$CuSO_4.5H_2O$. The magnetic anisotropy of this crystal has been very extensively studied by Krishnan and Mookherji [43-45]. The anisotropies obey

TABLE 12

| Crystal | $\chi_1-\chi_2$ | $\chi_1-\chi_3$ | $\Psi°$ | Ref.[a] |
|---|---|---|---|---|
| $Cu(SO_4).R_2(SO_4).6H_2O$ | | | | |
| R = K | 367 | 74 | -77.8 | 5 |
| | 322.3 | 72.4 | -77.4 | 30, 162 |
| NH$_4$ | 300 | 62 | +77 | 5 |
| | 285.7 | 67.6 | +75 | 30, 162 |
| Rb | 350 | 100 | -70.7 | 8 |
| | 329.7 | 94.2 | -69 | 30, 162 |
| Tl | 283 | 71 | -72.6 | 8 |
| | 302.5 | 84.9 | -72.1 | 30, 162 |
| Cs | 339 | 114 | -71.9 | 8 |
| | 329.7 | 111.2 | -71.8 | 30, 162 |
| $Cu(SeO_4).R_2(SeO_4).6H_2O$ | | | | |
| R = K | 367 | 95 | -53.5 | 8 |
| | 340.3 | 111.8 | +74.9 | 30, 162 |
| NH$_4$ | 253 | 118 | +66.7 | 8 |
| | 322.6 | 111.6 | +68.0 | 30, 162 |
| Rb | 354 | 78 | -73.3 | 8 |
| | 326 | 77.1 | -78.8 | 30, 162 |
| Tl | 328 | 22 | -72.3 | 8 |
| | 305.1 | 71.3 | -77.6 | 30, 162 |

Curie law very closely. The principal susceptibilities in their range of study can be represented by,

$$\chi_{||} = 0.572/(T + 4.5) \quad \text{and} \quad \chi_{\perp} = 0.399/(T - 2.0).$$

It may be noted here that the crystal is nearly uniaxial as far as its magnetic properties are concerned. The crystal shows anisotropy of the same order as the Tutton salts. Some other aspects of this crystal have been discussed in Sections II, A and II, C.

$CuSeO_4.5H_2O$. Mookherji and Tin [165] have reported the data on this crystal at room temperature. The principal anisotropies are very similar to that of $CuSO_4.5H_2O$.

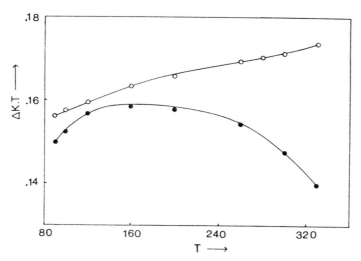

FIG. 43. Temperature variation (°K) of molecular anisotropy of copper ammonium sulfate (solid circles) and copper potassium sulfate (hollow circles) hexahydrates (Ref. 162).

**Copper Acetate Monohydrate:** The single crystals of this well-known dimeric compound [166-168] belong to a monoclinic system with two exchange-coupled copper dimers per unit cell. Its magnetic anisotropy was first reported by Guha [95] who showed that its average and principal susceptibilities show maxima at about 260°K. The variation of its principal anisotropies with temperature is shown in Fig. 44. Mookherji and Mathur [169] have remeasured its anisotropy in the liquid nitrogen temperature range. They find that the exchange integral is isotropic [170]. Explanation of its anisotropy data on $C_{2v}$ symmetry has been recently attempted by Bose et al. [171].

An interesting point is that although the average moment of this compound at room temperature is well below the "normal" value, its magnetic anisotropy is of the same order as that of magnetically dilute copper compounds.

**Copper Proprionate Monohydrate:** The compound shows magnetic behavior similar to copper acetate monohydrate [172]. Exchange integral is isotropic.

**Copper Formate Tetrahydrate:** The single crystal of this compound belongs to a monoclinic system and has a very unique layer structure with copper ions arrayed in sheets parallel to the (001) plane [173]. The measurement of magnetic anisotropy and principal susceptibilities is consistent with the theory of two dimensional antiferromagnetism [174]. Both $K_{||}$ and $K_{\perp}$ obey a Curie-Weiss law with almost the same Weiss constant, $\theta = -165°$ (see Fig. 45). Although the Cu-Cu distance within each layer of the crystal is

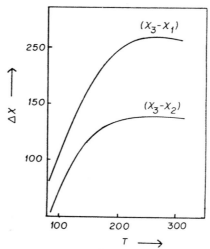

FIG. 44. Temperature variation (°K) of principal crystalline anisotropies of copper acetate monohydrate (Ref. 170).

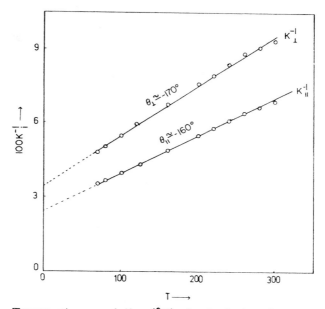

FIG. 45. Temperature variation (°K) of principal molecular susceptibilities of copper format tetrahydrate.

about 5.8 Å, the appreciable super-exchange interaction has been explained as due to the delocalized $\pi$-electrons associated with the resonating O - CH = O $\rightleftharpoons$ O = CH - O anion systems [175].

Measurement of its principal susceptibilities at temperatures below 80°K reveals some very interesting features [176, 177]. At about 62°K, all the three susceptibilities pass through broad maxima which has been suggested as indicative of a short range spin ordering. At about 17°K a very sharp maximum is observed both in $\chi_1$ and $\chi_3$, below which weak ferromagnetism develops. The transition at 17°K has been ascribed as due to some sort of cooperative phenomenon. It may be mentioned here that the (001) plane is almost isotropic down to about 60°K, below which a very large anisotropy develops in this plane. This is consistent with the expected temperature variation of $K_{||}$ and $K_{\perp}$ in this case [178, 179].

(b)  Tetrahedrally Co-ordinated Compounds. The magnetic anisotropy of tetrahedrally co-ordinated copper compounds has been recently the subject of some study. In the tetrahedral ligand field, the free ion ground term of the copper ion, $^2D$, is split into a doublet and a triplet, the triplet lying some 10,000 cm$^{-1}$ below the doublet. The axial field removes partly the orbital degeneracy of the ground triplet ($^2T_2$) and splits this level into an orbital singlet and doublet; the singlet has been observed to be lying lowest in all the compounds studied so far. The Stark pattern is shown in Fig. 46. Since the ground term is an orbital singlet, anisotropy is expected to be large. The splitting of the $^2T$ level in the axial field is observed to be very large in most of the compounds. We shall discuss below the behavior of some compounds.

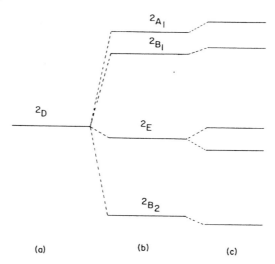

FIG. 46.  Stark pattern of tetrahedrally coordinated Cu(II) ion: (a) Free-ion, (b) Total field (Tetrahedral + Axial), (c) Spin-orbit coupling.

$Cs_2CuCl_4$: The X-ray structural analysis of this crystal by Helmholz and Kruh [180] shows that the crystal belongs to the orthorhombic system with two magnetically inequivalent molecules in the unit cell. Each copper ion is at the center of a tetrahedron formed by four chlorine atoms at its vertices at an average distance of 2.22 Å from the copper ion. The $[CuCl_4]^{2-}$ molecule is a tetragonal sphenoid having a fourfold inversion symmetry about one of the lines through central copper ion joining the middle points of two pairs of chlorine atoms. The tetrahedron is observed to be "squashed" along the tetragonal axis.

At 27°C, Mitra [181] reports,

$$\chi_c - \chi_a = 156, \qquad \chi_c - \chi_b = 341,$$

which yields $K_{||} - K_{\perp} = 525$, indicating that the anisotropy is about 30% of the average susceptibility, and is of the same order as that of octahedral copper compounds. The temperature variation of the principal moments [182] are shown in Fig. 47. The small temperature dependence of $\mu_{||}$ may probably be due to the high frequency term which is appreciable because of the small energy separation between the ground and excited states. It may partly also arise from the exchange interaction observed in this crystal from the ESR studies [183].

The fitting of the magnetic data [184] with the theory and the single crystal polarized spectra [185] indicate that the axial field splitting is very large here, $\Delta = 5000$ cm$^{-1}$, and is thus comparable to the overall cubic field splitting which is only about 9000 cm$^{-1}$. The metal-ligand bond is observed to be fairly covalent.

Gerloch [186] has recently shown that a calculation based on only $^2T_2$ level would lead to the wrong sign of anisotropy. This is an interesting

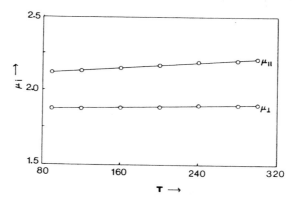

FIG. 47. Variation of principal molecular moments of $Cs_2CuCl_4$ with temperature (°K) (Ref. 147).

result and points out the usefulness of the technique. The average moment can, however, be fitted on the theory based on $^2T_2$ configuration alone.

$Cs_2CuBr_4$: Figgis et al. [187] have recently studied this crystal in the liquid nitrogen temperature range. Its behavior is similar to that of $Cs_2CuCl_4$. Lahiri et al. [188] have also studied its magnetic anisotropy.

Tetramethyl Ammonium Copper Halides: Lahiri et al. [188] have studied the anisotropy of $[(CH_3)_4N]_2CuCl_4$ and $[(CH_3)_4N]_2CuBr_4$ crystals down to liquid nitrogen temperature. The crystals also belong to an orthorhombic system and have distorted tetrahedral environment [189]. The behavior of $[(CH_3)_4N]_2CuBr_4$ is anomalous and the temperature variation of anisotropies is complicated.

Schiff-base Compounds: Figgis et al. [187] have recently reported data on some Schiff-base compounds. The temperature variation of principal moments of isopropyl derivative is rather puzzling. The increase in $\mu\|$ with the fall of temperature is difficult to understand.

(c) Five Co-ordinated Compounds. The magnetic anisotropy of some square pyramidal compounds have been studied. Under a $C_{4v}$ symmetry, the ordering of the energy levels is shown in Fig. 48. The ground state is an orbital singlet ($dx^2-y2$) and is far removed from the excited states. The magnetic anisotropy is therefore expected to be small and principal susceptibilities obeying a Curie law.

Gregson and Mitra [13] have recently reported detailed magnetic data on a series of copper dialkyldithiocarbamates. These crystals belong to monoclinic system with two sets of inequivalent molecules in the unit cell. The detailed X-ray data [191] on the diethyl derivative show that the compound is a dimer and is surrounded by an approximately square array of four sulfur atoms, with the average Cu-S distance of 2.31 Å, while the fifth

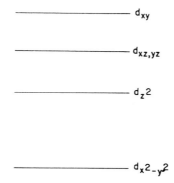

FIG. 48. Ordering of energy levels in copper dialkyldithiocarbamates (Ref. 13).

S comes from the second Cu(dedTC) unit as shown in Fig. 49. The Cu ion is thus in a square pyramidal environment with approximately $C_{4v}$ symmetry.

The room temperature values of the principal anisotropies are collected in Table 13. The anisotropies are all very low and lie between 6 and 10% of the average susceptibility. This indicates very large ligand-field energies. Indeed the quantitative analysis of the results indicated $\Delta E(^2B_1-^2B_2)$ and $\Delta E(^2B_1-^2E)$ to be, respectively, of the order of 40,000 and 25,000 $cm^{-1}$. $^2E$ is observed to be lying below $^2B_2$. The work on these compounds also reveals the importance of the magnetic anisotropy measurement in understanding the electronic structure of these low symmetry compounds.

The ligand field parameters calculated show, as expected, a very strong covalent metal-ligand bonding. The spin-orbit coupling and orbital reduction parameters show some anisotropy. The interaction within the dimer is weak.

(d)  Dodecahedrally Coordinated Compounds. According to the prediction of crystal field theory, the sign of the cubic field coefficient for tetrahedral and dodecahedral coordination is the same with the cubic field separation in the latter being twice that of the former. Hence the ordering of energy levels in the dodecahedrally coordinated compounds is expected to be similar to the tetrahedral compounds. Randie [193] has done some theoretical calculations on dodecahedral compounds. For copper compounds, he deduces that the ground state is $^2B_1$.

The only compound studied in this category is copper calcium acetate hexahydrate. The X-ray structure analysis [194] of this compound shows that the crystal belongs to the tetragonal system (space group I4/m) with four molecules in the unit cell. The copper atom occupies the $\overline{4}(S_4)$ site and is coordinated to a puckered square of four oxygen atoms. Four more oxygen atoms, two above and two below the puckered square complete the

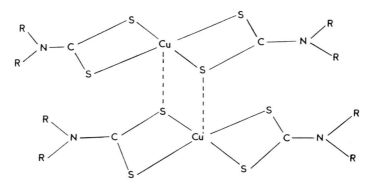

FIG. 49. Molecular geometry of copper dialkyldithiocarbamates.

TABLE 13

| Crystal | Temp. (°K) | $\chi_1 - \chi_2$ | $\chi_1 - \chi_3$ | $\Psi^\circ$ | Ref. |
|---------|-----------|-------------------|-------------------|--------------|------|
| Diethyl | 300 | 61.8 | 35.5 | -9.3 | 13 |
| Di-n-propyl | 300 | 15.4 | -9.8 | -50.2 | 13 |
| Di-n-butyl | 300 | 47.7 | -20.1 | -50.0 | 13 |

dodecahedral coordination about the copper ion (see Fig. 50). The symmetry of the polyhedron is approximately $D_{2d}$. The acetate ions act as bridging ligands between the copper and calcium atoms forming polymeric chains of alternate metal ions parallel to the c-axis of the crystal. It is then obvious that the symmetry axes of the molecules are parallel, and hence the crystalline magnetic ellipsoid is identical with the molecular ellipsoid (i.e. $\chi_{||} = K_{||}$, $\chi_\perp = K_\perp$).

Gregson and Mitra [195] have studied the magnetic anisotropy of this crystal down to about 90°K. At room temperature (292.0°K), $\chi_{||} - \chi_\perp = 475$, indicating that the anisotropy is about 30% of the mean susceptibility and is of the same order as observed in tetrahedral or octahedral copper compounds. The metal-ligand bond is observed to be almost ionic. The data indicate a small exchange interaction in this compound and are observed to be consistent with the theory of one dimensional antiferromagnetism [196].

## C. Rare Earth Compounds

As mentioned in the Introduction, a very detailed and systematic study of magnetic anisotropy has been done on the rare earth compounds. The paramagnetic behavior of these compounds is due to the unpaired 4f electrons which are shielded from the perturbation effect of the ligands by the outer-lying s- and p-electrons. There are two main differences in the problem of rare earth ions as compared to the transition metal ions. First, the shielding of the 4f electrons by the 5s and 5p electrons reduces considerably the overlap of the 4f electronic wavefunction with the ligand. A consequence of this fact is that the electrostatic ionic model appears to be reasonably more applicable to the problems of rare earth than to the transition metal compounds. It also imparts a great similarity in the physicochemical properties of these compounds. Second, in the transition metal compounds the perturbations which act upon a set of d electrons to remove the degeneracy are arranged in the order

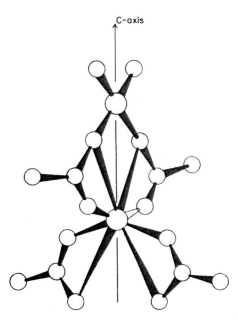

FIG. 50. Molecular structure of copper calcium acetate hexahydrate.

$$e^2/\gamma \sim V_{cubic} \overset{>}{} \lambda \quad . \quad \underline{L.S.} \sim V_{axial} \overset{>}{} kT.$$

That is, the interelectronic repulsions $(e^2/\gamma)$ are of the same order of magnitude as the cubic ligand field of an octahedral complex $(V_{cubic})$. These two quantities are, however, much larger than the effect of axial field $(V_{axial})$ and spin-orbit coupling $(\lambda \, L.S.)$. In the rare earth series ion the spin-orbit coupling coefficients lie between about 600 $cm^{-1}$ and 2500 $cm^{-1}$, and the order is,

$$e^2/\gamma > \lambda \, \underline{L.S.} > V_{cubic} \sim kT.$$

Hence, the ligand field in the rare earth ions serves to lift the $(2J + 1)$-fold degeneracy of the states of the free ion by an amount which is comparable to kT at room temperature, but which is much smaller than the separation between the states.

The fact that the 4f electrons are well shielded from the effects of their immediate environments tempts one to the conclusion that the magnetic behavior of these compounds may well approximate that of free ions.

Indeed the average moments of most rare earth compounds, except the special case of $Eu^{3+}$ and $Sm^{3+}$, are well explained on the basis of the simple expression,

$$\mu_{eff} = g\left[J\left(J+1\right)\right]^{\frac{1}{2}} \tag{83}$$

deduced for free ions. The free ion model will predict a zero magnetic anisotropy. From the measurements on some rare earth double sulfates, ethylsulfates and chlorides, Krishnan and Mookherji [8], however, discovered that most of the rare earth compounds are highly anisotropic. Their observation predicted two important conclusions: (a) that the effect of crystal field is very important in these compounds, and (b) that the symmetry of the crystal field in these compounds is not definitely cubic. The latter finding contradicted the earlier observation of Spedding and co-workers [197-199] from the optical spectra of these compounds that the symmetry of the crystal field in these compounds is cubic. Later ESR studies on the ethylsulfates [200,201] supported the findings of Krishnan and Mookherji.

The Hamiltonians for the crystal field potential of tetragonal and trigonal symmetry can be written in the following forms:

$$V_{Tet.} = A_2^{\circ}(3z^2 - \gamma^2) + A_4^{\circ}(35z^2 - 30\gamma^2 z^2 - 3\gamma^2)$$

$$+ A_6^{\circ}(231z^6 - 315z^4\gamma^2 + 105z^2 - 5\gamma^2)$$

$$+ A_4^4(x^4 - 6x^2 y^2 + y^4) + A_6^4(112z^2 - \gamma^2)(x^4 + y^4 - 6x^2 y^2)$$

$$V_{Trig.} = A_2^{\circ}(3z^2 - \gamma^2) + A_4^{\circ}(35z^4 - 30\gamma^2 z^2 + 3\gamma^2)$$

$$+ A_6^{\circ}(231z^4 - 315z^4\gamma^4 + 105z^2 - 5\gamma^6)$$

$$+ A_6^6(x^6 - 15x^4 y^4 + 15x^2 y^6 - y^6)$$

The notation $A_n^m$ is equal to the Elliot and Stevens [201] $A_n^m r^n$. Here the $A_n^m$ are the crystal field parameters which are evaluated from the experimental data. As we have remarked earlier about the similarity of the behavior of these compounds, one would also expect the crystal field parameters $A_n^m$ to vary smoothly through the series. But such is not the case. Elliot and Stevens proposed the variation of

$$A_n^m \alpha (Z - 55)^{-n}/4 \qquad (84)$$

where Z is the atomic number. A log-log plot of $A_n^m$ against (Z-55) for the ethyl sulfate or anhydrous chlorides does not yield the same slope which Eq. 84 demands. Jørgensen et al. [202], however, consider that the parameters $A_n^m$ have no real physical meaning and they just keep track of the energy differences between the seven f orbitals. It may be mentioned here that the values of $A_n^m$ can be calculated fairly accurately from optical absorption, fluorescence and ESR spectra, besides, of course, from the magnetic anisotropy.

Below, we shall discuss the data of each configuration separately.

### 1. $f^1$ ($Ce^{3+}$)

In this configuration magnetic anisotropy of some trivalent cerium compounds has been studied. The free ion ground term of Ce(III) ion is $^2F_{5/2}$. Next level $^2F_{7/2}$ is about 2250 cm$^{-1}$ above it. The spin-orbit coupling parameter for the free ion is about 643 cm$^{-1}$. At ordinary temperatures the effect of excited $^7F_{7/2}$ level is, therefore, neglected. The ground term $^2F_{5/2}$ is split into three Kramers doublets by the rhombic and cubic field with the separation comparable to kT. The detailed theory of magnetic susceptibility and anisotropy has been worked out by Elliot and Stevens [201].

Cerium Ethyl Sulphate. The magnetic anisotropy of $Ce(C_2H_5SO_4)_3.9H_2O$ was measured by Krishnan and Mookherji [8] and by Fereday and Wiersma [203]. The structure of the rare earth ethyl sulfates has been discussed in section III A, and the readers are referred to it. At room temperature (30°C),

$$\chi_{||} - \chi_{\perp} = 150,$$

amounting to about 8% of the average susceptibility. Fereday and Wiersma have reported the anisotropy data on this crystal down to about 14°K. The variation of anisotropy and $K_i^{-1}$ with temperature is shown graphically in Fig. 51. The room temperature principal susceptibilities are those reported by Krishnan and Mookherji, and the low temperature data are of Fereday and Wiersma. At about 14°K, the anisotropy increases to about 117% of the mean susceptibility, indicating thus the importance of anisotropic part of the crystal field. The principal susceptibilities obey Curie-Weiss law.

The low temperature anisotropy data of Fereday and Wiersma have been very satisfactorily explained by Elliot and Stevens [201] by taking the symmetry of the crystal field to be $C_{3h}$. The agreement with the room

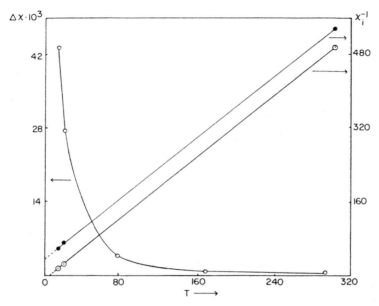

FIG. 51. Temperature variation of magnetic anisotropy and principal susceptibilities of cerium(III) ethylsulphate. Symbols: O, $\Delta\chi$ ; •, $\chi^{-1}$ ; and $\theta$, $\chi_{||}^{-1}$.

temperature data is rather poor, probably due to the neglect of excited $^2F_{7/2}$ term. Following set of parameters was deduced by them:

$$A_2^0 = -8.4 \text{ cm}^{-1} \qquad A_4^0 = -41.5 \text{ cm}^{-1}$$

$$A_6^0 = -94 \text{ cm}^{-1} \qquad A_6^6 = +1121 \text{ cm}^{-1}$$

**Cerium Double Nitrates.** The anisotropy of $Ce_2Mg_3(NO_3)_{12}.24H_2O$ and $Ce_2Zn_3(NO_3)_{12}.24H_2O$ crystals was measured by Krishnan and Mookherji [8]. These crystals belong to an isomorphous series of rhombohedral space group with one molecule in the unit cell. The room temperature anisotropy data are collected in Table 14. The crystals are fairly aniso-tropic about 18% of the average susceptibility. Mookherji [204] has reported the data at low temperatures. The temperature variation of its principal susceptibilities is very complicated (see Fig. 52). No explanation of this anomalous behavior has been offered. But see reference [336].

**Cerium Ammonium Sulfate.** The anisotropy of $Ce(NH_4)(SO_4)_2.4H_2O$ has also been reported by Krishnan and Mookherji [8]. The crystal belongs to the monoclinic system. At 30°C the anisotropies are:

$$\chi_1 - \chi_2 = 136, \quad \chi_1 - \chi_3 = 83 \quad \psi = 77.3°.$$

The anisotropy of the crystal at room temperature is comparable to that of the ethyl sulfate. The anisotropy increases at low temperatures in a similar way and becomes about 20% at about 100°K. The principal suscepti-bilities obey a Curie-Weiss law with very large Weiss constants, $\theta_i$ lying between 135° and 213° [204]. The origin of these large values of $\theta$ is diffi-cult to understand, particularly in view of the small values of $\theta$ in the ethyl sulfate crystal.

TABLE 14

| Crystal | Temp. (°K) | $\chi_\perp - \chi_{||}$ | Ref. |
|---|---|---|---|
| $Ce_2Zn_3(NO_3)_{12}.24H_2O$ | 303 | 650 | 8 |
| $Ce_2Mg_3(NO_3)_{12}.24H_2O$ | 303 | 637 | 8 |
|  | 300 | 640 | 204 |

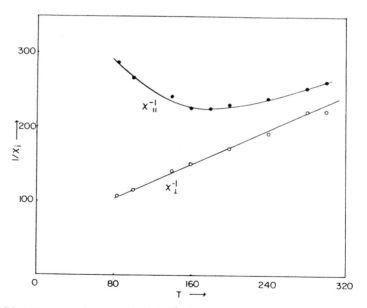

FIG. 52. Temperature variation (°K) of principal susceptibilities of $Ce_2Mg_3(NO_3)_{12}.24H_2O$ (Ref. 204).

## 2. $f^2$ $(Pr^{3+})$

In this configuration, praseodymium (III) ethyl sulfate, double nitrate, and octahydrated sulfate have been studied over a temperature range.

The free ion ground term of the $Pr^{3+}$ ion is $^3H_4$. The next excited term $^3H_5$ is about 2086 cm$^{-1}$ higher. The free ion spin-orbit coupling parameter is about 800 cm$^{-1}$, and so the effect of excited $^3H_5$ has been neglected in the theoretical calculation. In $C_{3h}$ symmetry the ground term $^3H_4$ splits into three singlets and three doublets; in the praseodymium ethyl sulfate a doublet is observed to be lowest.

Praseodymium Ethyl Sulfate: The magnetic anisotropy of this crystal has been very intensively studied. At room temperature the anisotropy is very small (Table 15), amounting to about 3% of the mean susceptibility. The variation of its anisotropy with temperature is interesting. Krishnan et al. [9] and Mookherji [205] measured its anisotropy down to about 80°K while Fereday and Wiersma [203] and Van den Handel [206] reported the data down to about 14°K. The anisotropy, at first, increases as the temperature falls and reaches maximum at about 200°K, below which it decreases very sharply and becomes zero at about 140°K. As the temperature is further lowered the anisotropy becomes negative ($\chi_\perp > \chi_{||}$) and then increases very rapidly such that at about 14°K it becomes about 50% of its average susceptibility (see Fig. 53). The principal susceptibilities, however, obey approximately a Curie-Weiss law.

The temperature variation of anisotropy of this crystal has been explained, on a quantitative basis, by Neogy and Mookherji [207]. Using the following values of the crystal field parameters,

$$A_2^0 = 50 \text{ cm}^{-1} \qquad A_4^0 = -100 \text{ cm}^{-1}$$

$$A_6^0 = -48 \text{ cm}^{-1} \qquad A_6^6 = +660 \text{ cm}^{-1}$$

they are able to explain the behavior of anisotropy in the liquid nitrogen temperature range. The above parameters yield that the ground $^3H_4$ term splits under $C_{3h}$ symmetry into the following energy levels:

$$W_1 = -155.2 \text{ (D)} \qquad W_2 = -138.6 \text{ (S)} \qquad W_3 = +51.6 \text{ (D)}$$

$$W_4 = +53.8 \text{ (D)} \qquad W_5 = +64.2 \text{ (S)} \qquad W_6 = +174.1 \text{ (S)}$$

Here (S) and (D) imply a singlet and a doublet, respectively. It is observed that the overall width of the multiplets is comparable to kT at T=300°.

Praseodymium Double Nitrates: $Pr_2Mg_3(NO_3)_{12}.24H_2O$ and $Pr_2Zn_3(NO_3)_{12}$ $.24H_2O$ have been studied by Krishnan and Mookherji [8]. At room temperature (see Table 16), the crystals are very feebly anisotropic, thus suggesting

TABLE 15

Data on $Pr_2(C_2H_5SO_4)_6 \cdot 18H_2O$

| Temp. (°K) | $\chi_{||} - \chi_{\perp}$ | Ref. |
|---|---|---|
| 303 | 241 | 8 |
| 300 | 245 | 205 |
| 290.9 | 252 | 206 |
| 290.0 | 318 | 203 |

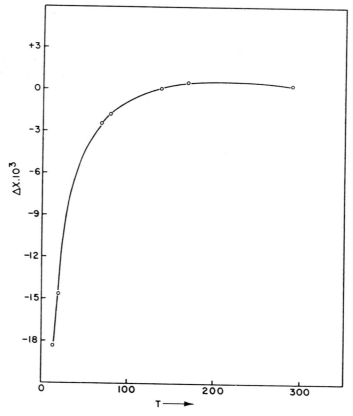

FIG. 53. Variation of the magnetic anisotropy of $Pr(C_2H_5SO_4)_3 \cdot 9H_2O$ (Ref. 203).

TABLE 16

| Crystal | Temp. (°K) | $\chi_\perp - \chi_{||}$ | Ref. |
|---------|------------|--------------------------|------|
| $Pr_2Zn_3(NO_3)_{12}.24H_2O$ | 303 | 653 | 8 |
| $Pr_2Mg_3(NO_3)_{12}.24H\ O$ | 303 | 632 | 8 |
| | 300 | 630 | 205 |

the similarity of the nature of its crystal field to the ethyl sulfate. However, the temperature variation of anisotropy of $Pr_2Mg_3(NO_3)_{12}.24H_2O$ down to 80°K reveals, contrary to the ethyl sulfate, quite a "normal" behavior. The relative ordering and separation of energy levels in this crystal are thus expected to be very much different from the ethyl sulfate.

$Pr_2(SO_4)_3.8H_2O$. At room temperature (30°K), Krishnan and Mookherji [8] reported,

$$\chi_1 - \chi_2 = 783, \quad \chi_1 - \chi_3 = -975, \quad \psi = +26°$$

which gives, $K_{||} - K_\perp = 1270$, indicating that the anisotropy of the molecule is very high, about 26%. At 100°K, the anisotropy increases to 57%. The large value of anisotropy of the octahydrated sulfate at room temperature is contrary to the nearly isotropic behavior of the ethyl sulfate and double nitrate. This indicates that the nature and symmetry of the crystal field in this compound is perhaps different from the ethyl sulfate and double nitrate.

The temperature variation of its anisotropy is, like the double nitrate, quite normal.

## 3. $f^3$ ($Nd^{3+}$)

The ground term of free $Nd^{3+}$ ion is $^4I_{9/2}$, and the next term $^4I_{11/2}$ is about 1800 cm$^{-1}$ above the ground term. The free ion spin-orbit coupling parameter is 900 cm$^{-1}$. In the crystal field calculation of magnetic susceptibility and anisotropy the effect of this excited term has, however, been neglected by all workers. In an axial field of $C_{3h}$ symmetry, the ground term $^4I_{9/2}$ is split into five Kramers doublets with a total multiplet width comparable to kT.

Neodymium Ethyl Sulfate: Magnetic anisotropy of this crystal has been studied by a large number of workers. The crystal shows very small anisotropy at room temperature (see Table 16a). Mookherji [208], and

TABLE 16a

Data on $Nd_2(C_2H_5SO_4)_6 \cdot 18H_2O$

| Temperature ($^\circ$K) | $\chi_{||} - \chi_{\perp}$ | Ref. |
|---|---|---|
| 303 | 600 | 8 |
| 290 | 778 | 203 |
| 291.5 | 764 | 210 |
| 300 | 597.6 | 216, 209 |

Mookherji and Mookherji [209] have reported the data down to about 85°K. Fereday and Wiersma [203] and Van del Handel and Hupse [210] have reported the anisotropy and principal susceptibility data down to about 14°K. Unfortunately the agreement between the data reported by these workers, particularly at low temperatures, is rather poor (see Fig. 54). Mahalnabis and Chachra [211] have deduced following crystal field parameters using the experimental data of Mookherji and Mookherji [209]:

$$A_2^0 = 45 \text{ cm}^{-1} \qquad A_4^0 = -100 \text{ cm}^{-1}$$

$$A_6^0 = -37.7 \text{ cm}^{-1} \qquad A_6^6 = 660 \text{ cm}^{-1}$$

This set of parameters agrees well with that deduced from ESR data [200] except the value of $A_2^0$, for which the ESR study yields $A_2^0 = -15 \text{ cm}^{-1}$. It is also interesting to compare these values with those extrapolated from the cerium ethyl sulfate using Eq. 84:

$$A_2^0 = -12 \text{ cm}^{-1} \qquad A_6^0 = -43 \text{ cm}^{-1}$$

$$A_4^0 = -24 \text{ cm}^{-1} \qquad A_6^6 = 545 \text{ cm}^{-1}$$

The agreement is very satisfactory, and thus it lends support to the validity of Eq. 84.

$Nd_2(SO_4)_3 \cdot 8H_2O$: The anisotropy of this crystal is much larger than the ethyl sulfate. At room temperature, Krishnan and Mookherji [8] report:

$$\chi_1 - \chi_2 = 806, \qquad \chi_1 - \chi_3 = -187, \qquad \psi = 14.6$$

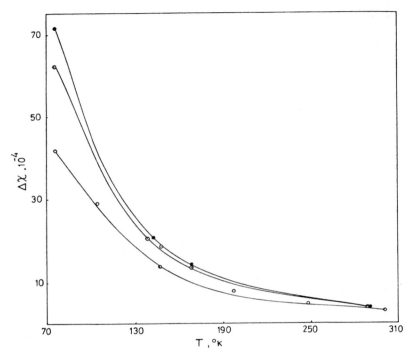

FIG. 54. Temperature variation of magnetic anisotropy of neodymium ethyl sulfate. Symbols: O, (Ref. 209); ⊙, (Ref. 210); and ●, (Ref. 253).

which gives $K||-K_\perp = 900$, indicating that the molecular anisotropy is about 11% of the average susceptibility. Mookherji [208] has measured its anisotropy down to about 80°K. Jackson's measurement [212] of its principal susceptibilities extends down to about 14°K. Here again the agreement between the data of these two workers is poor. Chachra [213] has deduced following set of the crystal field parameters from the data of Mookherji [208] using the symmetry of the crystal field to be tetragonal:

$$A_2^0 = 250 \text{ cm}^{-1} \qquad A_4^0 = 97.0 \text{ cm}^{-1}$$

$$A_6^0 = -70 \text{ cm}^{-1} \qquad A_4^4 = 600 \text{ cm}^{-1}$$

The quality of fitting obtained with these parameters is demonstrated in Fig. 55.

Neodymium Double Nitrates: $Nd_2Mg_3(NO_3)_{12}.24H_2O$ and $Nd_2ZN_3(NO_3)_{12}.24H_2O$ have been studied. At room temperature the anisotropy is about

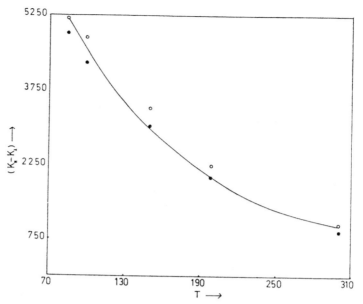

FIG. 55. Temperature variation of molecular anisotropy of neodymium sulphate octahydrate. Solid line is the theoretical curve of Chachra [213]. Circles are the experimental values—solid circles represent those of Jackson [212] and the hollow ones of Mookherji [208].

8% of the average susceptibility [8]. Mookherji [208] has measured its anisotropy down to about 85°K. The anisotropy is observed to increase in this temperature range by about 10 times.

## 4. $f^4$ (Pm$^{3+}$)

Trivalent promethium does not occur in nature and it is extremely radioactive, and so it has not been studied.

## 5. $f^5$ (Sm$^{3+}$)

The free-ion ground term of the trivalent samarium ion is $^6H_{5/2}$. The next excited term $^6H_{7/2}$ is only 993 cm$^{-1}$ above it. This separation is in fact less than its spin-orbit coupling parameter, about 1200 cm$^{-1}$. The closeness of the excited $^6H_{7/2}$ term makes its behavior, in a sense, similar to Eu$^{3+}$ compounds (discussed later).

Samarium Ethyl Sulfate: The magnetic anisotropy of this crystal has been studied down to about 90°K. At 30°K, Krishnan and Mookherji [8] report,

$$\chi_{||} - \chi_{\perp} = 172,$$

which is about 8% of the average susceptibility. Temperature variation of its anisotropy is shown in Fig. 56. Its close resemblance to the praseodymium ethyl sulfate is clear. No theoretical explanation of its anisotropy data has been attempted, although Frank [213a] has discussed its magnetic behavior. However the following set of parameters has been suggested from the ESR data [201]:

| | |
|---|---|
| $A_2^0 = -13 \text{ cm}^{-1}$ | $A_4^0 = -25 \text{ cm}^{-1}$ |
| $A_6^0 = -36 \text{ cm}^{-1}$ | $A_6^6 = 390 \text{ cm}^{-1}$ |

Samarium Double Nitrates. At room temperature, $Sm_2Mg_3(NO_3)_{12} \cdot 24H_2O$ and $Sm_2Zn_3(NO_3)_{12} \cdot 24H_2O$ both show anisotropy of the same order as the ethyl sulfate [8]. The temperature variation of anisotropy [208] of $Sm_2Mg_3(NO_3)_{12} \cdot 24H_2O$ is also very similar to the ethyl sulfate (see Fig. 56).

## 6. $f^6$ ($Eu^{3+}$)

The compounds of trivalent europium are probably the most interesting in the rare earth series. The free-ion ground term of $Eu^{3+}$ ion is $^7F_0$, and is thus non-magnetic. The free ion excited terms (J = 1, 2) are about 300 cm$^{-1}$ and 1000 cm$^{-1}$ above the ground term. As a result, unlike other ions, the free-ion multiplet intervals are comparable to kT in this case. This is the reason why the compounds of Eu(III) are fully paramagnetic at room temperature, although the magnetic moment in the first order is zero. Detailed theoretical calculation of susceptibility considering axial field has not been yet done; but Frank [213a] has done simple calculation taking only the cubic field.

The ground term of the Eu(III) being non-magnetic, the crystal field parameters are very insensitive to the mean moment. Study of magnetic anisotropy or principal susceptibilities would be rewarding in this regard. Europium ethyl sulfate is very suitable for such studies, since an extensive amount of supplementary data exist on this crystal.

Europium sulfate octahydrate has been studied between 300 and 85°K [214]. At 300°K,

$$\chi_1 - \chi_2 = 759, \quad \chi_3 - \chi_2 = 641, \quad K_{||} - K_{\perp} = 700$$

indicating that the molecular anisotropy is about 16% of the average susceptibility. At 85°K, this increases to about 66%. The temperature variation of principal anisotropies show, particularly at lower temperatures,

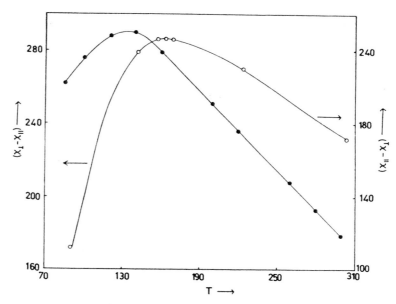

FIG. 56. Temperature variation of magnetic anisotropy of $Sm_2(C_2H_5SO_4)_6$ .18$H_2O$ (open circles) and $Sm_2Mg_3(NO_3)_{12}$.24$H_2O$ (solid circles).

the effect of Boltzmann depopulation from the paramagnetic levels (J = 1 and J = 2) (see Fig. 57). It will be interesting to extend the measurement below the liquid nitrogen temperatures.

## 7. $f^7$ ($Gd^{3+}$)

The free-ion ground term of the trivalent gadolinium ion is $^8S$. The excited terms are far removed, and hence the behavior of the gadolinium compounds is expected to be similar to the manganous Tutton salts (see Section III B5). The anisotropy of the crystals is expected to be extremely small and follow a $1/T^2$ law of temperature variation.

Magnetic anisotropy of $Gd_2(SO_4)_3$.8$H_2O$ crystals has been measured by Krishnan et al. [9]. At 30°C, they find

$$\chi_1 - \chi_2 = 750, \quad \chi_3 - \chi_2 = 745 \quad \psi = +15.5°$$

The crystal thus shows an anisotropy about 1.4% of the average susceptibility. This value is a little larger than that expected for S- state ions. The compound used by Krishnan et al. was not, however, very pure; they report the presence of 0.5% samarium and 1% terbium as impurities. Since the terbium compounds show very large anisotropy (see later) as

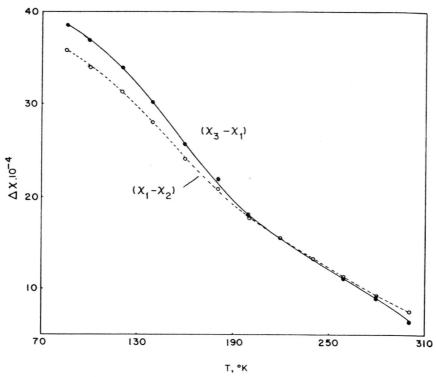

FIG. 57. Temperature variation of principal crystal anisotropies of europium sulfate octahydrate (Ref. 214).

compared to the nearly isotropic nature of the samarium compounds, the slightly larger value of anisotropy of the gadolinium sulphate may be largely due to the terbium impurities. Krishnan and Banerji [215] repeated the measurement using a much purer sample and obtained,

$$\chi_1 - \chi_2 = 36 \quad , \quad \chi_3 - \chi_1 = 16 \quad \psi = +17°$$

The anisotropy is now only about 0.07% and is very close to the value observed for manganous Tutton salts.

Krishnan et al. [9] have also reported the temperature variation of anisotropy of the above compound. The principal anisotropies obey $\frac{1}{T^2}$ law very closely.

## 8. $f^8$ (Tb$^{3+}$)

Free-ion ground term of the trivalent terbium ion is $^7F_6$. Next term $^7F_5$ is about 2043 cm$^{-1}$ above the ground term. The free-ion spin orbit coupling parameter is 1620 cm$^{-1}$ and is hence comparable to the free-ion miltiplet width. In the crystal field of cubic symmetry the ground term $^7F_6$ splits into six components having an overall separation of the ground-manifold of about 118 cm$^{-1}$ only.

Magnetic anisotropy of terbium ethyl sulfate has been reported by Mookherji [216]. At 300°K,

$$\chi_{||} - \chi_{\perp} = 9877$$

The anisotropy of the crystal is thus very large, about 24% of the average susceptibility. It is interesting that the terbium ethyl sulfate shows largest anisotropy among all the rare earth ethyl sulfates. The anisotropy of this crystal is indeed comparable to the octahydrated rare earth compounds. The large anisotropy of the crystal at higher temperatures may be due to the considerable contribution from the close-lying excited $^7F_5$ term through spin-orbit coupling. No theoretical analysis of the magnetic anisotropy of this crystal has so far been reported. However following crystal field parameters have been deduced from the optical absorption spectra [217]:

$$A_2^0 = 92 \text{ cm}^{-1} \qquad A_4^0 = -40 \text{ cm}^{-1}$$
$$A_6^0 = -30 \text{ cm}^{-1} \qquad A_6^6 = 290 \text{ cm}^{-1}$$

## 9. $f^9$ (Dy$^{3+}$)

The free-ion ground term of the trivalent dysprosium ion is $^6H_{15/2}$. The next excited term is about 3400 cm$^{-1}$ above the ground term. The free-ion spin-orbit coupling parameter is very large, about 1820 cm$^{-1}$. Under the influence of an axial crystal field, the ground term $^6H_{15/2}$ splits into eight Kramers doublets.

Dy$_2$(SO$_4$)$_3$.8H$_2$O. Neogy and Mookherji [218] have reported the anisotropy data on this crystal between 300° and 100°K. At 300°K,

$$\chi_1 - \chi_2 = 8748 \quad , \quad \chi_3 - \chi_2 = 12260, \quad K_{||} - K_\perp = 10500.$$

The anisotropy of the molecule is thus about 23%, and is of the same order as that of other rare earth sulfates. The molecular anisotropy increases by about 5.5 times between 300° and 100°K. Neogy and Mookherji have also carried out detailed theoretical calculation assuming the $Dy^{3+}$ ion to be in the crystal field of tetragonal symmetry. Following set of parameters is deduced by them (in $cm^{-1}$):

$$A_2^0 = 123 \qquad A_4^0 = -63$$

$$A_6^0 = -60 \qquad A_4^4 = 420.$$

The ground term $^6H_{15/2}$ splits into eight Kramers doublets with a total width of about 272 $cm^{-1}$. An interesting point of their calculation is that they calculate $g_\perp > g_{||}$, while experimentally they find $K_{||} > K_\perp$. The discrepancy is rationalized by the fact that there is an unusually large contribution to $K_{||}$ by the second order term, although the first order contribution to $K_\perp$ is slightly lower than that to $K_{||}$.

Dysprosium Ethyl Sulfate: Cooke et al. [219] have reported the temperature variation of principal susceptibilities of this crystal between 20° and 1°K. The principal susceptibilities in this temperature range are represented by,

$$\chi_{||} = \frac{(10.9 \pm 0.5)}{T} [ 1 + \frac{0.135 \pm 0.02}{T} ]$$

and

$$\chi_\perp = \frac{(0.03 \pm .01)}{T} [ 1.05_3 \pm 0.02 ].$$

## 10. $f^{10}$ ($Ho^{3+}$)

The free-ion ground term of $Ho^{3+}$ is $^5I_8$. The next excited term $^5I_7$ is about 5000 $cm^{-1}$ above the ground term. The free-ion spin-orbit coupling is about 2080 $cm^{-1}$.

The anisotropy of $Ho(C_2H_5SO_4)_3.12H_2O$ has been measured down to about 20°K [216, 220]. At 300°K,

$$\chi_{||} - \chi_\perp = 4953$$

which is about 11% of the average susceptibility. This value increases to about 13.7% at 85°K [216]. However, between 300° and 20°K, the anisotropy increases by 123 times [220].

No theoretical calculation based on the above data has been attempted.

## 11. $f^{11}$ ($Er^{3+}$)

The free ion ground term of trivalent erbium ion is $^4I_{15/2}$. The next excited term $^4I_{13/2}$ is about 6500 cm$^{-1}$ above the ground term. The spin-orbit coupling parameter is 2360 cm$^{-1}$. Under a crystal field of axial symmetry, the ground term splits into eight Kramers doublets - with an overall separation comparable to kT.

Erbium Sulfate Octahydrate: Neogy and Mookherji [221] have studied the anisotropy of this crystal between 300° and 85°K. At 300°K,

$$\chi_1 - \chi_2 = 603, \quad \chi_3 - \chi_2 = 7625, \quad K_{||} - K_{\perp} = 4130.$$

The crystal thus shows a moderate molecular anisotropy of about 12%, and is much less anisotropic than other octahydrated rare earth sulphates. At 100°K, the anisotropy increases to about 31%. Assuming the symmetry of the crystal field to be tetragonal, Neogy and Mookherji deduce the following parameters (in cm$^{-1}$) from their data.

$$A_2^0 = -73.0 \qquad A_4^0 = -10.7 \qquad A_4^4 = 400.0$$

$$A_6^0 = -53 \qquad A_6^4 = 19.0 \ .$$

The width of the splitting of the ground term in the axial field is calculated to be about 307 cm$^{-1}$.

Erbium Ethyl Sulfate: The crystal has been studied by a large number of workers. The anisotropy of this crystal was first measured by Fereday and Wiersma [203] down to about 14°K. However the measurement of all later workers established that the data of Fereday and Weirsma were seriously in error. The room temperature data on this crystal is collected in Table 17.

The low temperature measurement on this crystal by Mookherji [222] revealed a very interesting feature. The anisotropy of the crystal increases smoothly in the temperature range of measurement. However he observed a very large and continuous change (about 75°) in the setting angle of the crystal with the temperature (see Fig. 58). Since the crystal is supposed to belong to hexagonal system, the setting angle is expected to remain constant with temperature. The crystal is, therefore, believed to be undergoing some sort of gradual phase transformation which lowers its symmetry. It may be recalled that Saha [92] observed a similar phase transition in chromium potassium oxalate trihydrate (see Fig. 15).

TABLE 17

| Temperature (°K) | $\chi_\perp - \chi_{\parallel}$ | Ref. |
|---|---|---|
| 303 | +6590 | 8 |
| 303 | -2590 | 203 |
| 300 | +6202 | 216 |
| 290 | +5634 | 220 |

[a]Data on $Er_2(C_2H_5SO_4)_6 \cdot 18H_2O$

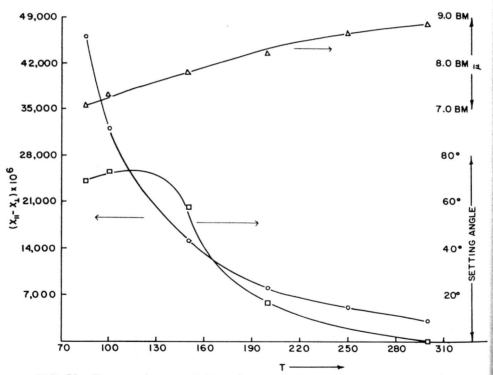

FIG. 58. Temperature variation of average moment, magnetic anisotropy and setting angle of erbium ethylsulfate [222]. Symbols: $\triangle$, average moment; O, magnetic anisotropy; $\square$, setting angle.

## 12.  $f^{12}$ ($Tm^{3+}$)

No experimental magnetic anisotropy data have so far been reported on the compounds of thulium ion.

## 13.  $f^{13}$ ($Yb^{3+}$)

The free trivalent ytterbium ion possessing $4f^{13}$ configuration has just one electron less than is required to complete the f shell.  Thus the free-ion ground term of $Yb^{3+}$ ion is $^2F_{7/2}$.  The next excited term $^2F_{5/2}$ is about 10,300 $cm^{-1}$ above the ground term.  The spin-orbit coupling parameter is about 2924 $cm^{-1}$.  The effect of the excited term has, therefore, been neglected in the calculation of susceptibility.  In the crystal field of axial symmetry, the ground $^2F_{7/2}$ term is split into four Kramers doublets with a total multiplet width comparable to kT at ordinary temperature.

Neogy (223) has reported the data on $Yb_2(SO_4)_3.8H_2O$ down to 100°K.  At the room temperature, the crystal is highly anisotropic like other rare-earth octahydrates, about 27% of the average susceptibility.  Assuming the symmetry of the crystal field to be tetragonal, he has deduced the following set of parameters from his data.

$$A_2^0 = -100 \text{ cm}^{-1} \qquad A_4^0 = -10 \text{ cm}^{-1} \qquad A_6^0 = -10 \text{ cm}^{-1}$$

$$A_4^4 = +350 \text{ cm}^{-1} \qquad A_6^4 = 0 \ .$$

No data have been reported on the ytterbium ethyl sulfate.

### D.  Diamagnetic Inorganic and Organometallic Compounds

In this sub-section, we shall discuss some simple inorganic compounds, metallocenes and phthalocyanines.

### 1.  Simple Inorganic Compounds

Magnetic anisotropy of several sulfates, nitrates, carbonates, chlorates, bromates etc. have been studied at room temperature.  We shall discuss them individually.

(a) Sulfates.  The anisotropy of some sulfates of divalent Ba, Ca, and Sr has been studied by Krishnan et al. [4].  The anisotropy is observed to be very small (see Table 18).  This is in conformity with the x-ray structural analysis which shows that the four oxygen atoms of the $[SO_4]^{2-}$ group are disposed more or less symmetrically about the sulfur atom, and are situated at the corners of a tetrahedron with the S atom at its center.  The observed

TABLE 18

| Crystal | $\chi_b - \chi_c$ | $\chi_c - \chi_a$ | Ref. |
|---------|---------------|---------------|------|
| BaSO$_4$ | 0.72 | 0.62 | 4 |
| SrSO$_4$ | 0.74 | 0.99 | 4 |
| CaSO$_4$ | 0.25 | -0.65 | 4 |
| CaSO$_4$.2H$_2$O | $\chi_1 - \chi_2 = 1.0$ | $\chi_1 - \chi_3 = 0.6$ | 4 |
| (Selenite) | | $\psi = 0$ | |

isotropic nature of the sulfates is thus a direct consequence of its structure.

(b) Nitrates and Carbonates. The anisotropy of several nitrates and carbonates have been studied at room temperature. The data are summarized in Table 19. A striking feature of their behavior is the large anisotropy exhibited by them, as contrasted with the nearly isotropic nature of the corresponding sulphates. Krishnan and Raman [225], and Krishnan et al [4] have attributed this large anisotropy to the intrinsic anisotropy of the $NO_3^-$ and $CO_3^=$ groups which have nearly planar structures. The X-ray and optical-birefringent studies support this observation.

(c) Chlorates and Bromates. Several chlorates and bromates have been studied. They exhibit large anisotropy similar to nitrates and carbonates, but of opposite sign (see Table 20). This has been explained by Krishnan et al. [4] as due to the fact that $ClO_3^-$ ions are not planar, but pyramidal with $Cl^-$ at the apex. Recently Tandon [224] has attempted an explanation of their negative anisotropy on the basis of Van Vleck's theory of polyatomic molecules [20].

(d) Double chlorides. Simon and Toussaint [31] have studied anisotropy of $K_2PtCl_4$ and $K_2PdCl_4$ crystals belonging to hexagonal system. The data are collected in Table 21. The small difference in the anisotropies of this compound may be due to the difference in the strength of chemical bonding, although other explanations can not be ruled out.

(e) Hydrated tetracyanides. $K_2Ni(CN)_4$.H$_2$O: The crystal belongs to the monoclinic system with 16 molecules in the unit cell. At room temperature, Rogers [51] reports:

$$\chi_1 - \chi_2 = 20.3, \quad \chi_1 - \chi_3 = 16.4, \quad \psi = 0.$$

Calculation of molecular anisotropy is not, however, possible in this case because of the large number of molecules in the unit cell with unknown orientations.

TABLE 19

| Crystal | $\Delta\chi$ | | Ref. |
|---|---|---|---|
| $NaNO_3$ | $\chi_\perp - \chi_{||}$ | $= 4.8$ | 4 |
| $KNO_3$ | $\chi_a - \chi_c$ | $= 4.8$ | 4 |
| | | $6.98$ | 36 |
| | $\chi_b - \chi_c$ | $= 4.8$ | 4 |
| | | $7.03$ | 36 |
| $CaCO_3$ (calcite) | $\chi_\perp - \chi_{||}$ | $= 4.0$ | 4 |
| | | $\simeq 4.0$ | 3 |
| $CaCO_3$ (aragonite) | $\chi_a - \chi_c$ | $= 4.0$ | 4 |
| | $\chi_b - \chi_c$ | $= 4.2$ | |
| $SrCO_3$ (strontianite) | $\chi_a - \chi_c$ | $= 4.8$ | 4 |
| | $\chi_b - \chi_c$ | $= 4.8$ | |
| $BaCO_3$ (witherite) | $\chi_a - \chi_c$ | $= 4.9$ | 4 |
| | $\chi_b - \chi_c$ | $= 5.0$ | |
| $(NH_4)NO_3$ | $\chi_\perp - \chi_{||}$ | $= 5.4$ | 224 |
| $AgNO_3$ | $\chi_\perp - \chi_{||}$ | $= 5.45$ | 224 |
| $NH_4HCO_3$ | $\chi_\perp - \chi_{||}$ | $= 5.0$ | 224 |
| $KHCO_3$ | $\chi_\perp - \chi_{||}$ | $= 5.0$ | 224 |

$BaM''(CN)_4.4H_2O$: Simon and Toussaint [31] have reported the anisotropy of a number of tetracyanides with the general formula $BaM''(CN)_4.4H_2O$ where M'' stands for divalent nickel, platinum and palladium. These crystals belong to the monoclinic system with four molecules in the unit cell. The data on anisotropies are collected in Table 22. The anisotropy is observed to increase inversely with the atomic number of M''.

$CaM''(CN)_4.5H_2O$: The anisotropy of the above compounds (with M'' = Ni, Pt and Pd) has also been studied by Simon and Toussaint [31]. The crystals belong to the orthorhombic system. The anisotropy data are collected in Table 23. Here again the anisotropy is inversely proportional to the atomic number of M''. At point of another interest is the close similarity of the

TABLE 20

| Crystal | $\chi_{||} - \chi_{\perp}$ | Ref. |
|---------|---------------------|------|
| $KClO_3$ | 3.8 | 4 |
|  | 5.54 | 224 |
| $NaClO_3$ | 5.54 | 224 |
| $Ba(ClO_3)_2$ | 5.54 | 224 |
| $Sr(ClO_3)_2$ | 5.54 | 224 |
| $KBrO_3$ | 5.18 | 224 |

TABLE 21

| Crystal | $\chi_{\perp} - \chi_{||}$ | Ref. |
|---------|---------------------|------|
| $K_2PtCl_4$ | 41.4 | 31 |
| $K_2PdCl_4$ | 50.3 | 31 |

TABLE 22

| Crystal | $\chi_1 - \chi_2$ | $\chi_1 - \chi_3$ | $\chi_3 - \chi_2$ | $\Psi$ | Ref. |
|---------|---------|---------|---------|----|------|
| $BaPt(CN)_4.4H_2O$ | 11.4 | 11.3 | 0.1 | $3°$ | 31 |
| $BaPd(CN)_4.4H_2O$ | 18.4 | 17.7 | 0.7 | $2°$ | 31 |
| $BaNi(CN)_4.4H_2O$ | 26.4 | 25.8 | 0.6 | $3°$ | 31 |

TABLE 23

| Crystal | $\chi_c - \chi_b$ | $\chi_c - \chi_a$ | $\chi_a - \chi_b$ | Ref. |
|---------|---------|---------|---------|------|
| $CaPt(CN)_4.5H_2O$ | 10.5 | 10.4 | 0.4 | 31 |
| $CaPd(CN)_4.5H_2O$ | 17.1 | 16.9 | 0.2 | 31 |
| $CaNi(CN)_4.5H_2O$ | 25.4 | 26.3 | 0.9 | 31 |

anisotropies for tetra- and penta-hydrated compounds. It appears that the diamagnetic anisotropy is rather insensitive to the small changes in the packing of the crystal lattice.

(f) KMnO4. The magnetic anisotropy of this crystal was measured by Mookherji [226] at room temperature. The anisotropies are very feeble. Guha [95] and Singhal [227] have reported its anisotropy data at low temperatures. A small temperature variation in its principal anisotropies was observed by Guha. Singhal, however, did not observe any such variation. The feeble paramagnetism exhibited by the compound may be of the high frequency type.

(g) Chromates and Dichromates. Magnetic anisotropy of several chromates have been studied over a temperature range. The anisotropies are observed to be very small (see Table 24). A small dependence of anisotropy on temperature was observed [228]. The anisotropies of the three compounds studied between 360° and 80°K can be represented by three constant formulas as below:

$Na_2CrO_4.4H_2O$:

$$\Delta K = 1.36 + \frac{28.0}{T} - \frac{1200}{T^2} \qquad ,$$

$(NH_4)_2CrO_4$:

$$\Delta K = 1.20 + \frac{60.0}{T} - \frac{3600}{T^2}$$

$K_2CrO_4$:

$$\Delta K = 1.17 + \frac{68.5}{T} - \frac{3900}{T^2}$$

Singhal and Mookherji [229] have also studied the temperature variation of dichromates. The anisotropies can be represented by,

$(NH_4)_2Cr_2O_7$: $\qquad \Delta \chi = 17.83 + \frac{682.5}{T} + \frac{35100}{T^2}$

and $Na_2Cr_2O_7.2H_2O$: $\qquad \Delta \chi = 11.5 + \frac{542.0}{T} + \frac{27600}{T^2}$

## 2. Metallocenes

Magnetic anisotropy measurements on ferrocene, osmocene and ruthenocene has yielded some very interesting results. We shall discuss them individually.

TABLE 24

| Crystal | $\chi_1 - \chi_2$ | $\chi_1 - \chi_3$ | $\overline{\mu}$ | Ref. |
|---------|-------------------|-------------------|------------------|------|
| $Na_2CrO_4.4H_2O$ | 1.38 | -0.11 | 0.62 | 228 |
| $(NH_4)_2CrO_4$ | 1.30 | -0.12 | 0.54 | 228 |
| $K_2CrO_4$ | $\chi_a - \chi_b = 0.10$ | $\chi_c - \chi_a = 1.30$ | 0.54 | 228 |

[a]The data refer to 300°K

(a) Ferrocene. The nature of chemical bonding in ferrocene has been the subject of long standing controversy. Many conflicting and contradictory models of its chemical bonding and molecular orbital calculations have been proposed [230-239]. For example, it has been contended by some workers that all the $\pi$-electrons on the cyclopentadienyl rings are donated to the iron atoms, while some workers have supported a single "metal-to-ring" bond.

It has been speculated that a knowledge of the number of $\pi$-electrons circulating on the cyclopentadienyl rings can be obtained from the measurement of its diamagnetic anisotropy. The problem of the diamagnetic anisotropy of such organo-metallic compounds is, in a sense, similar to that of the aromatic hydrocarbons (for details, see Section III, F). The simple Larmor-Langevin relationship as modified by Pauling [240] has been successful to a great extent in calculating the anisotropy:

$$\Delta K = \frac{Ne^2}{4mc^2} \sum_n r^2 \qquad (85)$$

where n is the number of $\pi$-electrons and $(r^2)_{av}$ is the mean-square distance of the circulating electrons from the orbital axis. Pauling assumed that $(r^2)_{av} = R^2$, R being the distance from the center of the ring to the carbon nuclei.

The single crystals of ferrocene belongs to monoclinic space group $(P2_1/a)$ with two molecules in the unit cell. At room temperature its anisotropies are [241-244]:

$$\chi_1 - \chi_2 = 37.0 \quad , \quad \chi_1 - \chi_3 = 40.9 \quad , \quad \chi_2 - \chi_3 = 3.88.$$

Assuming that $K_1 = K_2 = K_\perp$ and $K_3 = K_{||}$ the molecular anisotropy is calculated to be [244],

$$K_\perp - K_{||} = 75.$$

The earlier calculation of its molecular anisotropy [242, 243] appears to be in error.

The problem of diamagnetic anisotropy of a compound like ferrocene is not as straight forward as for aromatic hydrocarbons. This is because of the anisotropy in the high frequency paramagnetism which arises from the interaction of the ground $^1A_1$ state with the excited states. This contribution in ferrocene is, fortunately, zero. Then allowing for the paramagnetic anisotropy of the two cyclopentadienyl rings, as outlined by Dorffman [245], the corrected anisotropy of ferrocene becomes:

$$K_\perp - K_{||} = 57.$$

Application of Eq. 85 now gives n ≃ 9.0, i.e. about 4.5 electrons per ring. This value of $\pi$-electron supports Moffitt's [235] single $d_\pi$-$p_\pi$ bonding model which would predict 4 electrons per ring. The present result is also in agreement with the MO calculations of many workers [237-239] and the valence-bond model of Pauling [236]. It, however, does not agree with the model of complete donation of all $\pi$-electrons to the iron atom [230-233].

(b) Ruthenocene and Osmocene. Mulay and co-workers [246, 247] have recently studied the diamagnetic anisotropy of this compound at room temperature. The data of crystal susceptibilities are collected in Table 25. The anisotropies are very small. Using Eq. 85, n is found to be about 3.1 for ruthenocene and 2.6 for osmocene, whereas for ferrocene, as discussed above, n is about 4.6.

The decreasing $\pi$-electron density in the series—ferrocene, ruthenocene and osmocene—explains the lighter bonding and decreasing electrophilic reactivity reported on a qualitative basis for this series by Rausch and co-workers [248]. The number of $\pi$-electrons withdrawn from each ring (6 - n) shows an appropriate positive relation with the decrease in the divalent charge (2-$Z_{eff}$) on Fe, Ru, Os, where $Z_{eff}$ is the final charge on the metal atom calculated by Batsanov [249] for these metallocenes according to the molecular orbital theory.

## 3. Transition Metal Phthalocyanines

Transition metal phthalocyanines provide a very good series of compounds for diamagnetic anisotropy studies. They have an almost circular large area of circulating $\pi$-electrons, and thus provide good approximation to Eq. 85.

Martin and Mitra [303] have measured the diamagnetic anisotropy of nickel and zinc phthalocyanines at room temperature. The result shows

TABLE 25

| Crystal | $\chi_i$ | $K_i$ | $\Delta K$ | n | Ref. |
|---------|----------|-------|-----------|---|------|
| Ruthenocene | $\chi_a = -137$ | $K_1 = -148$ | | | |
| | $\chi_b = -148$ | $K_2 = -125$ | 40 | 3.1 | 242a |
| | $\chi_c = -165$ | $K_3 = -177$ | | | |
| Osmocene | $\chi_a = -183$ | $K_1 = -189$ | | | |
| | $\chi_b = -189$ | $K_2 = -175$ | 33 | 2.6 | 242a |
| | $\chi_c = -207$ | $K_3 = -215$ | | | |

that the area of the $\pi$-ring system of the Zinc phthalocyanine is larger than the corresponding area of nickel (II) phlhalocyanine.

### E.  Semiconductors and Minerals

Magnetic anisotropy of a large number of semiconductors and minerals has been studied at room and low temperatures. We shall discuss some of them below.

### Graphite

This is perhaps one of the first semiconductors to be studied very extensively. The crystal belongs to the hexagonal system with a perfect basal cleavage. The carbon atoms in it are arranged in layers parallel to the basal plane, the atoms in each layer forming a regular hexagonal network. The separation between adjacent layers is about 3.4Å, which is much larger than the distance between adjacent atoms in the same layer, namely 1.42Å. Thus the binding between the adjacent layers is very weak.

The measurement of magnetic anisotropy of graphite by Krishnan et al. [251-256] shows that the crystal exhibits a very large diamagnetic anisotropy which is almost wholly along the normal to the basal plane, i.e., $\chi_\perp = -0.5 \times 10^{-6}$, $\chi_{||} = -21.5 \times 10^{-6}$. This indicates that the mobility of the conduction electrons is particularly confined to the basal plane. This is exactly what is observed by electrical conductivity measurements which show a kind of superconductivity in the basal plane (about 10,000 times greater than the conductivity in the normal plane). Temperature variation of anisotropy and principal susceptibilities is interesting. Between 1200° and about 90°K, the variation of anisotropy is shown in Fig. 59. $\chi_\perp$ was observed to remain constant with temperature while $\chi_{||}$ was highly

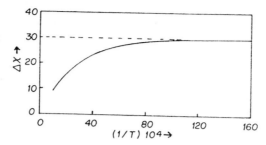

FIG. 59.  Temperature variation of magnetic anisotropy of graphite.

temperature dependent.  Recent measurements by Poquet et al. [257] substantiate the work of Krishnan et al.

Krishnan et al. [258,256] have demonstrated that the observed temperature variation of its diamagnetic anisotropy and principal susceptibility can be very satisfactorily explained on the basis of a free electron gas model. We shall briefly discuss it here.

An electron gas possesses, besides its spin paramagnetism, an appreciable diamagnetism superposed on it due to the quantized orbital motions of the electrons in the magnetic field.  For a free electron gas both the diamagnetic and the paramagnetic susceptibilities are easily calculated.  The diamagnetic susceptibility per unit volume of the gas is given by

$$Kd = - \frac{n\beta^2 F'(\eta)}{3\kappa T F(\eta)}$$

where n is the number of electrons per unit volume, $\beta$ is the Bohr magneton,

$$F(\eta) = \int_0^\infty \frac{x^{\frac{1}{2}} \cdot dx}{e^{x-\eta}+1} ,$$

$$F'(\eta) = \frac{\partial}{\partial \eta} F(\eta) = \frac{1}{2} \int_0^\infty \frac{x^{-\frac{1}{2}}}{e^{x-\eta}+1} \cdot dx$$

$$\eta = \zeta/\kappa T$$

$\zeta$ being the thermodynamic potential per electron.  To the same approximation, the paramagnetic susceptibility per unit volume is given by

$$Kp = \frac{n\beta^2 F'(\eta)}{\kappa T F(\eta)}$$

and hence Kp = -3Kd. The resultant susceptibility, K, is

$$K = K_p + K_d,$$

and is hence paramagnetic.

At very high temperatures, $T \gg T_0$ (where, $T_0$ is the degeneracy temperature), $F'/F$ tends to reach asymptotically the value 1, and the two susceptibilities will then conform to the Curie law,

$$Kd = -\frac{n\beta^2}{3\kappa T} \quad , \quad Kp = \frac{n\beta^2}{\kappa T}$$

At low temperatures, $T \ll T_0$, the two susceptibilities reduce to the temperature independent values:

$$Kd = -\frac{n\beta^2}{2\kappa To} \quad , \quad \kappa p = \frac{3n\beta^2}{2\kappa To}$$

The above model thus explains very successfully the temperature variation of anisotropy at high temperatures, and its constant value at lower temperatures. An interesting inference of this study is that tne conduction electrons in graphite are probably very nearly free in their motion. This makes graphite a suitable substance to verify the theories of electronic conduction due to free electrons.

## Molybdenite

Molybdenite crystals belong to the hexagonal system with perfect basal cleavage. The molybdenum and sulfur atoms in it are arranged in parallel layers, each layer of molybdenum atoms being sandwiched between two layers of sulfur atoms and these three layers form a composite layer by the repetition of which the whole structure is built. Datta [259] measured the magnetic anisotropy of this crystal at room and low temperatures. The room temperature values are:

$$\chi_\perp = -44.3, \quad \chi_{||} = -87.1$$

which yield $\chi_\perp - \chi_{||} = 42.9$. Like graphite it shows large anisotropy which is directed along the hexagonal axis of the crystal. The temperature variation of the magnetic anisotropy is also similar to graphite. At high temperature anisotropy follows a Curie law, $\chi_\perp - \chi_{||} = \frac{0.36}{T}$, whereas at low temperature it assumes a constant value $-48$. The presence of metallic electrons having their motion restricted in the basal plane, has been suggested from the magnetic data.

### α Silicon Carbide

Silicon carbide occurs in several polymorphic modifications. The $\beta$-SiC is cubic while the $\alpha$-form are either hexagonal or rhombohedral. These modifications are obtained by the different stacking sequences of closest packed plane layers formed by silicon and carbon atoms for the three possible stacking positions, the different types being classified according to the number of layers needed to complete the stacking sequences. Pure silicon carbide is transparent and almost colorless. Commercial samples show different colors presumably because of the presence of foreign impurities. Sigamony [260] reported the anisotropy measurement on the green variety. Recently Das [261] has done measurements on a large variety of commercial samples between 90° and 1000°K. The anisotropy is very much smaller as compared to graphite or molybdenite, and differ very slightly for various samples. Both $\chi_{||}$ and $\chi_{\perp}$ vary with temperature at higher temperatures; however, anisotropy is observed to remain practically constant over the whole range of temperature.

### Braunite

The chemical composition of braunite is generally represented by the formula $3Mn_2O_3.MnSiO_3$, according to which one of the Mn atoms is divalent and the remaining six are trivalent. An alternative formula $MnMnO_3$ in preference to $Mn_2O_3$ has also been suggested on chemical grounds. According to this formula, there are four divalent and three tetravalent Mn atoms in the braunite; none of them is trivalent. Measurements show that the anisotropy of this crystal is extremely small [262]

$$\chi_{\perp} - \chi_{||} = 0.364 \times 10^{-6},$$

thus supporting the formula $3\,MnMnO_3.MnSiO_3$.

### Manganite

It has been suggested that Mn atoms in this crystal is in divalent and tetravalent state. Presence of trivalent Mn atom has also been proposed. Krishnan and Banerji [263] measured the magnetic anisotropy of this crystal at room temperature. The anisotropies are very small:

$$\chi_c - \chi_a = 4.0$$

$$\chi_c - \chi_b = 3.0$$

thus suggesting that the Mn atom in the crystal can not be trivalent, since trivalent Mn compounds are expected to be highly anisotropic.

## Rhodochrosite (MnCO$_3$)

This crystal has been studied by a number of workers. Dupouy [264] measured the magnetic anisotropy of this crystal over a range of temperature and found,

$$\chi_{||} = \frac{387}{T-70} \quad \text{and} \quad \chi_\perp = \frac{387}{T-53} .$$

This disagrees with the measurement of Krishnan and Banerji [265] who obtained

$$\overline{\chi} = \frac{3.81}{T+13} ,$$

$$\text{and} \quad \Delta\chi = 7.2.$$

The anisotropy is thus very small, about 0.06% of the average susceptibility. This is in agreement with the Mn$^{2+}$ ion being in S- state. Measurement of Sucksmith [266] supports this conclusion.

## Siderite (FeCO$_3$)

This crystal has also been studied by several workers. Foex [267] studied the temperature variation of its principal susceptibilities and obtained,

$$\chi_{||} = \frac{C}{T-60} \quad \text{and} \quad \chi_\perp = \frac{C}{T-103} .$$

Krishnan et al. [5] reported its magnetic anisotropy at room temperature (303°K):

$$\chi_{||} - \chi_\perp = 6092.$$

Bizette and Tsai [268] measured its principal susceptibilities down to helium temperatures and reported an antiferromagnetic ordering at about 35°K. Mookherji and Mathur [269] has recently studied its magnetic anisotropy between 90° and 500°K. The compound is paramagnetic. A broad maximum in the susceptibility vs temperature curve was observed around 300°K.

## Tungstenite (WS$_2$)

The structure of this crystal is similar to molybdenite. Its magnetic anisotropy was studied by Datta and Roy Chowdhury [270] between 80° and 700°K. The crystal is observed to be paramagnetic. At 30°C,

$$\chi_{||} = 4917 \quad , \quad \chi_{\perp} = 6314.$$

The temperature variation of its principal moments show maxima in this temperature range.

## F. High Polymers

Magnetic anisotropy of natural and synthetic rubbers has been the subject of much study by many workers. Cotton-Feytis [271-274] first made some interesting observations on commercial rubber and observed that vulcanized rubber exhibited magnetic anisotropy. The effects of compression, stretching and hot and cold working upon crude rubber were investigated. The anisotropy was observed to increase in the early stages of elongation, but at about 400 to 500% elongation, the anisotropy tended towards a limit. Selwood and co-workers [37, 275-277] made extensive measurements on a large number of natural and synthetic rubbers. Some of their results are shown in Fig. 60. It may be noted that the change in anisotropy of natural rubber, as observed by these workers, did not tend towards a limit as Cotton-Feytis reported.

The rubber can be considered as uniaxial crystals, with the unique axis

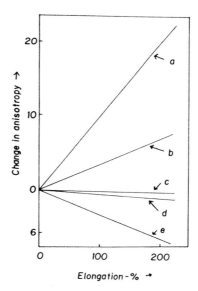

Fig. 60. Change in magnetic anisotropy of rubber versus percent elongation (determined by method of least square) (Ref. 276). a, natural rubber; b, polybutadiene X-453; c, polyethylene; d, polybutadiene XI; and e, Vistanex B-120.

parallel to the direction of stretching. X-ray studies on polyethylene indicate that the electron cloud of the $-CH_2$ groups is not spherical but is elongated in the plane of the carbon-hydrogen bonds, the plane normal to the chain axis. This distortion would lead to an increase in the mean square radii of the electronic orbits of the carbon atoms perpendicular to the chain axis, and hence it would give rise to a susceptibility parallel to the chain axis larger than perpendicular to it. Such an effect has been observed in polyethylene.

The figure shows that natural rubber exhibits a greater change in anisotropy with stretching than does Polybutadiene-X453. This has been attributed to the presence of larger numbers of side chains in the polybutadiene, which decreases the possibility of crystallization on stretching. The presence of unsaturated side chains will also increase the smaller principal susceptibility parallel to the chain axis, thus decreasing the anisotropy of the polymer.

Magnetic anisotropy of polystyrene has also been studied [277]. The maximum possible anisotropy for the unstretched and stretched polystyrene is about the same. However, the anisotropy of polyethylene terephthalate increases with the draw ratio [277].

Isihara et al. [279] have reported some measurements on natural rubber. The measurements carried out by varying the time of vulcanization showed that the anisotropy decreases with the time of vulcanization. Effect of temperature on the anisotropy of the natural rubber was also studied by them. An abrupt change in the anisotropy was observed at about 200°K, which was attributed to a second order phase transition. Above and below this temperature, the change in anisotropy with temperature was observed to be small.

### G. Aromatic and Aliphatic Hydrocarbons

The magnetic anisotropy of aromatic and aliphatic hydrocarbons has been very extensively studied both from experimental and theoretical view-points. In particular, the highly anisotropic character of benzene and its derivatives has provided, over the past few decades, considerable interest for experimental and theoretical investigations. It was first suggested by Raman and Krishnan [278], and by Raman [279, 280] that the large diamagnetic anisotropy of these compounds may be associated with the electrons moving in plane orbits of large area. Later on, Pauling [18] put this suggestion on a quantitative basis and calculated the anisotropy of a large number of aromatic hydrocarbons. He assumed that the $2p_z$ electrons (of the carbon atoms) are free to move, under the influence of the impressed fields, from one carbon atom to adjacent carbon atom. The idea of electrons moving in this way was not, of course, new, as it was indeed implicit in the theory of molecular orbitals applied earlier by Hückel [281-285] to aromatic hydrocarbons and unsaturated molecules. Pauling's semiclassical treatment was later substantiated by quantum mechanical treatment of London

[286-288]. London's method is a molecular orbital technique based on gauge-invariant atomic orbitals and is closely related to the Hückel's theory of $\pi$-electrons.

Pauling's model (Eq. 85) was observed to yield values which were in fair agreement with the experiment. It was, however, noticed that the values so calculated were in most cases, less than the experimentally observed values. For example, Pauling's calculation for benzene yields, $\Delta K = 49$, as against the experimental value of 59.6. Similar discrepancy is also observed for other compounds. London's method did not improve the agreement much further. McWeeny [289-291] has applied this method to a variety of hydrocarbons. The method has also been applied, with some extension, to porphyrins and heterocyclic systems by Berthier et al. [292], Abraham [293], and by Craig et al. [294].

Although the theory of London works well on relative values of diamagnetic anisotropy ($\Delta\chi/\Delta\chi_{Benzene}$) direct calculations on benzene give too small a value even if more refined methods of calculations are used [295, 296]. Lykos and Paar [297] have suggested that certain integrals in London's calculation were underestimated and that an enhancement factor has to be applied.

Pople [298-299] has recently suggested that only part of the anisotropy of benzene is due to the ring currents, about 30% being due to the local anisotropies comparable to those observed in other planar conjugated non-cyclic systems [300]. Singh [301] has indicated that due to the distortion in the $\underline{\sigma}$-electrons by the $\pi$-cloud, an additional correction is necessary. The small anisotropy observed in the plane of the molecule of these compounds is believed to be due to this distortion in the $\underline{\sigma}$-electrons.

The early studies on diamagnetic anisotropies of aromatic hydrocarbons were mainly directed towards the structural investigations. Some striking results about the structure of diphenyl, biphenyl, benzoquinone, chyrsene etc [4, 6, 302] were deduced from the magnetic measurements. It was, however, later realized [62] that the method was not general. The fact that a considerable measure of success attended the first attempts at such predictions, was due to these crystals being special cases. Nevertheless, for most of the large aromatic molecules, a perfectly general consideration of the crystalline diamagnetic anisotropy will give a clear indication as to the approximate orientation of the molecule. Lonsdale [19, 300] studied the diamagnetic anisotropy of some aliphatic and aromatic compounds from the viewpoint of chemical bonding. Recently some more work has been done in this direction [303, 304].

A full appreciation of the work done on aromatic and aliphatic hydrocarbons is not possible here. We shall therefore, discuss here only some compounds of chemical interest and tabulate in the end the data on a large number of compounds (Table 26).

TABLE 26[a]

| Crystal | $\chi_1-\chi_2$ | $\chi_1-\chi_3$ | $-K_1$ | $-K_2$ | $-K_3$ | $\Delta K$ | Ref. |
|---|---|---|---|---|---|---|---|
| Aromatic Compounds | | | | | | | |
| One-ring compounds | | | | | | | |
| Benzene | | | 37.3 | 37.3 | 91.2 | 53.9 | 4 |
| | | | | | | 59.7 | 305 |
| p-Benzoquinone | 40.2 | -1.2 | 24.3 | 28.7 | 67.1 | 40.6 | 6,62 |
| | 39.5 | 1.45 | 23.0 | 27.0 | 65.1 | 40.2 | 306 |
| Durene | 39.7 | 32.0 | 82.4 | 77.3 | 143.9 | 64.0 | 6,62 |
| Hexamethylbenzene | 1.6 | 62.7 | 101.1 | 102.7 | 163.8 | 61.9 | 6 |
| Hexachlorobenzene | 6.8 | 41.7 | 128 | 128 | 182 | 54 | 6 |
| p-Dichlorobenzene | 36.2 | 9.9 | - | - | - | - | 6 |
| | 36.4 | 10.3 | 78.3 | 50.3 | 120.2 | 55.9 | 396 |
| | 36.7 | 9.96 | 75.8 | 63.8 | 116.9 | 47.1 | 311 |
| p-Dibromobenzene | 32.2 | 9.1 | - | - | - | - | 6 |
| | 33.7 | 9.4 | 97.1 | 70.5 | 136.7 | 52.9 | 306 |
| p-Bromochlorobenzene | 35.0 | 9.7 | 87.6 | 59.9 | 129.0 | 55.2 | 306 |
| Chloranil | 22.3 | 2.3 | - | - | - | - | 302 |
| Tetrachloro-hydroquinone | 23.2 | 2.6 | 84.4 | 98.5 | 138.5 | 47.0 | 306 |
| | 25.7 | 25.9 | 103.0 | 114.3 | 144.9 | 36.3 | 306 |

| Compound | | | | | | | Ref. |
|---|---|---|---|---|---|---|---|
| 4(p-)Nitroaniline | 44.8 | 4.9 | 52.0 | 43.0 | 104.8 | 57.3 | 306 |
| p-Dinitrobenzene | 53.6 | 41.0 | - | - | - | - | 6 |
| m-Nitroaniline | 2.0 | 12.5 | - | - | - | - | 6 |
| Bromanil | 19.5 | -0.4 | - | - | - | - | 302 |
| s-Tetrabromobenzene | 40.2 | 2.8 | - | - | - | - | 302 |
| Isatin | 66.4 | 4.7 | 62.0 | 57.3 | 124.2 | 64.6 | 306 |
| Catechol | 30.5 | 27.8 | - | - | - | - | 6 |
| p-Acetotoluiol | 31.6 | 5.4 | - | - | - | - | 302 |
| Resorcinol | $\chi_c - \chi_a = 5.36$ $=5.4$ | $\chi_b - \chi_c = 13.17$ $=13.0$ | 49.2 | 49.2 | 103.2 | 54.0 | 312 302 |
| Phloroglucinol dihydrate | $\chi_c - \chi_a = 37.0$ $=37.1$ | $\chi_b - \chi_a = 36.0$ $=36.1$ | - | - | - | - | 302 |
| Acetanilide | $\chi_a - \chi_c = 35.6$ | $\chi_b - \chi_c = 26.6$ | 55.8 | 44.3 | 116.6 | 66.5 | 306 |
| Anthranilic Acid | $\chi_b - \chi_a = 53.2$ | $\chi_c - \chi_a = 47.2$ | 57.7 | 58.8 | 120.5 | 62.5 | 306 |
| Sodium acid phthalate | $\chi_b - \chi_a = 36.3$ | $\chi_c - \chi_a = 40.3$ | 59.1 | 76.5 | 134.1 | 66.3 | 306 |
| Potassium acid phthalate | $\chi_b - \chi_a = 35.6$ | $\chi_c - \chi_a = 41.7$ | 66.4 | 87.9 | 143.1 | 66.1 | 306 |
| m-Dinitrobenzene | $\chi_a - \chi_b = 13.7$ | $\chi_a - \chi_c = 62.2$ | - | - | - | - | 6 |
| s-Trinitrobenzene | $\chi_a - \chi_b = 6.9$ | $\chi_a - \chi_c = 13.4$ | 64 | 38 | 106 | 55 | 302 |
| p-Toluidine | $\chi_a - \chi_b = 24.4$ | $\chi_a - \chi_c = 1.3$ | - | - | - | - | 302 |

(continued)

TABLE 26 (Continued)

| Crystal | $\chi_1-\chi_2$ | $\chi_1-\chi_3$ | $-K_1$ | $-K_2$ | $-K_3$ | $\Delta K$ | Ref. |
|---|---|---|---|---|---|---|---|
| Hydroquinol | $\chi_\parallel-\chi_\perp=1.4$ | - | - | - | - | - | 6 |
| Two-ring Compounds | | | | | | | |
| Naphthalene | 122 | 29.3 | 39.4 | 43.0 | 187.2 | 104.8 | 4 |
| | 90.4 | 20.6 | 56.1 | 53.9 | 169.0 | 114 | 62 |
| | - | - | 54.7 | 52.6 | 173.5 | 119.8 | 313 |
| Biphenyl | 83.1 | 35.5 | 66.9 | 66.9 | 174.7 | 107.8 | 6 |
| | 77.5 | 33.5 | 67.7 | 61.7 | 183.7 | 119.1 | 306 |
| Dibenzyl | 83.1 | 35.5 | 90.7 | 86.7 | 205.5 | 116.8 | 4,307 |
| Azobenzene | 42.7 | -4.4 | 64 | 66 | 172 | 107 | 4 |
| | 48.6 | -4.9 | - | - | - | - | 302 |
| $\beta$-Naphthol | 86.0 | 18.1 | 46.6 | 50.2 | 194.4 | 146.0 | 4 |
| | 85.3 | 17.3 | 63.9 | 51.9 | 175.2 | 117.3 | 306 |
| 1-Naphthoic Acid | 19.8 | 106.4 | 70.5 | 58.9 | 192.5 | 127.8 | 306 |
| Stilbene | 56.6 | -0.8 | - | - | - | - | 4 |
| | 46.4 | -0.8 | - | - | - | - | 302 |
| Quinhydrone | 54.4 | 16.1 | - | - | - | 70.5 | 302 |
| 4,4'-Dichlorodiphenyl | 35.1 | -25.4 | - | - | - | - | 6 |

| Compound | | | | | | | Ref. |
|---|---|---|---|---|---|---|---|
| 4,4'-Dibromodiphenyl | 35.9 | −22.0 | – | – | – | – | 6 |
| Dimesityl | 27.5 | 3.8 | – | – | – | – | 6 |
| Diphenic Acid | 17.8 | 18.0 | – | – | – | – | 6 |
| 1,4-Naphthoquinone | 87.9 | 16.9 | – | – | – | – | 6 |
| $\alpha$-Naphthol | 51.9 | 61.9 | – | – | – | – | 6 |
| Fluorene | 84.0 | 37.0 | 72.6 | 72.6 | 193.6 | 121.0 | 6 |
| Naphthazarin | 59.0 | 12.9 | – | – | – | – | 302 |
| | 63.0 | – | – | – | – | – | 51 |
| Naphthalene tetrachloride | 62.4 | 3.9 | – | – | – | – | 302 |
| Diphenylamine | 46.8 | 3.9 | – | – | – | – | 302 |
| Tolane | 50.9 | 2.0 | – | – | – | – | 302 |
| Benzidine | 50.9 | 1.4 | 81.5 | 67.8 | 198.5 | 123.9 | 300 |
| | 19.0 | 51.5 | – | – | – | – | 302 |
| Acenaphthene | $\chi_b - \chi_c = 73.5$ | $\chi_a - \chi_c = 28.5$ | 72.0 | 70.5 | 185.5 | 114.25 | 4 |
| | $= 69.4$ | $= 26.0$ | | | | | 306 |
| Benzophenone | $\chi_b - \chi_c = 60.7$ | $\chi_a - \chi_b = 0.6$ | – | – | – | – | 4 |
| Salot | $\chi_a - \chi_c = 3.5$ | $\chi_b - \chi_c = 61.8$ | – | – | – | – | 4 |
| Hydrazobenzene | $\chi_a - \chi_b = 15.0$ | $\chi_c - \chi_b = 48.5$ | – | – | – | – | 4 |
| O-Tolidine | $\chi_a - \chi_b = 82.9$ | $\chi_a - \chi_c = 3.3$ | – | – | – | – | 6 |

(continued)

TABLE 26 (Continued)

| Crystal | $\chi_1-\chi_2$ | $\chi_1-\chi_3$ | $-K_1$ | $-K_2$ | $-K_3$ | $\Delta K$ | Ref. |
|---|---|---|---|---|---|---|---|
| γ-Biphenyl | $\chi_a-\chi_b=34.9$ | $\chi_a-\chi_c=15.9$ | - | - | - | - | 302 |
| Fluorenone | $\chi_a-\chi_b=56.9$ | $\chi_a-\chi_c=25.2$ | - | - | - | - | 6 |
| Benzil | $\chi_\parallel-\chi_\perp=45.6$ | - | - | - | - | - | 4 |
| Fluorene Alcohol | $\chi_\parallel-\chi_\perp=50.8$ | - | - | - | - | - | 302 |
| Three-ring Compounds | | | | | | | |
| Anthracene | 187.3 | 46.0 | 45.9 | 52.7 | 272.5 | 223.2 | 4 |
| | 136.3 | 27.4 | 75.8 | 62.6 | 251.8 | 182.6 | 62 |
| Anthraquinone | 42.2 | 124.45 | 76.1 | 64.5 | 217.9 | 147.6 | 306 |
| Acridine | 147.5 | 110.0 | 61.4 | 70.5 | 238.0 | 172.1 | 306 |
| Terphenyl | 117 | 48.6 | 98 | 98 | 260 | 162 | 6 |
| Phenanthrene | 126 | 39.9 | 74 | 74 | 240 | 166 | 6 |
| Fluoranthene | 109 | 40.0 | - | - | - | - | 6 |
| Dihydroanthracene | 73.0 | 7.0 | - | - | - | - | 302 |
| Thianthrene | 42.4 | -8.3 | - | - | - | - | 302 |
| Triphenylmethane | $\chi_b-\chi_a=24.4$ | $\chi_c-\chi_a=7.3$ | - | - | - | - | 302 |
| Alizarin | $\chi_b-\chi_c=81$ | $\chi_b-\chi_a=29$ | - | - | - | - | 302 |

### Four-ring Compounds

| Compound | | | | | | |
|---|---|---|---|---|---|---|
| Quaterphenyl | 168 | 70 | 129 | 129 | 345 | 216 | 6 |
| Chrysene | 170 | 48.1 | 88 | 88 | 306 | 218 | 6 |
| Pyrene | 97.5 | 125 | 80.6 | 80.6 | 303 | 222.4 | 6 |
| Naphthacene | 169.5 | 31.7 | - | - | - | - | 302 |

### Five-ring Compounds

| Compound | | | | | | |
|---|---|---|---|---|---|---|
| 1,2-Bezpyrene | 156.0 | 81.8 | 100 | 100 | 350 | 250 | 302 |
| Perylene | 131.3 | 60.9 | 80 | 80 | 320 | 240 | 302 |
| Dimethyl-dibenz-phthehanthrene | $\chi_c - \chi_a = 174$ | $\chi_b - \chi_a = 184$ | - | - | - | - | 302 |

### Aliphatic Compounds

| Compound | | | | | | |
|---|---|---|---|---|---|---|
| Oxalic acid dihydrate | 1.91 | 7.05 | 53.15 | 52.73 | 62.40 | 9.46 | 300 |
| Barbituric acid dihydrate | $\chi_a - \chi_b = 17.5$ | $\chi_c - \chi_b 18.4$ | 75.3 | 70.0 | 90.6 | 17.95 | 306 |
| N-Chloro-succinimide | $\chi_a - \chi_b = 12.91$ | $\chi_c - \chi_b = 9.96$ | 64.50 | 51.76 | 76.96 | 18.83 | 306 |
| N-Bromo-succinimide | $\chi_a - \chi_b = 13.07$ | $\chi_c - \chi_b = 9.03$ | 76.03 | 61.57 | 87.31 | 18.51 | 306 |
| Parabanic acid | 8.34 | 15.22 | 28.6 | 27.5 | 49.9 | 21.9 | 306 |
| Ammonium hydrogen d-tartrate | $\chi_a - \chi_b = 7.68$ | $\chi_a - \chi_c = 6.19$ | 37.41 | 38.57 | 46.80 | 8.81 | 306 |
| Potassium hydrogen d-tartrate | $\chi_a - \chi_b = 6.69$ | $\chi_a - \chi_c 6.15$ | 38.31 | 39.04 | 47.26 | 8.59 | 306 |

(continued)

TABLE 26 (Continued)

| Crystal | $\chi_1-\chi_2$ | $\chi_1-\chi_3$ | $-K_1$ | $-K_2$ | $-K_3$ | $\Delta K$ | Ref. |
|---|---|---|---|---|---|---|---|
| Chloroacetamide | 4.73 | 2.045 | 51.70 | 48.74 | 53.37 | 3.15 | 306 |
| Glycene | 1.59 | 6.17 | - | - | - | - | 36 |
| Urea | $\chi_c-\chi_a=2.57$ | | 31.84 | 31.84 | 36.98 | 5.14 | 314 |
| | | $\chi_\parallel-\chi_\perp=2.45$ | | | | | 336 |
| Urea nitrate | 7.05 | -1.09 | - | - | - | - | 314 |
| Urea oxalate | 12.54 | -3.27 | - | - | - | - | 314 |
| Thiourea | $\chi_c-\chi_b=2.52$ | $\chi_c-\chi_a=2.75$ | - | - | - | - | 36 |

[a] $K_1$ and $K_2$ lie in the molecular plane along the long and the short axes, while $K_3$ is perpendicular to the plane of the molecule.

## Metal-free Phthalocyanine:

Martin and Mitra [303] have recently measured the diamagnetic anisotropy of the single crystals of metal-free phthalocyanine. The phthalocyanine molecule contains a central C-N ring system of 18 $\pi$-electrons, which is very nearly circular. The system thus provides a very poor approximation to Eq. 85. From the measurement of crystal anisotropies, the molecular anisotropy due to the C-N ring is deduced to be: $K_\perp - K_{||} = 365$. Using the geometrical area of the C-N ring, eq. 85 gives $K_\perp - K_{||} = 470$.

## Cyanuric Compounds:

Magnetic anisotropy of some cyanuric compounds has been reported by Lonsdale [19]. These compounds are strongly anisotropic, though less than benzene or its derivatives. Cyanuric triazide, $C_3N_3(N_3)_3$, molecule is planar [309], and its molecular anisotropy is $\Delta K = 21.8$. For cyanuric trichloride, $\Delta K = 31.3$. Each of these electrons have six resonating electrons. Assuming these six electrons to occupy molecules orbits of approximately the same area as that of the cyanuric molecule itself, the anisotropy should be almost twice as much as is experimentally observed. On the other hand, if the electrons are assumed to be paired, forming double bonds, very little anisotropy should exist. The experimental data, therefore, indicates that the $\pi$-electrons do occupy molecular orbits, but with a strong tendency to the formation of double bonds in the molecule; in other words, there is an alternation of electron density round the nucleus, but perhaps not as great as would be expected for alternate double and single bonds between C and N atoms.

## Oxalic Acid Dihydrate:

The molecule is planar and the C - C distance in the oxalate group is about 1.44Å [310]. This indicates the presence of doublebond properties, presumably due to the conjugation of double bonds in adjacent carboxyl groups. The principal susceptibilities of the molecule are [300]

$$K_1 = -53.13$$

$$K_2 = -52.73$$

$$K_3 = -62.40,$$

$\Delta K$ being 9.47. The anisotropy is large compared with that usually expected for aliphatic molecules, and is about twice the value found for $CO_3^=$ group [4]. The large anisotropy of oxalates and carbonates indicates the presence of closed chains is not necessary for large anisotropy, since these molecules do not have closed chains of conjugated bonds. This observation has been of great help in many later calculations of diamagnetic anisotropy [298, 299].

## H. Miscellaneous

We shall discuss here briefly the work done on some compounds which could not be categorized in earlier subsections.

### Metals and alloys

Magnetic anisotropy of several metals (e.g. Bi, Sb, Tl, I etc.) has been studied in some details. John [315] has reported the temperature variation of the anisotropy of Bi. At 27°C, he observes:

$$\chi_{||} = 1.05$$

$$\chi_{\perp} = -1.45$$

As the temperature is raised, the anisotropy decreases linearly reaching a value of 0.09 at 260°C, beyond which it becomes practically constant right up to its melting point. Jones [316] has discussed the shape of its Fermi surface and shows that $\chi_{||}$ is mainly due to those parts of the Fermi surface which overlaps into the second main zone in the region of the [110] planes, while $\chi_{\perp}$ arises from the parts of the Fermi surface lying within the first zone. It has been pointed out that the increase or decrease in the number of electrons per atom would, respectively, decrease or increase the ratio $\chi_{\perp}/\chi_{||}$. Such variation in the Fermi distribution surface is observed by dissolving in the Bi small quantities of Sn, Pb, Te or Se.

Behavior of Sb is somewhat similar. At 290°K, Shoenberg and Uddin [317] observe,

$$\chi_{||} = -0.55 \quad , \quad \chi_{\perp} = -1.42$$

Diamagnetic anisotropy of Th single crystal has been reported by Rao and Subramanian [318]. At 30°C,

$$\chi_{||} = -0.407 \quad , \quad \chi_{\perp} = -0.163.$$

As the temperature is increased, $\chi_{\perp}$ remains constant while $\chi_{||}$ decreases, indicating the movements of atoms when the closed-packed hexagonal type is changed into the face centered cubic form.

The anisotropy of iodine crystal has been reported by Rao and Venkataramiah [319]. At room temperature,

$$\chi_{||} = -0.389 \quad , \quad \chi_{\perp} = -0.331.$$

Magnetic anisotropy of FeGe has been studied [320]. At 412°K, the spins appear to be aligned parallel to the c-axis, with ferromagnetic ordering within each c-plane.

## Naturally occurring substances

Magnetic anisotropy of some naturally occurring substances, like shells of Placuna placenta, Pinna bicolor, Meretrix casta, Vulsella rugosa, Mactra special, and Turbinella perium has been studied [321]. In all cases, the crystalline character has been established.

The anisotropy of the nacre of Turbo, Haliotis, Mytilus, Viridis has been measured and their crystalline structure has been established [322].

## Wood

Nilakanta [323] has studied the magnetic anisotropy of various specimens of wood and $\alpha$-cellulose (see Table 27). It has been found that the crystalline element in wood is definitely $\alpha$-cellulose, and that lignin and hemicellulose make no contribution to the magnetic anisotropy, suggesting either a random orientation if they are crystalline or amorphous structure. Anisotropy measurement also indicates that the direction of maximum diamagnetic susceptibility in the cellulose molecule is along the length of the chain.

## Biological molecules:

Magnetic anisotropy of some conjugated molecules of biological interests has been measured [324]. The results are summarized in Table 28.

TABLE 27

| Wood specimen | $\Delta\chi$ | $\alpha$-Cellulose % | Lignin % |
|---|---|---|---|
| 1. After extraction with alcohol-benzene and boiling water | 0.39 | 37 | 33 |
| 2. After extraction with 5% NaOH | 0.40 | 40 | 45 |
| 3. After partial chlorination | 0.68 | 59 | 15 |
| 4. $\alpha$-cellulose | 0.92 | 100 | – |

TABLE 28

| Crystal | $\frac{\Delta\chi}{\Delta\chi_{Benzene}}$ | Crystal | $\frac{\Delta\chi}{\Delta\chi_{Benzene}}$ |
|---|---|---|---|
| Adenine | 1.120 | Uracile | 0.107 |
| Guanine | 0.617 | Imidazole | 0.385 |
| Hypoanthine | 0.704 | Indole | 1.597 |
| Xanthine | 0.436 | Riboflavin | 1.660 |
| Acid Urique | 0.240 | 2-Amino-4-hydroxy pteridine | 1.093 |
| Cytosine | 0.283 | | |

Glass fibres:

The anisotropy of glass fibres spun from $NaPO_3$ and $B_2O_2$ melts has been investigated [325]. The values of $\Delta\chi/M$ (M being the molecular wt.) for $NaPO_3$ fibres being 0.0187 and 0.02788. If the glass is annealed for 24 hours at 375°C, the anisotropy vanishes. The anisotropy of boron oxide glass fibres is much larger than that of the sodium phosphate. The anisotropy in both cases is supposed to arise from the parallel arrangement of structural chains, in metaphosphate glass by infinite $(PO_4)$ chains.

Besides the above, references 326-329 may be included in the studies on transition metal paramagnetic compounds.

## IV. ACKNOWLEDGMENTS

It is a pleasure to thank Professor R. L. Martin for his constant encouragement and interest during the preparation of this article. Thanks are also due to Dr. C. G. Barraclough with whom the author had some very profitable discussions. The help of Mr. A. K. Gregson in the preparation of the manuscript is also gratefully acknowledged.

## ADDENDUM

Most of this article was completed in early 1969. Since then a large amount of work has been accomplished. We list below some interesting papers which developed from this more recent work, together with one or two papers which were not included earlier in this chapter.

1. Recently measurements on some high-spin five-coordinated nickel (II) and cobalt (II) compounds have been reported [330] between 300-90°K. The metal atoms in these compounds have nearly square pyramidal configuration.

Measurements on $[Ni(Ph_2MeAsO)_4NO_3]^+NO_3^-$ and $[Ni(Ph_2\ Me\ AsO)_4\ ClO_4]^+ClO_4$ show that these crystals are highly anisotropic in contrast to the feebly anisotropic character of the octahedral nickel (11) compounds. The average magnetic moments of these compounds, particularly of the chlorate complex, are rather high and lie in the range 3.4-3.5 B.M. The experimental results on the dinitrate complex have been analyzed in detail. In $C_{4\vartheta}$ symmetry the ground state is found to be $^3B_1$. Using the crystal field parameters deduced from polarized crystal spectra, a good fit is obtained with the experimental results of principal magnetic moments. A zero-field splitting of about 24 $cm^{-1}$ of the ground state $^3B_1$ was assumed. The excited states $^3E$, $^3A_2$, $^3B_2$ and $^3E$ are found to lie at 3890, 8330, 9140 and 11, 570 $cm^{-1}$ respectively from the ground state.

$[Co(Ph_2\ Me\ AsO)_4\ NO_3]^+NO_3^-$ has a very high magnetic moment of $\bar{\mu} = 5.07$ B.M. at room temperature. The single crystals of the compound are extremely anisotropic, the anisotropy being about 60% and 95% of the average susceptibility at 300 and 90°K respectively. $\mu_\perp$ is observed to remain nearly constant while $\mu_{||}$ changes significantly in the temperature range 300-90°K. The ground state in $C_{4\vartheta}$ symmetry is $^4A_2$ and the excited states $^4E$, $^4B_2$, $^4E$, $^4B_1$ are found to be at 650, 2030, 6120 and 10620 $cm^{-1}$, respectively, above the ground state. The presence of these low-lying states, which is common with low-symmetry complexes, appears to be responsible for the high magnetic moment and magnetic anisotropy of the complex.

2. Magnetic anisotropy of antipyrene complexes of ytterbium (III) and cerium (III) has been reported [331] between 300-90°K. The results have been interpreted within free-ion $f^{1,13}$ basis perturbed by spin-orbit coupling and a crystal field of $D_{3d}$ symmetry.

3. Paramagnetic anisotropies of some high-spin ferric complexes, like ferric acetylacetonate and ferric potassium oxalate trihydrate, have been measured between 300-90°K [332]. As expected, the anisotropies are very feeble and compare well with previous determinations [7]. Zerofield splitting parameter (D) is estimated to be very small and negative. No correction for the shape and diamagnetic anisotropy appears to have been made, but these corrections are very important in the present case.

4. Uenoyama et al [333] have reported a detailed magnetic anisotropy studies on acid ferrimyoglobin and ferrimyoglobin fluoride single crystal. The work is a beautiful application of this technique to deduce the zerofield splitting parameter (D) from the magnetic anisotropy studies. The myoglobin was extracted from sperm whale and purified in the usual way. Measurements were made between liquid nitrogen and helium temperatures. The magnetic anisotropy of these complexes in this temperature range is very large and the corrections mentioned earlier for high-spin iron (111) complexes are not very important.

The ferric ion in these compounds has a formal spin state of S = 5/2. The ground $^6A_{1g}$ term is split into three Kramers' doublets, Ms = $\pm \frac{1}{2}$,

± 3/2 and ± 5/2 with an overall splitting of 6D. ESR studies on the single crystals of these compounds have shown that D is very large ($\geq$ 5 cm$^{-1}$) and is positive (Ms = ± 1/2 lying lowest). In such a case, the appropriate spin-Hamiltonian can be written as

$$H = D \left\{ S_z^2 - \frac{1}{3} S (S + 1) \right\},$$

neglecting the rhombic term. Here z axis is taken perpendicular to the haem plane. The magnetic anisotropy in our usual notation is then given by

$$(\mu_\perp^2 - \mu_{||}^2) = \frac{3}{1 + e^{-2x} + e^{-6x}} \left\{ 8(1 + \frac{1}{x}) - (9 + \frac{11}{x})e^{-2x} \right.$$

$$\left. -5 (5 + \frac{1}{2x})e^{-6x} \right\}$$

where x = D/kT.

From their measurements, Uenoyama et al deduced D = 10.5 cm$^{-1}$ for acid ferrimyoglobin and D = 6.5 cm$^{-1}$ for ferrimyoglobin fluoride. These values are in good agreement with those from ESR studies.

5. Recently a series of compounds with the general formula $K_2 M(II) F_4$ has been discovered which appear to be ideal two-dimensional antiferro-magnets. Principal susceptibilities of $K_2 Ni F_4$, $K_2 Co F_4$ and $K_2 Fe F_4$ have been measured over a range of temperature [334, 335]. The crystals show characteristic behavior of such an interacting system. $K_2 Cu F_4$ is found to be a ferromagnet at low temperature.

6. Principal susceptibilities of the double nitrates of Ce(III), Pr(III) and Nd(III) have been measured down to 4.4°K [336].

7. Principal magnetic susceptibilities of guadinium vanadium(III) sulfate hexahydrate has been determined [337] between 1.5-20°K. Zerofield splitting parameter is deduced to be D = 3.74 cm$^{-1}$ which agrees well with ESR results.

8. Magnetic properties of single crystals of two "square" planar cobalt (II) compounds were studied between 300°-90°K. Both of these crystals show very similar magnetic behaviors, but very different from the planar cobalt (II) phthalocyanine described earlier (see page 265).

Cobalt (II) dithioacetylacetonate provides an interesting example where the cobalt (II) ion is bonded to four sulfur atoms of the dithioacetylacetonate ligands in very nearly a square planar geometry. The compound is low-spin (S = $\frac{1}{2}$) with $\overline{\mu}$ = 2.27 B.M. at room temperature which remains nearly temperature independent down to 90°K. Measurements on its single crystals reveal very characteristic features which are summarized below [16]:

|          | 300°K       | 90°K        |
|----------|-------------|-------------|
| $\mu_x$  | 2.93 B.M.   | 2.80 B.M.   |
| $\mu_y$  | 1.89        | 1.92        |
| $\mu_z$  | 1.77        | 1.55        |

Here x and y axes lie in the plane of the molecule and z axis is perpendicular to it. Note the large anisotropy in the (x-y) plane in contrast to the isotropic nature of this plane in cobalt (II) phthalocyanine (page 265). This unique behavior of the cobalt (II) dithioacetylacetonate has been rationalized [16] by assuming that the unpaired electron lies in the $d_z2$ orbital. A crystal field calculation on this basis indicates a large splitting of the $d_{xz,yz}$ orbitals with $d_z2\ d_{yz}$ = 1900 cm$^{-1}$ and $d_z2\ d_{xz}$ = 10,000 cm$^{-1}$.

Similar magnetic behaviors are observed [338] in another "square" planar cobalt (II) compound, trans-dimesitylbis(diethylphenyl phosphine) cobalt (II).

9. Recently very detailed measurements and analysis of the magnetic anisotropy studies on tetrahedrally coordinated nickel (II) compounds have been reported [339, 340]. The ground term in the tetrahedral ligand field is an orbital triplet which splits into a singlet and a doublet in the axial part of the ligand field. Hence the crystals of tetrahedrally coordinated nickel (II) compounds are expected to be fairly anisotropic as contrasted to the very feebly anisotropic character of octahedrally coordinated nickel (II) compounds. The results on the "tetrahedral" nickel (II) compounds are in general agreement with this expectation. One of the compounds, tetraethylammonium tetrachloronickelate (II), shows a very interesting phase transition around 218°K which is similar to that observed in cobalt fluosilicate (see page 262).

10. Paramagnetic anisotropy of binuclear copper (II) acetate monohydrate has been recently reinvestigated [341]. It is observed that the magnetic anisotropy data of all the previous workers [95, 169-171] are seriously in error, perhaps because of their erroneous identification of the crystal axes.

The recent measurement [341] gives following values at 300°K:

$$\chi_1 - \chi_2 = 200, \quad \chi_1 - \chi_3 = 114$$

A comparison of the above values with the earlier rusults summarized in Fig. 44 reveals the gross errors in previous measurements. The new data shows that the magnetic anisotropy of this binuclear compound is very sensitive to the anisotropy in high frequency paramagnetism, its contribution at 80°K and 300°K being about 70% and 25% of the total measured anisotropy respectively. This particular feature of this compound was utilized [341] in assigning its controversial electronic spectrum in terms of following ligand field transitions:

$$11,000 \text{ cm}^{-2} \quad : \quad d_{Z^2} \longleftarrow d_{X^2-Y^2}$$

$$14,400 \text{ cm}^{-1} \quad : \quad d_{XZ,YZ} \longleftarrow d_{X^2-Y^2}$$

$$17,000 \text{ cm}^{-1} \quad : \quad d_{XY} \longleftarrow d_{X^2-Y^2}$$

The new data conform very closely to the Bleaney-Bowers theory with constant values of ligand field parameters and exchange integral.

11. Gerloch and Quested [342] have recently reinvestigated the principal susceptibilities of $Co(NH_4)_2 (SO_4)_2.6H_2O$. They observe that the ligand field axes in this compound do not coincide with the crystallographic symmetry axes of the $(Co,6H_2O)$ system. This is an important result which could be true in several other systems with low symmetry.

12. Paramagnetic anisotropy of ferrous ammonium sulfate hexahydrate has been reinvestigated [343]. It is observed that the sign of its molecular anisotropy $(K_{||}>K_{\perp})$ deduced by previous workers [117, 121] is in error. The new measurements show that $K_{\perp}>K_{||}$ and hence $^5B_2$ lies below $^5E$ (see Fig. 27). The principal susceptibilities and their temperature variation can be explained quite well by a temperature-independent set of parameters. The previous necessity of invoking temperature dependence of the ligand field parameters disappears once correct sign of the anisotropy is assumed.

13. Principal susceptibilities of $K_2[Cu(H_2O)_6] (SO_4)_2$ have been again analyzed within molecular orbital approximation [344].

14. Paramagnetic anisotropy of $Co(bipy)_3 Br_2.6H_2O$ has been studied and correlated with N.M.R. contact shift studies [345].

15. A review article on paramagnetic anisotropy has recently appeared [346], which lists most of the references till 1970.

## REFERENCES

1. L. F. Bates, <u>Modern Magnetism</u>, Cambridge University Press, London, 1961, Chapter IV.

2. F. Stenger, <u>Wied. Ann. (Ann. Physik.)</u>, <u>20</u>, 304 (1883); <u>35</u>, 331 (1888).

3. C. G. W. König, <u>Wied. Ann. (Ann. Physik.)</u>, <u>31</u>, 273 (1887).

4. K. S. Krishnan, B. C. Guha, and S. Banerjee, <u>Phil. Trans. Roy. Soc.</u>, A231, 235 (1933).

5. K. S. Krishnan, S. Banerjee, and N. C. Chakravorty. <u>Phil. Trans. Roy. Soc.</u>, A232, 99 (1933).

6. K. S. Krishnan and S. Banerjee, Phil. Trans. Roy. Soc., A234, 265 (1935).

7. K. S. Krishnan and S. Banerjee, Phil. Trans. Roy. Soc., A235, 343 (1936).

8. K. S. Krishnan and A. Mookherji, Phil. Trans. Roy. Soc., A237, 135 (1938).

9. K. S. Krishnan, A. Mookherji, and A. Bose, Phil. Trans. Roy. Soc., A238, 125 (1939).

10. J. W. Stout and M. Griffel, J. Chem. Phys., 18. 1449 (1950).

11. M. Griffel and J. W. Stout, J. Chem. Phys., 18, 1455 (1950).

12. J. W. Stout and L. M. Matarrese, Rev. Mod. Phys., 25, 338 (1953).

13. A. K. Gregson and S. Mitra, J. Chem. Phys., 49, 3693 (1968).

14. C. G. Barraclough, R. L. Martin, S. Mitra, and R. C. Sherwood, J. Chem. Phys., 55, 1638 (1970).

15. R. L. Martin and S. Mitra, Chem. Phys. Letters, 3, 183 (1969).

16. A. K. Gregson, R. L. Martin and S. Mitra, Chem. Phys. Letters, 5, 310 (1970).

17. R. L. Martin and S. Mitra, Inorg. Chem., 9, 182 (1970).

18. L. Pauling, J. Chem. Phys., 4, 673 (1936).

19. K. Lonsdale, Proc. Roy. Soc., A159, 149 (1937).

20. J. H. Van Vleck, The Theory of Electric and Magnetic Susceptibilities, Oxford University Press, London, 1932.

21. C. J. Ballhausen, Introduction to Ligand Field Theory, McGraw-Hill, New York, 1962.

22. L. E. Orgel, An Introduction to Transition-Metal Chemistry, 2nd ed., Methuen, London, 1966.

23. B. N. Figgis, Introduction to Ligand Fields, Interscience, New York, 1966.

24. B. N. Figgis and J. Lewis, Progress in Inorganic Chemistry, 6, Chapter II, Interscience, New York, 1964.

25. J. S. Griffith, The Theory of Transition Metal Ions, Cambridge Univ. Press, New York, 1961.

26. M. H. L. Pryce, Nuovo Cimento, 6 (suppl.), 817 (1957).

27. B. Bleaney and K. W. H. Stevens, Reports Progr. Phys., 16, 108 (1953).

28. J. F. Nye, The Physical Properties of Crystals, Clarendon Press, Oxford, 1957.

29. A. Mookherji, Indian J. Phys., 20, 9 (1946).

326                                   S. MITRA

30. S. Datta, Indian J. Phys., 28, 239 (1954).

31. J. Simon and J. Toussaint, Bull. Soc. Royale Sci. (Liege), 32, 881 (1963).

32. A. H. Cooke, R. Lazenby, and M. J. M. Leask, Proc. Phys. Soc. London, 85, 767 (1965).

33. K. Lonsdale, Reports Progress Phys., 4, 368 (1937).

34. R. Pointeau, and E. Poquet, Compt. Rend., 249, 546 (1959).

35. S. Datta, Indian J. Phys., 27, 155 (1953).

36. D. A. Gordon, Rev. Sci. Inst., 29, 929 (1958); J. Phys. Chem., 64, 273 (1960).

37. P. W. Selwood, J. A. Parodi, and A. Pace, Jr., J. Am. Chem. Soc., 72, 1269 (1950).

38. B. N. Figgis and J. Lewis, in Tech. Inorg. Chem., 4, (Interscience, New York, 1964).

39. P. Groth, Chem. Krist., 2, 419 (1906).

40. J. M. Baker, B. Bleaney, and K. D. Bowers, Proc. Phys. Soc. (London), 69, 1205 (1956).

41. L. C. Jackson, Proc. Roy. Soc., A140, 695 (1933).

42. A. K. Gregson and S. Mitra, J. Chem. Phys., 49, 3696 (1968).

43. K. S. Krishnan and A. Mookherji, Phys. Rev., 50, 860 (1936).

44. K. S. Krishnan and A. Mookherji, Phys. Rev., 54, 533 (1938).

45. K. S. Krishnan and A. Mookherji, Phys. Rev., 54, 841 (1938).

46. K. S. Krishnan and A. Mookherji, Nature, 140, 896 (1937).

47. A. Bose, Indian J. Phys., 21, 275 (1947).

48. N. B. Brandt., Pribory i Tekh. Eksperimenta, No. 3, 114 (1960); (see also, Chem. Abstract., 55, 6080 (1961)).

49. G. Schoffa, O. Ristau, and G. Mai, Exptl. Tech. Physik., 7, 217 (1959).

50. H. Zijlstra, Rev. Sci. Inst., 32, 634 (1961).

51. M. J. Rogers, J. Am. Chem. Soc., 69, 1506 (1947).

52. C. W. Fleischmann and A. G. Turnes, Rev. Sci. Inst., 37, 73 (1965).

53. H. H. Van Mal, J. Sci. Inst., 44, 446 (1967).

54. D. G. Thakurta and D. Mukhopadhyay, Indian J. Phys., 40, 69 (1966).

55. D. Neogy, S. Banerji, P. Kumar, and A. Mahalanabis, Indian J. Phys., 41, 744 (1967).

56. S. C. Mathur, Proc. Nat. Inst. Sci. India Pt.A, 26, 581 (1960).

57. U. S. Ghosh and R. N. Bagchi, Indian J. Phys., 36, 538 (1962).

58. D. Neogy, Indian J. Pure & Appl. Phys., 1, 123 (1963).

59. D. Neogy, Indian J. Pure & Appl. Phys., 1, 401 (1963).

60. J. K. Ghose, Indian J. Pure & Appl. Phys., 2, 94 (1964).

61. J. K. Ghose, Indian J. Pure & Appl. Phys., 4, 175 (1966).

62. K. Lonsdale and K. S. Krishnan, Proc. Roy. Soc., A156, 597 (1936).

63. U. S. Ghosh and S. Mitra, Indian J. Phys., 38, 19 (1964).

64. S. Mitra, P. K. Ghose, S. K. Dutta Roy, Indian J. Phys., 37, 552 (1963).

65. J. K. Ghose, Indian J. Phys., 40, 457 (1966).

66. S. Bhagavantanam, Crystal Symmetry and Physical Properties, Academic Press, London, 1966.

67. M. A. Lasheen and S. Tadros, Acta Cryst., A24, 287 (1968).

68. E. C. Stoner, Phil. Mag., 36, 803 (1945).

69. J. A. Osborne, Phys. Rev., 67, 351 (1945).

70. S. Uyeda, M. D. Fuller, J. C. Belshe, and R. W. Girdler, J. Geophysical Research, 68, 279 (1963).

71. P. Rhodes and G. Rowlands, Proc. Leeds Phil. Soc., 6, 191 (1954).

72. J. L. Snoek, Physica, 1, 649 (1934).

73. M. Majumdar, Indian J. Phys., 36, 111 (1962).

74. W. Hofmann, Z. Kristallogr., 78, 279 (1931).

75. N. W. Grimes, H. F. Kay, and M. W. Webb, Acta Cryst., 16, 823 (1963).

76. M. W. Webb, H. F. Kay, and N. W. Grimes, Acta Cryst., 18, 740 (1965).

77. L. Pauling, Z. Kristallogr., 72, 482 (1930).

78. W. C. Hamilton, Acta Cryst., 15, 353 (1962).

79. J. A. A. Ketelaar, Physica, 4, 619 (1937).

80. W. H. Zachariasen, J. Chem. Phys., 3, 197 (1935).

81. G. S. Bogle and V. Heine, Proc. Phys. Soc., A67, 734 (1954).

82. J. M. Robertson, J. Chem. Soc., 615 (1935); ibid., 1109 (1936); ibid., 219 (1937); ibid., 36 (1940).

83. A. K. Gregson and S. Mitra - (Unpublished, 1968).

84. R. P. Dodge, D. H. Templeton, and A. Zalkin, J. Chem. Phys., 35, 55 (1961).

85. J. Van den Handel and A. Siegert, Physica, 4, 871, (1937).

86. B. N. Figgis, J. Lewis, and F. Mabbs, J. Chem. Soc., 2480 (1960).

87. A. S. Chakravarty, Proc. Phys. Soc. (London), 74, 711 (1959).

88. W. H. Brumage, C. R. Quade, and C. C. Lin, Phys. Rev., 131, 949 (1963).

89. A. Bose, R. Chatterji, and R. Rai, Proc. Phys. Soc. (London), 83, 959 (1964).

90. W. T. Astbury, Proc. Roy. Soc., A112, 448 (1926).

91. J. N. van Niekerk and F. R. L. Schoening, Acta Cryst., 5, 196, 499 (1952).

92. P. R. Saha, Indian J. Phys., 41, 628 (1967).

93. I. I. Rabi, Phys. Rev., 29, 174 (1927).

94. L. C. Jackson, Proc. Roy. Soc., A140, 695 (1933).

95. B. C. Guha, Proc. Roy. Soc., A206, 353 (1951).

96. N. Kurty and F. Simon, Proc. Roy. Soc., A149, 152 (1935).

97. J. H. Van Vleck, J. Chem. Phys., 9, 85 (1941).

98. B. F. Hoskins, R. L. Martin, A. H. White, Nature, 211, 677 (1966).

99. R. L. Martin and A. H. White, Inorg. Chem., 6, 712 (1967).

100. R. L. Martin and S. Mitra - (Unpublished).

101. H. H. Wickman, A. M. Trozzolo, H. J. Williams, G. W. Hull, and F. R. Merritt, Phys. Rev., 155, 563 (1967).

102. A. P. Ginsberg, R. L. Martin, and R. C. Sherwood, Inorg. Chem., 7, 932 (1968).

103. P. W. Anderson, in Magnetism, Vol. 1, (Academi Press, New York, 1963).

104. R. L. Martin, New Pathways in Inorganic Chemistry (Cambridge Univ. Press, Chapter 9, 1968).

105. J. B. Howard, J. Chem. Phys., 3, 811 (1935).

106. H. Kamimura, J. Phys. Soc. Japan, 11, 1171 (1956).

107. A. Bose, A. S. Chakravarty, and R. Chatterji, Proc. Roy. Soc. A255, 145 (1960).

108. A. K. Gregson and S. Mitra, Chem. Phys. Letters, 3, 392 (1969).

109. P. Groth, Chem. Krist., 1, 418, 325 (1906).

110. V. Barkhatov, Acta Phys. Chim, URSS, 16, 23 (1942).

111. V. Barkhatov and H. Zhdanov, Acta Phys. Chim. URSS, 16, 43 (1942).

112. L. C. Jackson, Proc. Phys. Soc. (London), 50, 707 (1938).

113. B. Bleaney and M. C. M. O'Brien, Proc. Phys. Soc. (London), B69, 1216 (1956).

114. B. N. Figgis, Trans. Faraday Soc., 57, 204 (1961).

115. G. Schoffa, O. Ristau, and K. Ruckpaul, Zhur, Eksptl. i Theoret. Fiz., 35, 641 (1958).

116. B. N. Figgis, J. Lewis, F. E. Mabbs and G. A. Webb, J. Chem. Soc., 422 (1966).

117. A. Bose, A. S. Chakravarty, and R. Chatterji, Proc. Roy. Soc., A261, 207 (1961).

118. D. Palumbo, Nuovo Cimento, 8, 271 (1958).

119. A. Bose and R. Rai, Indian J. Phys., 39, 176 (1965).

120. H. Eicher, Z. Physik., 171, 582 (1963).

121. E. König and A. S. Chakravarty, Theoret, Chim. Acta (Berl.), 9, 151 (1967).

122. A. Bose, Indian J. Phys., 22, 483 (1948).

123. M. S. Joglekan, Z. Krist., 98, 411 (1938).

124. M. Majumdar and S. K. Datta, Indian J. Phys., 41, 590 (1967).

125. L. C. Jackson, Phil. Mag., 4, 269 (1959).

126. R. Ingalls, Phys. Rev., A133, 787 (1964).

126a. D. W. Dale, R. J. P. Williams, C. E. Johnson, and T. L. Thorp, J. Chem. Phys. 49, 3441 (1968).

127. C. G. Barraclough, R. L. Martin, S. Mitra, and R. C. Sherwood, J. Chem. Phys. 55, 1643 (1970).

128. J. W. Stout, M. I. Steinfeld, and M. Yuzuri, J. App. Phys., 39, 1141 (1968).

129. R. Kubo, Phys. Rev., 88, 568 (1952).

130. L. C. Jackson, Phil. Trans. Roy. Soc., A224, 1 (1923); A226, 107 (1927).

131. A. Bose, Indian J. Phys., 22, 276 (1948).

132. R. Schlapp and W. G. Penney, Phys. Rev., 42, 666 (1932).

133. N. Uryû, J. Phys. Soc. Japan, 11, 770 (1956).

134. A. Bose, A. S. Chakravarty and R. Chatterji, Proc. Roy. Soc., A261, 43 (1961).

135. J. N. van Niekerk and F. R. L. Shoening, Acta Cryst., 6, 609 (1953).

136. B. C. Guha, Nature, 184, 50 (1959).

137. A. Mookherji and S. C. Mathur, Physica, 31, 1547 (1965).

138. M. Majumdar and S. K. Datta, J. Chem. Phys., 42, 418 (1965).

139. A. Bose, L. C. Jackson and R. Rai, Indian J. Phys., 39, 7 (1965).

140. S. Ray, Indian J. Phys., 38, 176 (1964).

141. A. Ohtsubo, J. Phys. Soc. Japan, 20, 82 (1965).

142. S. K. Dutta Roy and B. Ghosh, J. Phys. Chem. Solids, 29, 1511 (1968).

143. J. H. Van Vleck, Phys. Rev., 41, 208 (1932).

144. J. H. Van Vleck, Disc. Faraday Soc., no.26, 96 (1958).

145. H. M. Powell and A. F. Wells, J. Chem. Soc., 359 (1935).

146. K. S. Krishnan and A. Mookherji, Phys. Rev., 51, 774 (1937).

147. A. Bose, S. Mitra, and R. Rai, Indian J. Phys., 39, 357 (1965).

148. A. Bose, R. Rai, S. Kuman, and S. Mitra, Physica, 32, 1437 (1966).

149. B. N. Figgis, M. Gerloch, and R. Mason, Proc. Roy. Soc., A279, 210 (1964).

150. K. D. Bowers and J. Owen, Rep. Prog. Phys., 18, 304 (1955).

151. F. A. Cotton, D. M. L. Goodgame and M. Goodgame, J. Am. Chem. Soc., 83, 4690 (1961).

152. R. P. van Stapele, H. G. Beljers, P. F. Bongers, and H. Zijlstra, J. Chem. Phys., 44, 3719 (1966).

153. M. A. Porai-Koshits, Trudy Inst. Krist, Akad. Nauk SSSR, 10, 117 (1954).

154. K. S. Krishnan and A. Mookherji, Phys. Rev., 51, 528 (1937).

155. S. Mitra, J. Chem. Phys., 49, 4724 (1968).

155a. G. S. Zdhanov and Z. V. Zvankova, Zh. fiz. Khim., 24, 1339 (1950).

156. A. Bose, S. C. Mitra, and S. Datta, Proc. Roy. Soc., A248, 153 (1958).

157. A. Bose and R. Chatterjee, Proc. Phys. Soc. (London), 82, 23 (1963).

157a. C. A. Beevers and C. M. Schwartz, Z. Krist., 19, 157 (1935).

158. A. Mookherji, Indian J. Phys., 20, 9 (1945).

159. C. A. Beevers and H. L. Lipson, Z. Krist., 83, 123 (1932).

160. B. Bhattacharya and M. Majumdar, Indian J. Phys., 40, 549 (1966).

161. U. Öpik and M. H. L. Pryce, Proc. Roy. Soc., A238, 425 (1957).

162. A. Bose, S. C. Mitra, and S. K. Datta, Proc. Roy. Soc., A239, 165 (1957).

163. A. Mookherji, Indian J. Phys., 19, 63 (1945).

164. A. Bose, Indian J. Phys., 22, 483 (1948).

165. A. Mookherji and M. T. Tin, Z. Krist., A101, 412 (1939).

166. J. N. van Niekerk and F. R. L. Schoening, Acta Cryst., 6, 227 (1953).

167. B. Bleaney and K. D. Bowers, Proc. Roy. Soc., A214, 451 (1952).

168. B. N. Figgis and R. L. Martin, J. Chem. Soc., 3837 (1956).

169. A. Mookherji and S. C. Mathur, J. Phys. Soc. Japan, 18, 977 (1963).

170. A. Mookherji and S. C. Mathur, Nature, 186, 370 (1962).

171. A. Bose, R. N. Bagchi and P. Sengupta, J. Appl. Phys., 39, 1146 (1968).

172. S. Mitra and P. Sengupta, Physica, 31, 363 (1965).

173. R. H. Kiriyama, H. Ibamoto, and K. Matsuo, Acta Cryst., 7, 482 (1954).

174. A. K. Gregson and S. Mitra, J. Chem. Phys., 51, 5226 (1969).

175. R. L. Martin and H. Watermann, J. Chem. Soc., 1359 (1959).

176. H. Kobayashi and T. Haseda, J. Phys. Soc. Japan, 18, 541 (1963).

177. R. B. Flippen and S. A. Friedberg, J. Chem. Phys., 38, 2652 (1963).

178. M. E. Fischer, Physica, 26, 618 (1960).

179. M. F. Sykes and M. E. Fischer, Phys. Rev. Letters, 1, 312 (1958).

180. W. C. Helmholz and R. F. Kruh, J. Am. Chem. Soc., 74, 1176 (1952).

181. S. Mitra, Indian J. Pure & Appl. Phys., 2, 333 (1964).

182. A. Bose, S. Mitra and R. Rai, Indian J. Phys., 39, 357 (1965).

183. M. Sharnoff, J. Chem. Phys., 41. 2203 (1964).

184. A. Bose, S. Lahiri and U. S. Ghosh, J. Phys. Chem. Solids, 26, 1747 (1965).

185. J. Ferguson, J. Chem. Phys., 40, 3406 (1964).

186. M. Gerloch, J. Chem. Soc., A2023 (1968).

187. B. N. Figgis, M. Gerloch, J. Lewis, and R. C. Slade, J. Chem. Soc., A2028 (1968).

188. S. Lahiri, D. Ghose, and D. Mukhopadhyay, Indian J. Phys., 40, 671 (1966).

189. B. Morosin and E. C. Lingafelter, J. Phys. Chem., 65, 50 (1961).

190. S. Mitra and A. K. Gregson - (Unpublished).

191. M. Bonamico, G. Dessy, A. Vaciago, and L. Zambonelli, Acta Cryst, 19. 619, 886 (1965).

192. C. Furlani, E. Cervone, F. Calzona, and B. Baldanza, Theoret. Chim. Acta, 7, 375 (1967).

193. M. Randie, J. Chem. Phys., 36, 2094 (1962).

194. D. A. Langs and C. R. Hare, Chem. Comm., 890 (1967).

195. A. K. Gregson and S. Mitra, J. Chem. Phys., 50, 2021 (1969).

196. M. E. Fischer, J. Math. Phys., 4, 124 (1963).

197. F. H. Spedding, J. Chem. Phys., 5, 316 (1937).

198. F. H. Spedding, J. P. Howe, and W. H. Keller, J. Chem. Phys., 5, 416 (1937).

199. F. H. Spedding, H. F. Hamlin, and G. C. Nutting, J. Chem. Phys., 5, 416 (1937).

200. R. J. Elliot and K. W. H. Stevens, Proc. Roy. Soc., A215, 437 (1952); A218, 553 (1953).

201. R. J. Elliot and K. W. H. Stevens, Proc. Roy. Soc., A219, 387 (1953).

202. C. K. Jørgensen, R. Pappalardo, and H. Schmidtke, J. Chem. Phys., 39, 1422 (1962).

203. R. A. Fereday and E. C. Wiersma, Physica, 2, 575 (1935).

204. A. Mookherji, Indian J. Phys., 23, 217 (1949).

205. A. Mookherji, Indian J. Phys., 23, 309 (1949).

206. J. H. Van den Handel, Physica, 8, 513 (1941).

207. D. Neogy and A. Mookherji, Indian J. Pure & Appl. Phys., 2, 28 (1964).

208. A. Mookherji, Indian J. Phys., 23, 410, 445 (1949).

209. T. Mookherji and A. Mookherji, Indian J. Pure & Appl. Phys., 4, 43 (1966).

210. J. H. Van den Handel and J. C. Hupse, Physica, 9, 225 (1942).

211. S. P. Chachra and A. Mahalenabis, Indian J. Pure & Appl. Phys., 6, 55 (1968).

212. L. C. Jackson, Proc. Roy. Soc., A170, 266 (1939).

213. S. P. Chachra, Indian J. Pure. Appl. Phys., 3, 459 (1965).

213a. A. Frank, Phys. Rev., 48, 771 (1935).

214. D. Neogy and A. Mookherji, Indian J. Phys., 39, 342 (1965).

215. K. S. Krishnan and S. Banerjee, Phys. Rev., 59, 770 (1941).

216. T. Mookherji, Curr. Sci., 33 (No.16), 487 (1964).

217. K. S. Thomak, S. Singh, and G. K. Dieke, J. Chem. Phys., 38, 2180 (1963).

218. D. Neogy and A. Mookherji, Physica, 31, 1325 (1965).

219. A. H. Cooke, D. J. Edmonds, F. R. McKim, and W. P. Wolf, Proc. Roy. Soc., A252, 246 (1959).

220. A. H. Cooke, R. Lazenby, and M. J. M. Leask, Proc. Phys. Soc., 85, 767 (1965).

221. D. Neogy and A. Mookherji, J. Phys. Soc. Japan, 20, 1332 (1965).

222. T. Mookherji, Indian J. Phys., 38, 587 (1964).

223. D. Neogy, Physica, 29, 974 (1963).

224. S. P. Tandon, Z. Physik. Chem., 228, 151 (1965).

225. K. S. Krishnan and C. V. Raman, Proc. Roy. Soc., A115, 549 (1927).

226. A. Mookherji, Indian J. Phys., 18, 187 (1944).

227. O. P. Singhal, Proc. Phys. Soc. London, 79, 389 (1962).

228. O. P. Singhal, and T. Mookherji, Proc. Phys. Soc. London, 81, 117 (1963).

229. O. P. Singhal and T. Mookherji, Proc. Rajasthan Acad. Sci., 9, Pt. 1-6 (1962); Chem. Abst., 59, 14701 (1963).

230. H. H. Jaffe, J. Chem. Phys., 21, 156 (1953).

213. E. O. Fischer, Angew Chem., 67, 475 (1955).

232. E. O. Fischer, Rec. trav. Chim., 75, 629 (1956).

233. E. Ruch, Rec. trav. Chim., 75, 638 (1956).

234. F. A. Cotton and G. Wilkinson, Z. Naturforsch., 96, 435 (1954).

235. W. Moffitt, J. Am. Chem. Soc., 76, 3386 (1954).

236. L. Pauling, The Nature of Chemical Bond, Cornell University Press, Ithaca, New York, 1960, 3rd ed.

237. E. M. Shustorovich and M. Ye. Dyatkina, Zhur. Neorg. Khim., 3 2721 (1958).

238. M. Ye. Dyatkina, Usp. Khim., 27, 1, 57 (1958).

239. J. P. Dahl and C. J. Ballhausen, Kgl. Danske Videnskab. Sekskab., Mat.-fys. Medd, No. 5, 33 (1961).

240. L. Pauling, J. Chem. Phys., 4, 673 (1936).

241. L. N. Mulay and Sr. M. E. Fox, J. Am. Chem. Soc., 84, 1308 (1962).

242. L. N. Mulay and Sr. M. E. Fox, J. Chem. Phys., 38, 760 (1963).

243. R. Mathis, M. Sweeney, and M. E. Fox, J. Chem. Phys., 41, 3652 (1964).

244. A. K. Gregson and S. Mitra - (Unpublished).

245. Ya. G. Dorfman, Diamagnetism and the Chemical Bond Edward Arnold Publishers, 1965.

246. L. N. Mulay and V. Withstandley, J. Chem. Phys., 43, 4522 (1965).

247. L. N. Mulay and I. L. Mulay, Anal. Chem., 38, 501R (1966).

248. M. D. Rauch, E. O. Fischer, and H. Grubert, J. Am. Chem. Soc., 82, 76 (1960).

249. S. S. Batsanov, Izv. Sibirsk. Otd. Akad. Nauk SSSR, No. 8, 110 (1962).

250. R. L. Martin and S. Mitra, J. Chem. Phys., 55, 1426 (1971).

251. K. S. Krishnan, Nature, 133, 174 (1934).

252. K. S. Krishnan and N. Ganguli, Nature, 139, 155 (1937).

253. K. S. Krishnan and N. Ganguli, Z. Krist., A100, 530 (1939).

254. K. S. Krishnan and N. Ganguli, Curr. Sci., 3, 472 (1935).

255. N. Ganguli, Phil. Mag., 21, 355 (1936).

256. K. S. Krishnan and N. Ganguli, Proc. Roy. Soc., A177, 168 (1941).

257. E. Poquet, N. Lumbroso, J. Hoarau, A. Marchand, A. Pacault, and D. E. Soule, J. Chim. Phys., 57, 866 (1960).

258. K. S. Krishnan and N. Ganguli, Nature, 145, 31 (1940).

259. A. K. Datta, Indian J. Phys., 18, 249.

260. A. Sigamony, Proc. Indian Acad. Sci., 19A, 377 (1944).

261. D. Das, Indian J. Phys., 41, 525 (1967).

262. K. S. Krishnan and S. Banerjee, Z. Krist., 101, 507 (1939).

263. K. S. Krishnan and S. Banerjee, Trans. Faraday Soc., 35, 385 (1939).

264. M. G. Dupouy, Ann. Phys. (Paris), 15, 495 (1931).

265. K. S. Krishnan and S. Banerjee, Z. Krist., 99, 499 (1938); (See also: Compt. Rend., 241, 369 (1955).

266. W. Sucksmith, Proc. Roy. Soc., A133, 179 (1931).

267. M. Foex, Ann. Phys. 16, 174 (1921).

268. H. Bizette and B. Tsai, C.R. Acad. Sci., 238, 1575 (1954).

269. A. Mookherji and S. C. Mathur, J. Phys. Soc. Japan, 20, 1336 (1965).

270. A. K. Dutta and B. C. Roy Chowdhury, Indian J. Phys., 23, 131 (1949).

271. E. Cotton-Feytis, Compt. Rend., 214, 485 (1942).

272. E. Cotton-Feytis, Compt. Rend., 215, 299 (1942).

273. E. Cotton-Feytis, Rev. gén. caout., 21, 27 (1944).

274. E. Cotton-Feytis, Rubber Chem. Technology, 18, 8 (1945).

275. E. M. Weir and P. W. Selwood, J. Am. Chem. Soc., 73, 3484 (1951).

276. E. W. Toor and P. W. Selwood, J. Am. Chem. Soc., 74, 2364 (1952).

277. P. W. Selwood, Magnetochemistry, Interscience, New York, 1956.

278. A. Isihara, H. Kusumoto, H. Nagano, and K. Oshima, J. Chem. Phys., 21, 1909 (1953).

278a. C. V. Raman and K. S. Krishnan, Proc. Roy. Soc., A113, 511 (1927).

279. C. V. Raman, Nature, 123, 945 (1929).

280. C. V. Raman, Nature, 124, 412 (1929).

281. E. Hückel, Z. Phys., 70, 204 (1931).

282. E. Hückel. Z. Phys., 76, 628 (1932).

283. E. Hückel, Z. Phys., 72, 310 (1931).

284. E. Hückel, Z. Phys., 83, 632 (1933).

285. E. Hückel, Int. Conf. Phys. Lond., 2, 35 (1934).

286. F. London, C.R. Acad. Sci. Paris, 205, 28 (1937).

287. F. London, J. Phys. Radium, 8, 397 (1937).

288. F. London, J. Chem. Phys., 5, 837 (1937).

289. R. McWeeny, Proc. Phys. Soc. (London), 64A, 261, 921 (1951).

290. R. McWeeny, Proc. Phys. Soc. (London), 65A, 839 (1952).

291. R. McWeeny, Proc. Phys. Soc. (London), 66A, 714 (1953).

292. G. Berthier, M. Mayot, B. Pullman, and A. Pullman, J. Phys. Radium, 13, 15 (1952).

293. R. J. Abraham. Mol. Phys., 4, 145 (1961).

294. D. P. Craig, M. L. Heffernan, R. Mason, and N. L. Paddock, J. Chem. Soc., 1376 (1961).

295. T. Itoh, K. Ohno, and H. Yoshizumi, J. Phys. Soc. Japan, 10, 103 (1955).

296. L. Caralp and J. Hoarau, J. Chim. Phys., 60, 889 (1963).

297. R. G. Paar and P. G. Lykos, J. Chem. Phys., 28, 361 (1958).

298. J. A. Pople, J. Chem. Phys., 38, 1276 (1963).

299. J. A. Pople, J. Chem. Phys., 41, 2559 (1964).

300. K. Lonsdale, J. Chem. Soc., 364 (1938).

301. L. Singh, Trans. Faraday Soc., 54, 1117 (1958).

302. S. Banerjee, Z. Kristallogr., A100, 216 (1938).

303. C. G. Barraclough, R. L. Martin, and S. Mitra, J. Chem. Phys., 55, 1427 (1971).

304. H. A. Allen and N. Muller, J. Chem. Phys., 48, 1626 (1968).

305. J. Hoarau, J. Joussot-Dubien, B. Lemanceau, N. Lumbroso, and A. Pacault, Cah. Phys., no. 74, 34 (1956).

306. M. A. Lasheen, Phil. Trans. Roy. Soc., A256, 357 (1964).

307. J. M. Robertson, Proc. Roy Soc., A150 348 (1935).

308. J. Donohue, J. Phys. Chem., 56, 502 (1952).

309. I. E. Knagg, Proc. Roy. Soc., A150, 576 (1935).

310. J. M. Robertson and I. Woodward, J. Chem. Soc., 1817 (1936).

311. B. Lamanceau, J. Chim. Phys., 56, 933 (1959).

312. K. Lonsdale, Nature, 137, 826 (1936).

313. N. Lumbroso and A. Pacault, C. R. Acad. Sci. Paris, 245, 686 (1957).

314. K. Lonsdale, Proc. Roy. Soc., A177, 272 (1941).

315. W. J. John, Z. Krist., 101, 337 (1939).

316. H. Jones, Proc. Roy. Soc., A144, 225 (1934); ibid, A147, 396 (1934).

317. D. Shoenberg and M. Z. Uddin, Proc. Camb. Phil. Soc., 32, 499 (1936).

318. S. Ramachandra Rao and K. C. Subramaniam, Nature, 136, 336 (1935).

319. S. Ramachandra Rao and H. V. Venkataramiah, Curr. Sci., 14, 19 (1945).

320. K. A. Blom, O. Beckman, and M. Richardson, Solid State Comm., 5, 977 (1967).

321. P. Nilakantan, Proc. Indian Acad. Sci., 4A, 542 (1936).

322. P. Nilakantan, Proc. Indian Acad. Sci., 2A, 621 (1935).

323. P. Nilakantan, Proc. Indian Acad. Sci., 7A, 38 (1938).

324. A. Veillard, B. Pullman, and G. Berthier, Compt. rend., 252, 2321 (1961).

325. B. K. Banerjee, Glastech. Ber., 33, 8 (1960).

326. A. Mookherji and R. B. Lal, Indian J. Pure & Appl. Phys., 3, 288 (1965).

327. A. Mookherji and R. B. Lal, Nature, 207, 853 (1965).

328. K. Hirakawa, T. Hashimoto, and K. Hirakawa, J. Phys. Soc. Japan, 16, 1934 (1961).

329. A. H. Heeger, O. Beckman, and A. M. Portis, Phys. Rev., 123, 1652 (1961).

330. M. Gerloch, J. Kohl, J. Lewis, and W. Urland, J. Chem. Soc. (A 1970) 3269, 3283.

331. M. Gerloch, and D. J. Mackey, J. Chem. Soc. (A 1970) 3030, 3040.

332. M. Gerloch, J. Lewis, and R. C. Slade, J. Chem. Soc. (A 1969) 1422.

333. H. Venoyama, T. Iizuka, H. Morimoto, and M. Kotani, Biochim. Biophys. Acta 160 (1968) 159

334. K. G. Srivastava, Phys. Letters 4 (1963) 55.

335. E. Legrand and A. Van den Bosch, Solid State Comm. 7 (1969) 1191.

336. K. H. Hellwege, S. H. Kwan, H. Lange, W. Rummel, W. Schembs, and B. Schneider, Z. Phys. 167 (1962) 487.

337. J. N. McElearney, R. W. Schwartz, S. Merchant, and R. L. Carlin, J. Chem. Phys. 55, 466 (1971).

338. R. B. Bentley, F. E. Mabbs, W. R. Smail, M. Gerloch, and J. Lewis, Chem. Comm., 119 (1969).

339. S. Lahiri, D. Mukhopadhyay, and D. Ghosh, Indian J. Phys., 42, 320 (1968).

340. M. Gerloch and R. C. Slade, J. Chem. Soc., A 1022 (1969).

341. A. K. Gregson, R. L. Martin, and S. Mitra, Proc. Roy. Soc. A320, 473 (1971).

342. M. Gerloch and P. N. Quested, J. Chem. Soc., A 2307 (1971).

343. A. K. Gregson and S. Mitra, Chem. Phys. Letters, 13, 313 (1972).

344. W. DeW. Horrocks, Inorg. Chem., 9, 690 (1970).

345. W. DeW. Horrocks and D. DeW. Hall, Coordination Chem. Revs., 6, 147 (1971).

Each name is followed by the page numbers on which it is mentioned.

Reference numbers appear in parentheses and the page where each reference is listed, is indicated by an underlined number.